Communications
in Computer and Information Science　　2202

AF148130

Rationale

The CCIS series is devoted to the publication of proceedings of computer science conferences. Its aim is to efficiently disseminate original research results in informatics in printed and electronic form. While the focus is on publication of peer-reviewed full papers presenting mature work, inclusion of reviewed short papers reporting on work in progress is welcome, too. Besides globally relevant meetings with internationally representative program committees guaranteeing a strict peer-reviewing and paper selection process, conferences run by societies or of high regional or national relevance are also considered for publication.

Topics

The topical scope of CCIS spans the entire spectrum of informatics ranging from foundational topics in the theory of computing to information and communications science and technology and a broad variety of interdisciplinary application fields.

Information for Volume Editors and Authors

Publication in CCIS is free of charge. No royalties are paid, however, we offer registered conference participants temporary free access to the online version of the conference proceedings on SpringerLink (http://link.springer.com) by means of an http referrer from the conference website and/or a number of complimentary printed copies, as specified in the official acceptance email of the event.

CCIS proceedings can be published in time for distribution at conferences or as postproceedings, and delivered in the form of printed books and/or electronically as USBs and/or e-content licenses for accessing proceedings at SpringerLink. Furthermore, CCIS proceedings are included in the CCIS electronic book series hosted in the SpringerLink digital library at http://link.springer.com/bookseries/7899. Conferences publishing in CCIS are allowed to use Online Conference Service (OCS) for managing the whole proceedings lifecycle (from submission and reviewing to preparing for publication) free of charge.

Publication process

The language of publication is exclusively English. Authors publishing in CCIS have to sign the Springer CCIS copyright transfer form, however, they are free to use their material published in CCIS for substantially changed, more elaborate subsequent publications elsewhere. For the preparation of the camera-ready papers/files, authors have to strictly adhere to the Springer CCIS Authors' Instructions and are strongly encouraged to use the CCIS LaTeX style files or templates.

Abstracting/Indexing

CCIS is abstracted/indexed in DBLP, Google Scholar, EI-Compendex, Mathematical Reviews, SCImago, Scopus. CCIS volumes are also submitted for the inclusion in ISI Proceedings.

How to start

To start the evaluation of your proposal for inclusion in the CCIS series, please send an e-mail to ccis@springer.com.

R. Geetha · Nhu-Ngoc Dao · Saeed Khalid
Editors

Advances in Artificial Intelligence and Machine Learning in Big Data Processing

First International Conference, AAIMB 2023
Chennai, India, August 17–18, 2023
Proceedings, Part-I

 Springer

Editors
R. Geetha (ID)
S.A. Engineering College
Chennai, Tamil Nadu, India

Nhu-Ngoc Dao (ID)
Sejong University
Seoul, Korea (Republic of)

Saeed Khalid (ID)
Bialystok University of Technology
Białystok, Poland

ISSN 1865-0929   ISSN 1865-0937 (electronic)
Communications in Computer and Information Science
ISBN 978-3-031-73064-1   ISBN 978-3-031-73065-8 (eBook)
https://doi.org/10.1007/978-3-031-73065-8

This Springer imprint is published by the registered company Springer Nature Switzerland AG
The registered company address is: Gewerbestrasse 11, 6330 Cham, Switzerland

If disposing of this product, please recycle the paper.

Preface

These two CCIS volumes constitutes the refereed proceedings of the International Conference on Advances in Artificial Intelligence and Machine Learning in Big Data Processing, AAIMB 2023, held at S.A. Engineering College, Chennai, India, during August 17–18, 2023. All the accepted papers were carefully double-blindly reviewed by national and international reviewers. They were organized in topical tracks as follows: Artificial Intelligence and Machine Learning, Intelligent Information Processing, Pattern Recognition and Analysis, Sentiment Analysis, Computational Optimization, Information Extraction and Prediction, Decision Making, Virtual reality, Natural Language Processing, AI Algorithms, Fuzzy logic & Neural System. Adaptive agents, AI Planning Strategies and Tools, Reinforcement Learning, Robotics, Computer Vision, Expert Systems, Pattern Recognition, Component Analysis, Social Network Analysis, Cloud-Based Services, Big Data Analytics, Data Mining Challenges with Big Data, Big Data Analytics and Big Data Governance, Big data Analytics and Applications Data, Text, Web Mining, & Visualization, Data & Knowledge Engineering, Techniques for Data Visualization, Data Stream Algorithms and Information Processing, For the AAIMB 2023 conference 183 papers were been received, out of which 51 full-length papers were accepted and presented at the conference.

We would like to acknowledge and thank the conference committee members and international reviewers who spent a lot of time on the arduous task of selecting the qualified papers from all the submissions. Their excellent professional advice enabled us to come out with this volume of proceedings of the conference. We thank all the participants and the management of S.A. Engineering College, who encouraged us morally and financially.

August 2023

R. Geetha
Nhu-Ngoc Dao
Saeed Khalid

Organization

General Chair

Geetha R.   S. A. Engineering College, Chennai, India

Program Committee Chairs

Anuradha M.   S. A. Engineering College, Chennai, India
Hemanand D.   S. A. Engineering College, Chennai, India
Nalini M.   S. A. Engineering College, Chennai, India
Kaliraj V.   S. A. Engineering College, Chennai, India
Nhu-Ngoc Dao   Sejong University, Seoul, Republic of Korea
Abdel-Badeeh M. Salem   Ain Shams University, Egypt
Nazir Ahmad   King Khalid University, Saudi Arabia

Steering Committee

Nhu-Ngoc Dao   HCMOU, Vietnam
Nazir Ahmad   K. A. Nizami Centre for Quranic Studies,
                                   Aligarh, India
Dinakaran K.   S.A. Engineering College, India
Thilagavathi N.   S.A. Engineering College, India
Muthumari Lakshmi S.   S.A. Engineering College, India
Abdel-Badeeh M. Salem   Ain Shams University, Egypt
Sheela Lourdusamy   Pentecost University, Ghana
Xavier N. Fernando   Ryerson University, Canada
Mokhtar M. M. Alihamid   University of El Imam El Mahdi, Aba Island,
                                   Sudan
Muhammad Ghalib   De Montfort University, Dubai, United Arab
                                   Emirates
Kindie Biredagn Nahato   Debre Berhan University, Debre Birhan, Ethiopia
Senthil Ramadoss   Majan University College, Oman
J. Jean Justus   St. Joseph's College of Engineering, India
Suresh Kumar S.   Rajalakshmi Engineering College, India
Sathish Kumar P. J.   Panimalar Engineering College, India

Chembian W. T.   Vel Tech HighTech Dr.Rangarajan Dr.Sakunthala
                                  Engineering College, India
Eswaramoorthy V.   Bannari Amman Institute of Technology, India
Subhashini P.   Vel Tech Multi Tech Dr.Rangarajan
                                  Dr.Sakunthala Engineering College, India
Gopalakrishnan B.   Bannari Amman Institute of Technology, India
Sundara Murthy S.   Bannari Amman Institute of Technology, India
Vanitha M.   Saveetha Engineering College, India
Sathya Priya J.   Velammal Engineering College, India
Ashok Kumar P. M.   KL University, India
Brindha Merin J.   B.S. Abdur Rahman Crescent Institute of
                                  Science & Technology, India
Mercy Paul Selvan   Sathyabama Institute of Science and Technology,
                                  India
Saraswathi S.   SSN College of Engineering, India
Bhavadharini R. M.   VIT, India

Program Committee

Siva Kumar S.   S. A. Engineering College, Chennai, India
Chitra Devi D.   S. A. Engineering College, Chennai, India
Balakrishnan C.   S. A. Engineering College, Chennai, India
Mani A.   S. A. Engineering College, Chennai, India
Balasubramanian M.   S. A. Engineering College, Chennai, India
Muthukumarasamy S.   S. A. Engineering College, Chennai, India

Additional Reviewers

Prasana Devi   Ananthi S. N.
Prasana Kumar   Julia Faith S.
Sangeetha J.   Veena T.
Aruna K. B.   Chithra D.
Thilagam T.   Krishnaveni S.
Shobana R.   Seetha A.
Belshia Jebamalar   Sutha Merlin
Abirami A.   Muralitharan G.
Kavitha M.   Aravinth Kumar R.
Ravanan R.   Chitti Babu B.
Susi K.   Jayavel R.
Lakshmipriya   Sureka V.
Mabel Rose R. A.   Sudha L.

Adaikkammai A.   Suntheya A. K.
Dharaniya R.   Keerthana D.
Lakshmi Priya S.   Senthil P.
Abdul Rahaman A.   Hima Vijayan
Selvaraj M. D.   Asha S.
Ram Prasad Padhy   Valarmathi K.
Gayathri S.   Usha N. S.
Arumai Shiney S.   Janshi Pragada
Prasanth K.   Lalithasree

Contents – Part I

Artificial Intelligence and Data Analytics

A Novel Revolutionizing Medical Surgery Procedures Using Mixed Reality ... 3
G. A. Senthil, J. Abinaya, K. Jency Oliviya, and R. Keerthana

Data Mining-Based Classification Algorithms for Predicting Mental Health 21
K. Vijay, P. T. S. Shahul Hameed, M. Bhavani, and M. Jaeyalakshmi

Topological Navigation of Path Planning Using a Hybrid Architecture
in Wheeled Mobile Robot 32
Vengatesan Arumugam and Vasudevan Algumalai

Abnormal Behaviour Detection in Surveillance Videos 45
*R. Vijayakumar, D. Sorna Shanthi, B. Bhuvaneswaran,
and M. Pragadeesh*

ISAApp – Image Based Smart Attendance Application 56
*Aritra Dutta, G. Suseela, G. Niranjana, Pushpita Boral, Pranav Gupta,
and Subha Bal Pal*

A Self-learning Ai-Based Information Leak Protection System 68
M. Jaeyalakshmi, P. Rohit Gangadhar, M. Srivatsan, and M. Bhavani

Deep Learning

Enhancing Abnormal Object Detection in Camera-Based Systems
Through Computer Vision and Deep Learning Techniques 81
*K. Veena, NagaHemanth Murari Alluguddu, A. Sai Simha Reddy,
A. Deepa, M. Selvi, and P. Kathambari*

Detection and Classification of Brain Tumor in Magnetic Resonance
Images Using CNN 97
S. Nivedha, A. Mani, and S. Muthukumarasamy

Diagnosis of Parkinson's Disease Using Machine Learning and Deep
Learning Techniques 111
*S. Sharanyaa, M. Sambath, A. Ganesh, A. Hammadh Ahmed,
and S. Ganesh*

A Survey on Deep Learning Based Human Activity Recognition System 124
  Ansu Liz Thomas and J. E. Judith

A Deep Learning Approach for Non - invasive Body Mass Index Calculation .. 135
  *S. Harish Nandhan, J. Remoon Zean, A. R. Mahi, R. Meena,
  and S. Mahalakshmi*

Early-Stage Detection of Alzheimer's Disease Using MRI Scans
with Deep Learning 147
  *R. Sarala, P. Bharath, S. Lakshman Raj, M. Selva Kumar,
  and M. D. Harish Srinivas*

Penguin Search Optimization with Deep Learning Based Cybersecurity
Malware Spectrogram Image Classification 158
  J. Jeyalakshmi, M. Santhiya, and R. Jegatha

Detection and Classification of Skin Disease Using CNN 171
  J. Jeyalakshmi, M. Santhiya, and M. Shobana

Estimation of Above Ground Biomass Using Machine Learning and Deep
Learning Algorithms: A Review 181
  S. Arumai Shiney and R. Geetha

URL Phishing Detection Using Deep Learning and Machine Learning
Techniques 197
  *R. Jegadeesan, Dava Srinivas, N. Sankar Ram, R. Janakiraman,
  M. Jhansi, C. H. Sanjana, N. Akshitha, and C. H. Saicharan*

Enhanced Disease Recognition and Classification in Black Gram Plant
Leaves Using Deep Learning 213
  K. Prasanth, P. Kabilamani, G. Sangar, V. Kaliraj, and V. Rajasekar

Ensemble Deep Learning Approach for Identification of DDOS Attack 225
  C. Balakrishnan and V. S. Prassana kumar

ROCLT: Enhanced Text Classifier for Sentiment on Imbalanced Multiclass
Tweet Data Using Hybrid Deep Learning Techniques 234
  M. Rameshraja and J. Arunadevi

Computer Vision to Animal Footprint Classification Based on Deep
Learning Model 246
  A. Rifana Fathima and K. Dhanalakshmi

Speech Emotion Recognition Using CNN Classifier Based on Deep
Learning Model 257
   M. Archana, D. Shanthi, and Pavan Kumar Vadrevu

Face Detection and Recognition for Criminal Identification System Using
Deep Learning 270
   K. Ramyadevi, M. Balasubramanian, G. Surya Prakash, and V. S. Tharun

Automated Essay Grading System for IELTS Using Bi-LSTM 280
   *Chandan Kumar Sangewar, Chinmay Pagey, Aman Kumar,
   and R. Krithiga*

Automized Quick Prediction of Skin Cancer Diagnosis by Enhanced Deep
Convolutional Neural Network 292
   *V. S. Jeyalakshmi, N. Bala Shunmugam, M. Kavitha,
   and D. Paulin Diana Dani*

Clustering Based Demand Prediction Using Long Short-Term Memory
(LSTM) in Retail Supply Chains 303
   S. Praveena and S. Prasanna Devi

Early Detection of Diabetic Retinopathy Using Deep Convoulutional
Neural Network 315
   K. Vijay, P. Krithiga, S. Kavirakesh, S. Swetha, and B. Vishal

Author Index 329

Contents – Part II

Artificial Intelligence and Data Analytics

Feature Fusion Based Bayesian Model Detection in Prognosis
of Glioma – A Survey 3
   K. H. Mohammed Sazzad, M. Nethra, S. Santhya,
   and A. Arnold Sylevester

AI Voice Assistant Using Python 20
   E. Kamalanaban, P. Selvarani, A. Aranidhi, Daniel Abraham,
   Andene Gowtham, and J. Jayaprakash

Research Output on Artificial Intelligence in India: A Scientometric Study 25
   S. Anbukkarasi, K. Sivasekaran, and M. Rethi

Analyzing Hand Gestures Using Object Detection and Processing It
into Local Language 37
   K. Sangeetha, V. S. Balaji, P. Kamalesh, and P. S. Anirudh Ganapathy

A Review of an Automated Model for Sexist Language Detection
and Replacement of Sexist Terms 45
   M. S. Shriram, S. Sushmitha, and Shravanthi Murugesan

Machine Learning

Detection of DoS Attack Using Machine Learning in Software Defined
Network 61
   M. A. Gunavathie, S. Kousalya, J. Celina Rekha, S. Rasikaranjani,
   and D. Mahalakshmi

Regression Model Approach Towards Concrete Compressive Strength
Prediction and Evaluation 72
   Vijayalakshmi G. V. Mahesh, CP Achyutha Gowda,
   Alla Vamsi Krishna, and Leti Manish Kumar

An Enhanced Artificial Neural Network Mode for Type 2 Diabetes
Classification Using SMOTE and SMOTE-Tomek with Effective Feature
Selection Methods 84
   E. Sabitha and M. Durgadevi

Machine Learning Based Traffic Congestion and Accident Prevention
Analysis 105
    A. Sathya Sofia, C. P. Thamil Selvi, S. Suganya, P. Francis Antony Selvi,
    and M. Shanthalakshmi

A Comparative Evaluation of Machine Learning Methods for Predicting
Chronic Kidney Disease 116
    K. Navaz, S. Yazhinian, N. Muthuvairavan Pillai, and N. Purushotham

Detecting Implicit Aspects of Customer Experience in the Hotel Industry
Using a Machine Learning Algorithm 126
    S. Jayanthi and S. S. Arumugam

Vehicle Insurance Claim Prediction 139
    V. Sureka, K. B. Aruna, L. Sudha, and A. K. Suntheya

Ensemble Learning Models for Detecting Spam Over Social Networks
Using RFE 150
    V. Saraswathi, A. Adaikkammai, Anitha Jebamani, D. Devi,
    and R. Radhika

An Intelligent Machine Learning Framework for Melanoma Classification
System: A Critique 165
    S. Sridevi, S. Gowthami, and K. Hemalatha

Online Network Intrusion Detection System for IOT Structure Using
Machine Learning Techniques 176
    K. Mahalakshmi and B. Jaison

An Analysis of Machine Learning Tools and Algorithms 195
    R. Sindhuja, R. Nanmaran, and A. Madhan Kumar

Ensemble Learning-Based Android Malware Detection 205
    V. Priya and A. Sathya Sofia

Handling Imbalanced Data for Credit Card Fraudulent Detection:
A Machine Learning Approach 220
    E. Sujatha, V. Umarani, K. S. Rekha, P. V. Gopirajan,
    and V. Manickavasagan

An Enhanced Learning Model Based on an Improved Random Forest
Classifier and an Integrated Attribute Selector for Healthcare Datasets 234
    S. Rajeashwari and K. Arunesh

A Systematic Literature Analysis of Scientometric Research in the Field
of Cloud-Based Services: A 2023 Update 250
  S. Karthika, S. Balachandran, and S. Sivankalai

Optimizing Coronary Illness Prediction Using Hyperparameter Tuning
Through Machine Learning 263
  M. G. Vaishnavi and D. Shanthi

Detection and Classification of Intracranial Tumor in Machine Learning
Using Fuzzy C-Means Algorithm 276
  R. Roobika, D. Shanthi, and N. Sivakamy

A Survey of Anomaly Detection in Video Surveillance 292
  N. Muthurasu and V. Rajasekar

Prediction and Comparison of ML Algorithm for Heart Disease 303
  G. Belshia Jebamalar, J.A. Adlin Layola, J. Rajalakshmi, T. Thilagam,
  M. Kausthuban, and R. Nareshraj

Comparative Analysis of Classifiers for Chat Classification: A Study
on Random Forest, Support Vector Machine, and Multinomial Naive
Bayes with Bag of Words and TF-IDF 314
  K. Hemakirthiga and J. Arunadevi

Author Index 327

Artificial Intelligence and Data Analytics

A Novel Revolutionizing Medical Surgery Procedures Using Mixed Reality

G. A. Senthil$^{(\boxtimes)}$, J. Abinaya, K. Jency Oliviya, and R. Keerthana

Department of IT, Agni College of Technology, Chennai, India
senthilga@gmail.com

Abstract. In medical field, human error is a major complicated and failures in all the surgical procedures. On the other hand, mixed reality enables the process of reconstructive surgeries. Mixed Reality (MR) or Hybrid Reality (HR) is a technology that combines elements of Virtual and Augmented Reality (VR and AR). This research paper discusses the implementation of Virtual Reality and Augmented Reality (AR) in medicine, specifically virtual surgery technology. Surgeons can utilize MR to construct accurate, comprehensive simulations of the human anatomy to plan and practice treatments before carrying them out on live patients. With the widespread adoption of technological advances in the medical sphere, the potential for MR advancement is limitless. This study covers the content and technological properties of mixed reality-based medical operation visualisation systems. It also discusses a few prevalent uses of mixed reality technology in healthcare settings. The Simulation we have also used Unity 3D, which is a gaming engine for creating Mixed Reality (MR) applications, such as those used in surgery. Real-time rendering, physics simulation, and cross-platform portability are just a few of the tools and capabilities that make Unity 3D an appealing alternative for MR creation.

Keywords: Virtual Surgery · Augmented Reality (AR) · Magnetic Resonance Imaging (MRI) · Virtual Reality (VR) · Mixed Reality (MR) · Head Mount Display · Unity 3D · Vuforia · HoloLens · Computed Tomography (CT) · Hybrid Reality (HR)

1 Introduction

This research describes a revolutionary strategy that uses mixed reality (MR) technology to revolutionize medical surgery operations. MR provides a tremendous opportunity to improve surgical precision, optimize patient outcomes, and alter the surgical scene by merging components of Virtual Reality (VR) and Augmented Reality (AR). The goal of this study is to investigate the potential of MR in revolutionizing medical surgery techniques [1]. The study investigates the influence of MR on many elements of surgical practice, including preoperative planning, intraoperative guiding, and postoperative monitoring, using a complete evaluation of literature, case studies, and practical implementations [2].

The concept of mixed reality is an innovation that enables the virtual and physical worlds to coexist, allowing things to be modified in real time implementation. Mixed

© The Author(s), under exclusive license to Springer Nature Switzerland AG 2025
R. Geetha et al. (Eds.): AAIMB 2023, CCIS 2202, pp. 3–20, 2025.
https://doi.org/10.1007/978-3-031-73065-8_1

reality devices can be utilised in surgical planning and training. The advent of MR will user in novel revolutionary advances in health professions education, healthcare research, interaction, and patient care [3]. The combination of Mixed Reality (MR) technology combines Virtual Reality, also known as VR, and Augmented Reality, also known as AR, technologies to provide a fully-immersive and dynamic experience. MR technology is being researched and developed for application in surgery in the medical industry [4]. Using VR and AR during surgery allows doctors to receive real-time information and direction, improving accuracy and lowering the possibility of problems [7–10].

The Uses of Mixed Reality in surgery offers the potential to improve surgical planning, physician training, and outcomes for patients. Before performing operations on actual patients, surgeons can rehearse procedures in a controlled setting by employing VR simulations [5]. This allows them to gather experience and sharpen their skills. During surgery, augmented reality (AR) technology can offer real-time information and advice, helping to increase accuracy and lower the chance of complications. Mixed Reality technology is becoming the most common frontline technological innovation for healthcare applications. Moreover, Mixed reality (MR) can be utilized to perform scans on patients [6]. MR technology may be utilized to construct virtual models of patients and their anatomical structures that can be seen and modified in three dimensions. One modern use of MR is in medical imaging, such as Computed Tomography (CT) and the use of Magnetic Resonance Imaging (MRI) [22, 23]. Mixed Reality technologies can be used to create a virtual representation of the human body of the patient or the specific anatomical feature under examination. This virtual model may be seen and altered across three dimensions, providing a more comprehensive and interactive examination of the anatomy of the patient's being examined [21].

Moreover, Mixed Reality Technology may be utilized in concert with other scanning technologies, such as ultrasound, to give real-time guidance during surgeries. Mixed Reality, for instance, that may be used to create a virtual representation of the patient's anatomy, which is subsequently overlaid onto a real-time ultrasound image to provide a more precise and comprehensive perspective of the anatomy. Ultimately, Mixed Reality has the potential to improve medical imaging and patient outcomes by delivering more accurate and complete information about the patient's anatomy. Further research is required to properly comprehend the advantages and drawbacks of MR in surgery, which is still in its infancy. Nonetheless, the potential for better patient outcomes, medical education, and training has encouraged researchers and healthcare professionals to investigate and develop this technology. It is necessary to conduct a complete study that presents a fair image and makes integrative comparisons between various cutting-edge visualisation approaches. Highly advanced medical 3D workstations, which are often used in radiology, are underutilised as tools to aid these emerging technologies [12]. The effectiveness of these workstations for cardiac computed tomography (CT) and magnetic resonance imaging (MRI) in evaluating congenital heart defects (CHD) will be examined in this paper [24]. Finally, we discuss technical challenges, medicinal uses for augmented reality, hybrid reality, virtual reality, and 3D printing, as well as unresolved problems and potential directions for the future of technology [16].

The study's primary findings include that Mixed Reality Simulations are preferred for exhibiting complicated CHD lesions, increasing depth perception, illustrating the spatial relationship between heart portions, educating about the condition, and enabling pre-operative planning [11]. The 3DPHM was ranked as the best instrument for enhancing depth perception, promoting patient communication, and functioning as a pathologi-cal learning aid. By offering extra useful data for the diagnostic evaluation of CHD patients, MR and 3DPHM both function as supplementary tools to the present image visualisation approach [13]. Mixed reality in surgery is currently under investigation and development, and it is not commonly employed in clinical practise. However, it is a promising technology that is projected to gain traction in the future [15] (Fig. 1 shows framework for mixed reality).

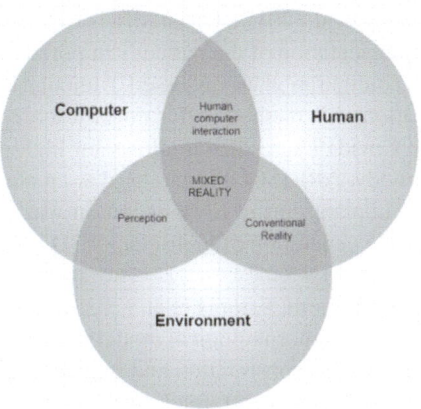

Fig. 1. Framework for Mixed Reality

The findings show that MR technology gives surgeons an immersive and interac-tive environment in which to see patient-specific anatomical components, allowing for perfect preoperative planning. The use of real-time imaging data and virtual overlays improves surgical navigation by aiding surgeons in identifying important structures and increasing surgical precision. MR allows for real-time guiding during surgical operations by superimposing virtual models over the patient's anatomy. This enhanced visualiza-tion improves surgeon perception, resulting in better tool placement, more accurate tissue manipulation, and fewer surgical mistakes. Furthermore, MR provides real-time feedback on vital signs and patient data, allowing for earlier intervention and better decision-making. Furthermore, MR transforms surgical training and education by pro-viding students with realistic virtual settings in which to practice complicated procedures. This immersive training promotes skill development, procedural competency, and learner confidence, resulting in better patient outcomes.

2   Related Work

Thomas M. Gregory et al. (2018), The orthopaedic surgeon was able to drag the 3-D reconstruction of the scapula to the visible part of the bone in three minutes while simultaneously having real-time access to the patient's medical information and the operational strategy. It was a typical reverse shoulder arthroplasty procedure. During this treatment, there was an issue with the vault's restricted bone supply (caused by the patient's anatomy, which had a Walch A2-type glenoid).

Jens T. Verhey et al. (2019), The integration review included studies on nasogastric tube insertion, facet joint injection, catheterization, needle guiding, and resection planning, however needle insertion was not defined as a surgical operation. Trainees now have access to realistic and highly engaging operational simulations without the need for supervision. With this access to multidimensional rebuilding based on patient imaging and the ability to communicate remotely with colleagues outside the operating room, the engaging in orthopaedic surgeon is becoming more adept at preoperative planning and intraoperative navigation without the use of fluoroscopy.

Charite–Berlin, Germany et al. (2017), On an MR-HMD, the surgeon's field of view was overlay with a 3D representation of the patient's important liver structures. During open hepatic surgery, this configuration was assessed. Although live surgeries have a long history in medical education, AR methods offer novel teaching tactics. This teaching strategy is critical in the Covid19 age since it allows for high-quality remote learning without endangering anybody. ARRISCOPE was used to create the visualizations. The installation of a cochlear implant during ENT operation was demonstrated using AR technology. This enabled the generation of high-quality 3D images as well as based on images measurements.

Janno Torop, University of Tartu, Estonia et al. (2022), A smart segmentation technique is utilised to generate a 3D morphological model with magnification and rotating effects for optimal CT and MRI reconstruction and upload it to the cloud platform for physical parameter adjustment, placement, rendering, and high-dimensional presentation. MR holograms give a new technology platform for telesurgery cooperation, greater visual assistance during treatments, improved precision and safety of screw insertion, and an easy communication tool between doctors and patients.

Rahul Prasanna Kumar et al. (2019), Resection (Liver Resection Planning Tool) uses deformable Bezier surfaces and a resection margin (safety indicator) to recreate liver operation planning. Assuming the tumour is spherical, the point-to-point distance between each point on the Bezier plane and the tumour's centre is computed to determine the resection margin. Clinical use-cases for mixed reality applications are generally acknowledged and provide deeper understanding of patient physiology as well as better planning for surgery. A method is provided for developing, viewing, and interacting with patient-specific 3D models in the HoloLens.

Jieun Han et al. (2019), To remove common terminologies, build synonym text sets, unify singular and plural of the same terms, and reflect academic compound nouns, data refinement and normalisation were carried out. The objective of this study's bibliometric assessment of worldwide studies is the understanding, diagnosis, and treatment of mental disorders utilising virtual and augmented reality (VR/AR) technology. The SciVerse Scopus database was used to gather relevant documents for the 1980–2021 research period. According to the findings, this field of study has witnessed tremendous growth since 2017, is predominantly carried out by academics in rich countries with cutting-edge digital technology, and obtains the greatest citations for publications on anxiety and phobias. Real-world uses suggest that AR/VR Technologies are accepted and simple to use, and that medical education is critical.

H Brun et al. (2018), Holographic manipulation tools have the potential to help surgical planning for heart defects that are congenital, with female and younger users being preferred. The 3D heart model was segmented from CTA images and visualised in HoloLens® for 36 members of the paediatric cardiac team, resulting in high anatomical recognition and diagnostic output.

3 Proposed Work

Medical professionals must be highly skilled and able to carry out intricate procedures with accuracy and precision. The margin for error during actual surgeries can be rather high, and training for these procedures can be time-consuming and expensive. By enabling more effective and efficient training and assisting surgeons during actual procedures, mixed reality technology offers a potential solution to these problems.

Also, by enabling more precise and accurate procedures, mixed reality technology has the potential to improve patient outcomes. By providing real-time information and recommendations, Mixed Reality Technology can assist to reduce the risk of complications and improve surgical results.

Last but not least, mixed reality technology has the potential to lower the cost of surgical procedures and medical training. Mixed Reality training can help to cut back on the requirement for pricey resources and equipment by enabling more effective and efficient training. The use of follow-up operations and additional medical care can be decreased with mixed reality technology since it improves surgical results.

In summary, mixed reality technology's futuristic benefits for the healthcare sector, such as greater learning, improved outcomes for patients, and cheaper costs, have prompted academics and healthcare professionals to examine and develop this technology for medical applications.

4 Methodology

Virtual and Augmented Reality Technology is integrated in surgical operations using mixed reality in medicine. It improves the surgeon's perception and gives crucial information during the procedure by fusing computer-generated images and data with real-time imaging of the surgical field. Here are a few techniques frequently employed in mixed reality-based medical surgery:

1. **Surgical Navigation:** Using patient-specific imaging data, such as CT scans or MRI, mixed reality can help with preoperative planning by building a 3D representation of the patient's anatomy. During the surgery, surgeons may then see the surgical site in real-time and manoeuvre through intricate anatomical systems with precision.
2. **Visualization of Augmented Reality (AR):** Using AR, virtual data is superimposed over the actual surgical setting. The field of vision of surgeons is filled with computer-generated pictures that are projected onto a transparent display, such as a head-mounted display (HMD). Anatomical structures, real-time data, surgical guides, or tool tracking information can all be included in these representations, which can help with direction and improve situational awareness.

3. **Virtual Reality (VR) Simulation**: Before performing actual surgery, surgeons can practice in a realistic, immersive virtual environment using virtual reality (VR) simulation. Using hand controllers or specialized surgical equipment, surgeons can model surgeries, manipulate virtual anatomy, and fine-tune their techniques. Surgery planning, surgical skill improvement, and error reduction are all benefits of VR simulations.

4. **3D Modelling and Printing:** Anatomical models or surgical guidance customized to a patient can be made using 3D printing and modelling technology combined with mixed reality. These models can be used by surgeons to prepare for surgery, practice difficult procedures, and increase accuracy while performing them.

5. **Intraoperative Imaging Integration:** Integrating intraoperative imaging into the surgeon's augmented view is possible with mixed reality systems, such as fluoroscopy or ultrasound. With the help of this integration, correct tumour resection, organ preservation, or implant placement is made possible by the real-time visualization of internal structures and the superimposition of imaging data onto the patient's anatomy.

6. **Telemedicine and Remote Collaboration:** Mixed Reality can help telemedicine and remote cooperation, enabling surgeons in various locations to interact and work together while doing surgeries. Using virtual annotations and real-time video streaming, specialists may direct and help surgeons in difficult instances, fostering knowledge exchange.

These techniques show how mixed reality has the potential to improve surgical accuracy, speed, and safety while also opening up new possibilities for surgical education and remote healthcare delivery. It's crucial to keep in mind that the field of mixed reality in medicine is still developing, and particular applications and procedures may differ based on the organization, the surgeon, and the available technology.

4.1 Simulation Steps

a. **Head-Mounted Displays (HMDs):** The aforementioned are head-worn devices that show virtual or augmented reality material, such as VR headsets. Built-in sensors such as cameras, accelerometers, and gyroscopes are frequently used to track the user's motions and head orientation.

b. **Hand-Held Controllers:** Such devices are gadgets that allow users to engage in a number of ways with Virtual or Augmented Reality Settings. They often include buttons, triggers, and thumb sticks that can be used to perform specific actions, such as moving objects or selecting options.

c. **Tracking Systems:** These are devices or systems that are used to track the user's movements and position in a virtual or augmented reality environment. They may include sensors such as cameras, infrared or ultrasonic sensors, or even specialized markers that are placed in the environment.

d. **Software:** This includes the programs and applications that run on the MR system and provide virtual or augmented reality content. These can include specialized development tools for creating MR experiences, as well as pre-made experiences such as games, simulations, or educational content.

e. **Hardware:** MR-enabled devices such as smartphones, tablets, and PCs, with specialized processors and graphics cards that can handle the computational demands of the VR and AR content.

4.2 Data Preparation

Among the most often used technologies for acquiring health-related data are:

1. Customer Relationship Management System (CRM) in which it generates information, handles generic data, and investigates a variety of issues.
2. Personal patient data is collected and analysed by electronic health record (EHR) systems to produce more profound conclusions.
3. Mobile applications are software programmes that run on mobile devices that link doctors to patients while also gathering and accessing data from many different databases.

The data collected from the patients by the above tools are converted into DICOM format which is a standard format for telemedicine.

The uncompressed DICOM information is far too huge to fit into a headset. As a result, DICOM data from the healthcare facility's PACS were quickly imported and pre-processed in a Cloud Unit (Azure Graphic Processing Unit - Microsoft Corp, Redmond, WA, USA). During the procedure, the DICOM data becomes accessible to the headsets via a specific radiological holographic programme that is connected to the Cloud Unit via Wi-Fi.

5 Mixed Reality

Figure 2 shows Mixed Reality processes information regarding reality using a set of sensors, cameras, and typically AI-enabled technologies to produce digitally augmented situations.

Whenever a person puts on a pair of mixed-reality glasses, the cameras and sensors in the spectacles communicate with a software programme that gathers as much information about the surroundings as possible, basically producing a virtual map of the actual world.

Using that map, MR technology may transmit holographic pictures along with information into the world.

To function properly, MR must be capable to monitor:

a. Interfaces and bounds of objects (through scene interpretation and geographical mapping).
b. An individual's posture and motions.
c. Tangible things and locales.
d. Environmental illumination and sound effects (to create authenticity).

Internet of Things, improved input sensing, and environmental perceptions enable Mixed Reality mechanisms to successfully blend the real and virtual worlds in ways that go transcend the fundamentals of Augmented Reality.

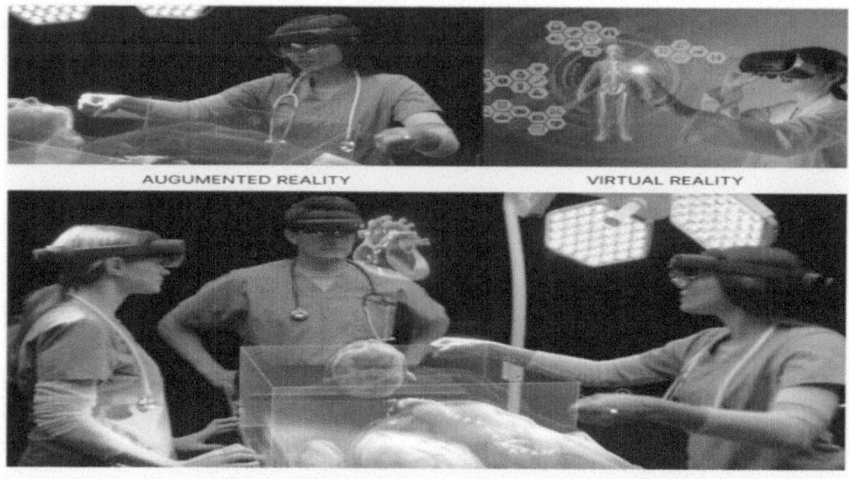

Fig. 2. Simulation of Augmented Reality and Virtual Reality in Surgical Process

5.1 Interaction and Visualization

We have made various interaction buttons infused to the device like rotate, zoom in, zoom out, move forward, move backward, and a lot more. With the air tap option, you may enable or disable all of these interaction methods. This makes the surgeon to better visualize the condition of the organ (Fig. 3 shown HoloLens Navigation tools).

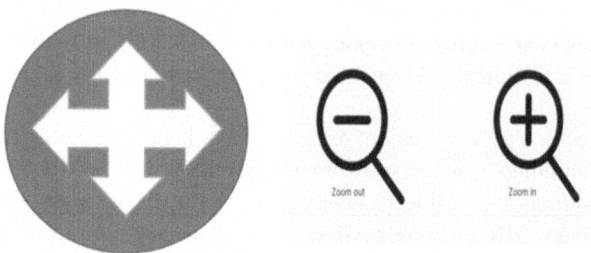

Fig. 3. HoloLens Navigation

Move, Rotate, and Scale: Using the aforementioned operations, the user may move, spin around, or scale a 3D surface model. Upon activation, the interactive buttons turn green.

MRI/CT Pictures (Medical Image Viewer): Seeing the CT/MR scans linked to the 3D models is a crucial interaction that is typically necessary. When this interaction is active, the matching CT/MRI image stack is loaded in a separate 2D viewer, and the user has the option to navigate throughout an image stack.

Chosen View: Since there are many models (such as lumps), only one portion may be shown at a time. This interaction enables for the separation of views.

DICOM (Digital Imaging and Communications in Medicine): Is a digital image format, non-proprietary data transmission protocol, and file structure for biological images and image-related data. Examples include CT scans, MRIs, and other types of imaging technology.

All the medical images of the patients are converted into a standard format (DICOM) and stored in the database called DICOM database which is above mentioned in the diagram. Also, there will be storing and retrieving of patient information between the main server and database (Fig. 4 highlights architecture diagram of medical surgery using mixed reality(MR) model view).

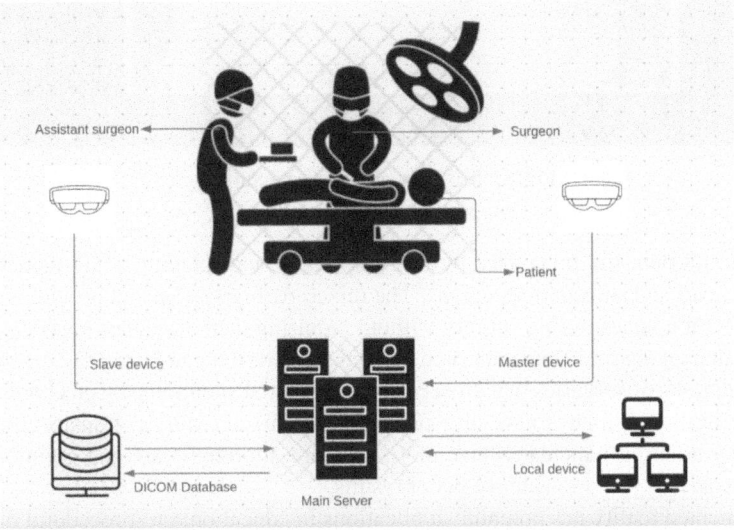

Fig. 4. Architecture Diagram of Medical Surgery using Mixed Reality

The local device is operated by the surgeon or assistant surgeon for the processing of data of the patient's information. The slave device operates at a remote location according to a command from the master device.

Now, let us see how to implement mixed reality in **Lung Surgery**.

During the surgery, real-time data from the patient, such as vital signs and imaging data, can be incorporated into the MR environment to provide the surgeon with real-time information and guidance.

Using the MR environment, the surgeon can perform the surgery while viewing the virtual lung model and real-time data. The MR environment can guide the location of the incision, the position of the surgical tools, and the placement of vital tissues such as blood arteries and nerves.

The MR environment may be utilised after surgery to monitor the patient's recuperation and offer real-time data on their status. This can help to identify potential complications early and allow for timely intervention (Fig. 5 visualize mixed reality in lung surgery).

The implementation of Mixed Reality in cardiac surgery is similar. The use of computer-generated graphics superimposed on the surgeon's field of vision during complex procedures is known as "mixed reality" in heart surgery. The surgeon can better

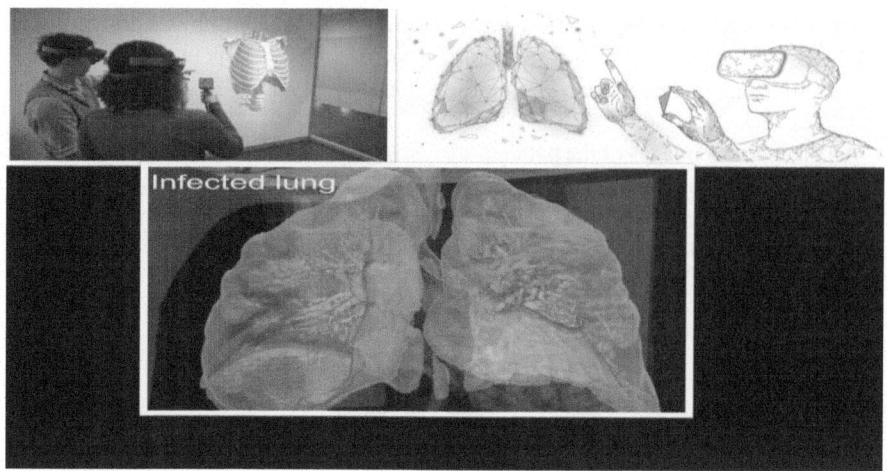

Fig. 5. Mixed Reality in Lung Surgery

visualize and plan the procedure by using a headgear displaying a 3D model of the patient's heart and surrounding organs. The mixed reality system can provide surgeons with constant updates during surgery without requiring them to glance away from their work by displaying real-time data such as vital signs or oxygen levels. By offering virtual simulations of surgeries that closely resemble actual situations, this technology can also help in the training of new surgeons. Overall, mixed reality is transforming heart surgery by assuring patients receive more accurate, thorough care and giving doctors more extensive support to achieve better results.

Augmented reality has potential applications in education, pre-procedural planning and simulation, and procedural counselling. Surgery or minimally invasive cardiac interventions benefit greatly from augmented reality. It was shown that percutaneous mitral valve repair and determining the ideal annuloplasty ring size (especially for surgeons with little expertise) could both benefit from augmented reality-enhanced transoesophageal echocardiography. Another study found that incorporating ECG-gated CT data in augmented reality allows for the use of less contrast agents during valve-in-valve surgery and transcatheter cardiac valve implantation. Using augmented reality, a transcatheter pacemaker was installed in a patient with congenitally corrected transposition of the great arteries. Surgeons have been helped by mixed reality utilizing HoloLens to better grasp the intricate morphologies of CHD [12].

5.2 Cardiac Surgery in Mixed Reality

In addition, mixed reality (MR) technology can be utilized to guide surgeons during cardiac surgery and give them a more accurate and thorough perspective of the patient's anatomy. The procedures for conducting cardiac surgery in mixed reality are as follows:

a. **Data Collection:** A 3D model of the patient's heart and surrounding structures is produced using the patient's medical imaging data, such as CT or MRI scans.

b. **MR Headset Setup:** A 3D model of the patient's heart and nearby structures can be seen in real-time, superimposed on the surgeon's actual image of the patient's body, thanks to the use of an MR headset.

c. **Surgical Planning:** The surgeon can carefully plan the surgical operation, including the positioning of incisions and the trajectory of surgical instruments, using the MR headset and other tools.

d. **Intraoperative Guidance:** The MR headset gives the surgeon real-time guidance and visualization during the procedure, enabling them to move more precisely and accurately across the patient's anatomy.

Using MR, surgeons can communicate with other specialists remotely and work together in real-time to enhance accuracy and results.

Heart surgery can benefit from the use of mixed reality in several ways, including a more precise and minimally invasive approach to surgery, a lower risk of complications, and quicker recovery times. But it's crucial to remember that, as with any medical operation (Fig. 6 exposes mixed reality in cardiac surgery).

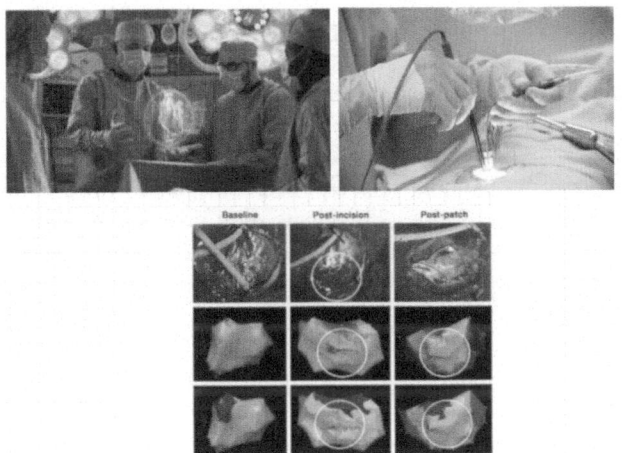

Fig. 6. Mixed Reality in Cardiac Surgery

5.3 Scanning for Breast Surgical

We created an interface for presenting supine breast MRI overlay on the patient using the HoloLens during planning for surgery to compare the location-specific accuracy of the HoloLens tumour representations to conventional localization approaches. In addition, a pilot trial in patients with palpable tumours is now underway [14]. We also conducted perceptual accuracy studies to compare the symmetry of 2D holograms presented on graph paper to known coordinates where the holograms should be displayed. We created a pipeline for presenting supine breast MRI overlay on the patient using the HoloLens during surgical planning to compare the spatial accuracy of the HoloLens tumour representations with conventional localization approaches. In order to compare the symmetry of 2D holograms projected on graph paper to known coordinates where the holograms should be displayed, we also conducted perceptual accuracy experiments. Because the surgeon utilises the predicted hologram's form to generate an outline of the tumour for our surgical planning application, it is critical to keep the tumour's shape as well as its placement in space. We conducted brief research in which six individuals were asked to sketch the contours of presented holograms on graph paper, as reported by Srinivasan et al. (2017). Early in the year, researchers evaluated our mixed-reality system's 2D perceptual accuracy [15].

Before donning the HoloLens and executing the alignment task, each participant had their interpupillary distance (IPD) measured. Subjects may be able to enhance this alignment by utilising voice commands to manipulate the rendered ArUco tag holograms to account for errors caused by biometric variation. Further voice commands produced holograms with thirty randomly positioned dots and eight more graphics.

5.4 Scanning for Brain Surgery

The preoperative planning of tumour localization with the HoloLens was compared to conventional neuronavigation in the operating theatre in patients who needed brain tumour surgery. Using magnetic resonance imaging and the HoloLens, 3-dimensional holograms of the patient's head and tumour were produced, which were then projected onto the actual patient's head. The 2-dimensional projection of the tumour boundaries on the patient's head skin that the neurosurgeon saw was created using both the HoloLens and neuronavigation. As the gold standard, neuronavigation was used to evaluate the precision of Hololens localization. 25 patients were involved in this investigation. In each instance, holograms were effectively made. The median difference between tumour localization using the HoloLens and conventional neuronavigation in nine patients was 0.4 cm (interquartile range: 0–0.8). With quantitative result evaluations, this progressive clinical investigation provides proof of conception for the practical viability of using the HoloLens for surgical planning of brain tumour surgery. To increase the accuracy of this wearable mixed-reality technology, more research is required. This prospective clinical investigation provides proof of concept for the practical use of Hololens for operating room brain cancer surgical planning and includes quantifiable outcome data. The accuracy and clinical usability of this smart mixed reality gadget must be improved [18].

5.5 Mixed-Reality Guiding for Depression Therapy with Brain Stimulation

More than 16 million adult Americans suffer from depression, and more than half do not improve with treatment. A significant antidepressant therapy known as transcranial magnetic stimulation (TMS) targets particular brain circuits in charge of mood and behaviours.

The relative position of the TMS coil directly affects both the efficacy and the danger of TMS, and correct neuronavigation can significantly lessen both impacts. In this post, we offer development tools for an original hybrid virtual reality neuronavigation system that would let a TMS practitioner visualise the patient's brain architecture superimposed right on the head. This is accomplished by combining tracking patients with brain magnetic resonance imaging (MRI) imagery to produce an improved and straightforward depiction.

Transcranial magnetic stimulation (TMS) is still a significant therapy option for depressive symptoms even if 52% of patients do not react to medication and 33% do not reach remission. Repetitive TMS (RTMS) has been proven to totally alleviate depression in 34% and 38% of patients by delivering trains of stimulation pulses to the dorsolateral prefrontal cortex (DLPFC). The two most popular techniques for establishing coils for the DLPFC are the "5 cm rule" and the Beam approach. The current research presents experimental outcomes from an innovative use of already-existing technology (MRI, TMS, Augmented Reality, and Tracking) for a simplified neuronavigation for depression TMS treatment. MR imaging, the Microsoft Hololens virtual reality headset, and the Intel Real sense camera are used to assess the patient's brain structure [19].

Technologies that project MRI images onto the head of the patient to direct TMS brain stimulation for depression. A holographic depiction of the TMS device and a 3D volumetric simulation of the subject's brain can be used to guide TMS coil location. The marker-based system works well as a proof-of-concept, but calibration using fiducial markers is necessary for more accurate marker placement. Face monitoring calculates the head posture from anatomical areas and updates the pose of the 3D head depiction in the Hololens display space. More head-to-hologram accuracy measurements are necessary to improve display accuracy. By linking the RGBD camera to a tiny portable PC, the setup would become wireless, allowing surgeons great flexibility and making internal organ location simpler [20].

5.6 MRI Before Surgery

The patient has a preoperative supine magnetic resonance imaging (MRI) the day prior surgery to identify the tumour using an intravenously administered gadolinium-based contrast agent. Six MR-visible fiducial markers are applied to the patient before the scan at various locations throughout the breast.

The positions of these biomarkers are documented during MR dataset post-processing, and the tumour and skin are partially automatically separated using active contour segmentation using ITKSNAP3. The MR pictures, marker placements, skin and tumours subdivision mesh, as well as other materials, are all imported into a Unity 3D project before it is constructed and posted as an "app" to the Microsoft HoloLens.

The use of transmitted visuals by doctors in locating a tumour, in our demonstration of a mixed reality medical application, it occurs during surgical planning for a breast lumpectomy. Every year, due to positive margins, hundreds of thousands of patients who have lumpectomies need a second procedure. This has a severe psychological impact and causes many women who are diagnosed with breast cancer to postpone therapy.

Preliminary study suggests that mixed reality could have an impact. We are at present attempting to improve the body registration of the holograms, with a focus on accounting for breast deformation changes caused by patient their position, as well as perceptual changes caused by the HoloLens' placement on users with a variety of biometric variations (such as eye spacing and position).

Mixed Reality has significant medical applications beyond breast surgery. Extensions of these approaches may also be useful for head and neck surgery, orthopaedics surgery, and surgeries on other solid organ tumors. Mixed reality has a lot of promise to change medicine and help a lot of patients in the future, and the surgeons at our institution are excited about adding it to their surgeries [17].

5.7 Graphical Representation

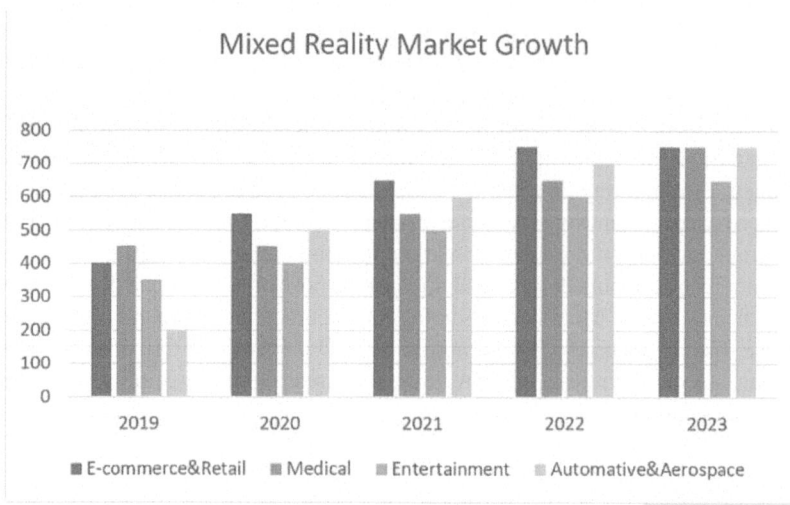

Fig. 7. Mixed Reality Market Growth

The above graph shows the market growth of mixed reality for the past years from 2019–2023. According to a Technology Research Future analysis on mixed reality (MR) development, the space would increase at a CAGR of 43.26% during 2022 and 2030. Ideas like holographic collaboration are becoming more feasible thanks to the development of MR technologies like the Meta Quest Pro. MR also paves the way for software like 3D and virtual modelling to reduce the resource and cost needs of innovation (Fig. 7 shows analysis of mixed reality market growth).

Once MR technologies are widely used, we can find that innovation in various other tech spheres also picks up speed. Teams will be able to work more freely with experts around the world because of MR. Workers can communicate with virtual blueprints and designs that don't require expensive processing thanks to devices.

The predictions estimate the CAGR for the MR industry to be 43.26%. The expansion is the result of more investments in platforms, software, and hardware. The new, more easily available technology is being introduced to the market by businesses like PICO, Microsoft, and metadata.

6 Result and Discussion

The outcomes of this research can provide significant insights on the acceptability of mixed reality technology and research trends in surgery. These results can aid physicians in assessing future mixed reality surgical medical research. These statistics may provide an academic context for choosing whether to embrace new surgical technologies. Hospitals may also periodically assess the maturity of mixed reality technology in surgery.

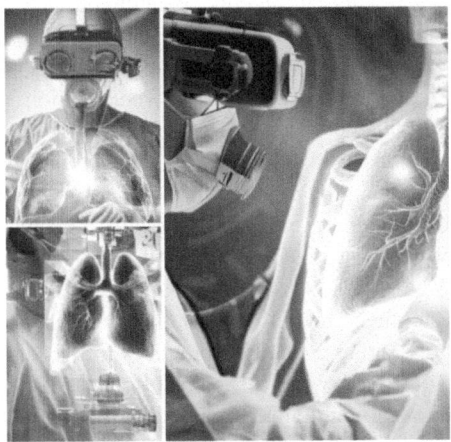

Fig. 8. 3D Modelling View of The Mixed Reality Surgery

Especially compared to standard surgical techniques, using a Mixed Reality headset could enhance outcomes for both the surgeon and the patient without jeopardising the treatment's safety. Because of 3-D holograms, the surgeon may observe the keyframes of the process as well as essential organs (nerves, arteries…) in real-time. This enhances procedural precision and safety, potentially reducing time and allowing for more exact implant placement. In particular, despite the reality that novice users could find it tedious, the technical installation of the surgical instrument and data preparation before to surgery only took a few minutes. Figure 8 shows 3D Modelling research reveals that mixed reality constitutes a paradigm change in the area of surgery. The combination of immersive visualisation, real-time feedback, and better training capabilities enables surgeons to conduct treatments with more precision and confidence. As MR technology advances and becomes more widely available, its disruptive influence on surgical practise is likely to grow, eventually enhancing patient care and outcomes.

7 Conclusion

In conclusion, mixed reality (MR) has emerged as a viable surgical technique. It has the potential to transform the way procedures are planned, conducted, and monitored to improve patient outcomes. Surgeons may use MR technology to generate virtual models of patients and their anatomical features that can be viewed and modified in three dimensions, allowing for a more thorough and interactive approach to surgical planning and simulation. MR can be utilized during surgery to overlay virtual models over real-time patient data, offering real-time advice on the location of incisions, instrument placement, and identification of essential anatomical features. This can assist to prevent problems, increasing accuracy, and improving patient safety. Furthermore, MR may be utilized to monitor the patient's health in real-time throughout surgery, giving crucial feedback to the surgical team. The use of MR in surgery is still in its early stages, and further research is necessary to properly understand its potential advantages and limitations. For the broad use of MR in surgical practice, challenges such as the expense of MR equipment, the requirement for specialized training for surgical teams, and regulatory issues must be addressed. Finally, magnetic resonance imaging (MRI) has the potential to revolutionize the area of surgery by improving visualization, planning, and guiding during treatments.

Finally, this study demonstrates the promise of mixed reality (MR) technology as a revolutionary and game-changing strategy in medical surgery operations. The combination of virtual and augmented reality has various advantages, including greater surgical precision, improved visualization, real-time assistance, and advanced training possibilities.

More research, development, and clinical trials are required to fully investigate the potential of MR in surgery and to create best practices for its safe and successful application. We should anticipate witnessing a growing effect of MR in the area of surgery as MR technology advances, leading to improved patient outcomes and enhanced surgical procedures.

8 Future Work

Future improvements should concentrate on standardization, cost-effectiveness analysis, long-term outcome assessment, accuracy, efficiency and interdisciplinary cooperation. These findings demonstrate how powerfully MR can change surgical procedures and elevate patient care using Hybrid Deep Learning Techniques in Artificial Intelligence. The Future Enhancement to implement in Hybrid Deep Learning Mixed Reality (HDLMR) Technology will be fuelled by ongoing enrich research and development, which will boost surgical accuracy, patient outcomes, and healthcare effectiveness.

References

1. Hu, H.Z., Feng, X.B., Shao, Z.W., Xie, M., Xu, S., Wu, X.H., Ye, Z.W.: Application and prospect of mixed reality technology in medical field. Curr. Med. Sci. **39**(1), 1–6 (2019). https://doi.org/10.1007/s11596-019-1992-8
2. Gao, Y., Tan, K., Sun, J., Jiang, T., Zou, X.W.: Application of mixed reality technology in visualization of medical operations. Chin. Med. Sci. J. **34**(2), 103–109 (2019). https://doi.org/10.24920/003564. PMID: 31315751
3. Gregory, T.M., Gregory, J., Sledge, J., Allard, R., Mir, O.: Surgery guided by mixed reality: presentation of a proof of concept. Acta Orthop. **89**(5), 480–483 (2018)

4. Dash, S., Shakyawar, S.K., Sharma, M., Kaushik, S.: Big data in healthcare: management, analysis, and prospects. J. Big Data **6**(1), 1–25 (2019)
5. Rudrapatna, V.A., Butte, A.J.: Opportunities and challenges in using real-world data for health care. J. Clin. Investig. **130**(2), 565–574 (2020)
6. Kumar, R.P., Pelanis, E., Bugge, R., Brun, H., Palomar, R., Aghayan, D.L., Fretland, Å.A., Edwin, B., Elle, O.J.: Use of mixed reality for surgery planning: assessment and development workflow. J. Biomed. Inform. **112**, 100077 (2020)
7. Gerup, J., Soerensen, C.B., Dieckmann, P.: Augmented reality and mixed reality for healthcare education beyond surgery: an integrative review. Int. J. Med. Educ. **11**, 1 (2020)
8. Sadeghi, A.H., El Mathari, S., Abjigitova, D., Maat, A.P., Taverne, Y.J., Bogers, A.J., Mahtab, E.A.: Current and future applications of virtual, augmented, and mixed reality in cardiothoracic surgery. Ann. Thorac. Surg. **113**(2), 681–691 (2022)
9. Lu, L., et al.: Applications of mixed reality technology in orthopedics surgery: a pilot study. Front Bioeng. Biotechnol. **22**(10), 740507 (2022). https://doi.org/10.3389/fbioe.2022.740 507.PMID:35273954;PMCID:PMC8902164
10. Kumar, R.P., Pelanis, E., Bugge, R., Brun, H., Palomar, R., Aghayan, D.L., Fretland, Å.A., Edwin, B., Elle, O.J.: Use of mixed reality for surgery planning: assessment and development workflow. J. Biomed. Inform. **112**, 100077 (2020). https://doi.org/10.1016/j.yjbinx. 2020.100077
11. Han, J., Kang, H.-J., Kim, M., Kwon, G.H.: Mapping the intellectual structure of research on surgery with mixed reality: bibliometric network analysis (2000–2019). J. Biomed. Inform. **109**, 103516 (2020). https://doi.org/10.1016/j.jbi.2020.103516
12. Goo, H.W., Park, S.J., Yoo, S.-J.: Advanced medical use of three-dimensional imaging in congenital heart disease- augmented reality, mixed reality, virtual reality and three-dimensional printing. Korean Soc. Radiol. **21**, 133 (2018). https://doi.org/10.3348/kjr.2019.0625
13. Lau, I., Gupta, A., Ihdayhid, A., Sun, Z.: Clinical applications of mixed reality and 3D printing in congenital heart disease. Biomolecules **12**, 1548 (2022). https://doi.org/10.3390/biom12 111548
14. Boetes, C., et al.: Breast tumors: comparative accuracy of MR imaging relative to mammography and the US for demonstrating extent. Radiology **197**(3), 743–747 (1995)
15. Carbonaro, L.A., Tannaphai, P., Trimboli, R.M., Verardi, N., Fedeli, M.P., Sardanelli, F.: Contrast-enhanced breast MRI: spatial displacement from prone to supine patient's position. Preliminary results. Eur. J. Radiol. **81**(6), e771–e774 (2012)
16. Carter, T., Tanner, C., Beechey-Newman, N., Barratt, D., Hawkes, D.: MR navigated breast surgery: Method and initial clinical experience. In: MICCAI 2008: 11th International Conference, Proceedings, Part II, pp. 356–363. International Conference on Medical Image Computing and Computer-Assisted Intervention, New York (2008)
17. Perkins, S.L., Lin, M.A., Srinivasan, S., Wheeler, A.J., Hargreaves, B.A., Daniel, B.L.: A mixed-reality system for breast surgical planning. In: IEEE International Symposium on Mixed and Augmented Reality (ISMAR-Adjunct) (2017). https://doi.org/10.1109/ismar-adj unct.2017.92
18. Incekara, F., Smits, M., Dirven, C., Vincent, A.: Clinical feasibility of a wearable mixed-reality device in neurosurgery. World Neurosurg. **118**, e422–e427 (2018)
19. Greener, R.: The Market Growth of Mixed Reality - XR Today: XR Today (2022). www.xrt oday.com/mixed-reality/the-market-growth-of-mixed-reality
20. Leuze, C., Yang, G., Hargreaves, B., Daniel, B., McNab, J.A.: Mixed-reality guidance for brain stimulation treatment of depression. In: IEEE International Symposium on Mixed and Augmented Reality Adjunct (ISMAR-Adjunct), Munich, Germany, pp. 377–380 (2018). https:// doi.org/10.1109/ISMAR-Adjunct.2018.00109

21. Roopa, D., Prabha, R., Senthil, G.A.: Revolutionizing education system with interactive augmented reality for quality education. Mater. Today Proc. **46**(9), 3860–3863 (2021). https://doi.org/10.1016/j.matpr.2021.02.294

22. Prabha, R., Senthil, G.A., Lazha, A., VijendraBabu, D., Roopa, D.: A novel computational rough set based feature extraction for heart disease analysis. In: I3CAC 2021: Proceedings of the First International Conference on Computing, Communication and Control System, I3CAC (2021), Bharath University, Chennai, India, p. 371. European Alliance for Innovation (2021). https://doi.org/10.4108/eai.7-6-2021.2308575

23. Prabha, M.R., Prabhu, R., Suganthi, S.U., Sridevi, S., Senthil, G.A., Babu, D.V.: Design of hybrid deep learning approach for Covid-19 infected lung image segmentation. J. Phys. Conf. Ser. **2040**(1), 012016 (2021). https://doi.org/10.1088/1742-6596/2040/1/012016

24. Prabha, R., Razmah, M., Veeramakali, T., Sridevi, S., Yashini, R.: Machine learning heart disease prediction using KNN and RTC algorithm. In: International Conference on Power, Energy, Control and Transmission Systems (ICPECTS), Chennai, India, pp. 1–5 (2022). https://doi.org/10.1109/ICPECTS56089.2022.10047501

Data Mining-Based Classification Algorithms for Predicting Mental Health

K. Vijay, P. T. S. Shahul Hameed[(✉)], M. Bhavani, and M. Jaeyalakshmi

Department of Computer Science and Engineering, Rajalakshmi Engineering College, Chennai,
India
shahulhameed.pts.2019.cse@rajalakshmi.edu.in

Abstract. The rapid and diverse development of society has led to increased pressures on living conditions, education, and employment, resulting in significant psychological challenges. Addressing the importance of improving mental health education for children has become a subject of widespread concern and crucial for society as a whole. This proposal aims to introduce a novel system based on a common optimization algorithm for evaluating mental health intelligence, addressing the limitations of existing procedures, such as low work efficiency and high misjudgment rates. By combining artificial neural network (ANN) algorithms and extended decision trees, a comprehensive optimization method is proposed. The research begins by examining the current state of intelligence measurement in mental health, utilizing data mining to gather information from mental health intelligence tests. The gathered data is then analyzed and categorized using compound learning algorithms. The system's effectiveness and superiority are assessed through a series of unique simulation experiments. This research not only addresses the shortcomings of current systems but also proposes innovative strategies for mental health intelligence scoring to enhance precision, effectiveness, and other essential characteristics. Ensuring stability and meeting the needs of users are also primary objectives of this system. Overall, this proposal strives to contribute to the advancement of mental health assessment methodologies and education, benefiting individuals and society as a whole.

Keywords: Decision Tree · Machine Learning · ANN · Data mining · Precision · Psychology · Health Intelligence · Mental Health

1 Introduction

People are under pressure to survive because of the quick development of modern society. As a result of the increased stress brought on by social competitiveness, an increasing number of people experience psychological issues, including emotional and behavioral issues. Every year, there are cases of mental illnesses and disorders [1]. There was a sharp increase in the number of people suffering from mental illness. Dealing with the psychological issues that people experience, which have a variety of reasons, is currently the most crucial responsibility. There are two approaches that need to be investigated: how to better the outcomes of mental health and how to offer mental health services to people. Modern technology is required to increase the level of mental work's scientific rigor. The establishment of a sophisticated information-based system for mental health assessment is essential [2–5]. As consumer mental health improved, so did data gathering and mental health data collection. Its job is incredibly one-sided and only partially applicable from a managerial standpoint. Scales or questionnaires are typically used in traditional mental health assessment approaches to gather data. The nature of the survey items and the reliability of the findings are subject to some restrictions. Since the creation of the Internet, network technology has become more frequently used for mental health assessments. Survey participants' propensity to self-evaluate in a more private setting can lessen psychological strain. Common Internet software monitors user status to varied degrees.

It is evident that internet rating systems are excellent instruments for evaluating college students' mental health [6]. An automated rating system for users of online forums was developed using a rating framework, a hierarchical analysis of user mental health data, and a variety of strategies. The process requires time and it was created as a web-based solution for counseling college students' mental health. To enable users to more effectively deliver psychological services, carefully assess the demands of users and segment the system into various functional groups. The recall rate is not that high, though [21–25].

1.1 Machine Learning

The growing use of machine learning in the area of mental health is an important growth trend. Machine learning is a method for automatically evaluating or predicting new and unknowable facts by finding patterns and regularities in enormous amounts of data and information. Suppose that, in line with Tom's official definition of machine learning, performance is used to assess how well a computer programme performs on a particular task if the programme improves the task's performance through the application of experience [7–11]. Text data is the only type of data that is currently appropriate for machine learning. Social media, clinical reviews and records, electronic medical records, and publications are a few of the sources. Particularly social media may offer a wealth of information on users' psychological and behavioral footprints. Machine learning-based automatic classification is more precise than manual classification [20]. Currently, it is frequently used to forecast emotional, cognitive, and addictive behavior data on behavior and physiology; this data comes mostly from eye tracking, wearable mobile devices, physiological multichannel recording devices, and expression analysis systems. Machine learning algorithms come in three different varieties:

1. Controlled class
2. Semi-supervised learning
3. Original investigation

Use supervised learning to ascertain correlations between data and features and labels, and training to assess new data. The model can forecast when new data will be labeled. For instance, based on brain imaging data, machine learning algorithms can automatically separate patients with traumatic brain injuries from healthy people and distinguish between various illnesses [12, 20]. The data mining-based mental health intelligent evaluation system that can efficiently advance system information management, address issues with out-of-date system resources, fully utilize the advantages of the internet, create a new internet environment, and effectively address psychological issues faced by contemporary individuals [18, 19].

2 Related Work

Data mining-based classification algorithms can be used to predict mental health outcomes based on various factors such as demographics, behaviors, and clinical history. Some of the commonly used classification algorithms in this area include decision trees, logistic regression, k-nearest neighbors (KNN), support vector machines (SVM), and neural networks. Decision trees are popular in mental health prediction because they are easy to interpret and provide insight into the most important predictors of mental health outcomes. Logistic regression is another commonly used algorithm that can be used to estimate the probability of a person developing a mental health condition based on specific risk factors. KNN and SVM are both effective at classifying individuals into groups based on their characteristics and can be used to identify high-risk groups for mental health conditions. Finally, neural networks are a powerful tool for analyzing complex datasets and can be used to identify patterns and relationships that may be difficult to detect using other algorithms.

The exact dataset being examined and the research issue being addressed will both have an impact on the algorithm that is selected? Before making a decision, it is crucial to carefully weigh the advantages and disadvantages of each algorithm. In order to guarantee the quality and dependability of the results, it is also crucial to make sure that the data utilised for analysis is representative of the population being studied and that the proper data pretreatment measures are followed.

Electronic medical records are more time-efficient than the currently-used conventional storage systems, but they introduce a single point of failure for patients due to their centralised data storage. Encryption algorithms are an integral part of the medical management system, ensuring that only authorised users may gain access to sensitive patient data [12, 13].

Traditional psychological research has a small sample size, making it difficult to conduct studies with successful predicted outcomes [17].

The results of certain research are inconclusive and unclear because of the poor quality of the data and the absence of covariate information. This issue also arises with value-based significance tests when trying to replicate study results. Machine learning cross-validation techniques have a tremendous deal of potential to increase the reproducibility of psychological investigations when used rigorously and consistently. Machine learning based on technology can build learning models from a lot of data [20–22].

The sudden and dramatic increase in the number of patients during the coronavirus (COVID-19) pandemic caused the closure of all educational institutions and the introduction of online teaching, e-learning, and virtual gatherings. Predicting QoE in online learning environments is crucial because of the significance of behavioural aspects of teaching and learning between instructors and students. In this study, we introduce a novel prediction model for employing data mining to uncover instructional and e-learning technologies in online classrooms. Effective QoE parameters in online classrooms are identified using association rules mining and supervised approaches. The experimental results demonstrated how the proposed prediction model has sufficient accuracy, precision, and recall for forecasting how students will behave during e-learning and online classes [23].

Several researchers are using machine learning to forecast the complicated psychological issues that humans may encounter. An anxiety condition or stress-related hunch. There are several features that can be used to predict suicide risk. On the social media site Weibo, they are building a suicide detector. This detector rates Weibo user suicides in real time using a multilayer perceptron algorithm. The probability prediction accuracy can be as high as 95%. Carpenter evaluated the potential for anxiety in kids between the ages of two and five using information from preschool psychiatric research and machine learning. That generalized anxiety disorder and separation anxiety had the highest predictive accuracy was 90% [24, 25].

3   Methodology

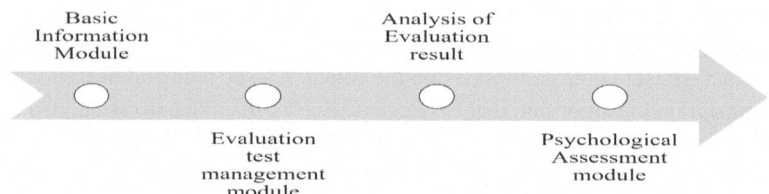

Fig. 1. This figure depicts the flow of program across various modules of the algorithm

Typically, machine learning does not create associations or make conclusions about unrelated things. It's crucial to design and pick specialized psychiatric interventions. In order to help people accurately comprehend their mental health while keeping the scientific and informational biases of counseling to promote mental health, this research investigates a decision tree algorithm. Creating an intelligent system for mental health assessment based on the tool gives a complete unbiased picture of mental health. The overall system methodology is illustrated in Fig. 1.

Data is collected from various people consisting of different age groups through google forms. Google forms with a series of 10 questions related to the individual's mental health has been created and distributed among the people. These questions contain both negative and positive aspects of life. By answering these questions one can find the happiness index of individuals. The questions have binary response yes or no using which we have created a dataset consisting of binary values 0 and 1 and the dataset was labelled manually according to the response given by the individuals. The collected dataset is given as input to the proposed algorithm.

Data sources are insufficiently diverse; the proposal's participants' on-campus behavioral data are only partially given because only network behavioral data were employed. The objective of the following phase will be to provide more information on campus network behavior in order to enhance the network behavioral data. Using behavioral data from the use of meal cards and library borrowing, perception models of students' mental states are created.

There are not enough feature dimensions. The examination of novel feature dimension creation methods follows. The feature dimension data now in use were created using just two regularity measures and a dependence measure. Future research will examine other intermediate variables based on network behavioral data. Its existence always improves the sample dataset's feature dimension.

Model browsing's size is insufficient. Only two machine learning classification models are used in this proposal. Using deep learning network structure models for model testing, the author's model's performance is assessed.

3.1 Decision Tree Algorithm

Using the data mining approach to psychological evaluation, the client software of this system produces evaluation findings. Once the user rating score data has been processed, a database has been created, the rating results have been analyzed using a decision tree algorithm, and ratings have been generated. The processes in the data mining procedure for psychological assessment include data extraction, cleaning, model selection, and output. Based on the user's responses, the system creates the first record. A mental health assessment dataset is obtained after preprocessing of data integration, extraction, cleansing, and transformation. The data mining process is illustrated in Fig. 2.

A decision tree algorithm is used in data mining techniques. Data classification uses mining findings to accurately classify user data. Decision tree algorithms are inductive learning algorithms that generate a collection of chaotic examples based on examples in order to establish categorization rules. Decision trees are typically used to solve categorization issues in two steps. The training and training sets are used to create a decision tree classification model, which is then used to categorize samples of unclassified categories. For a certain sample type, the classification process begins at the root node, ends

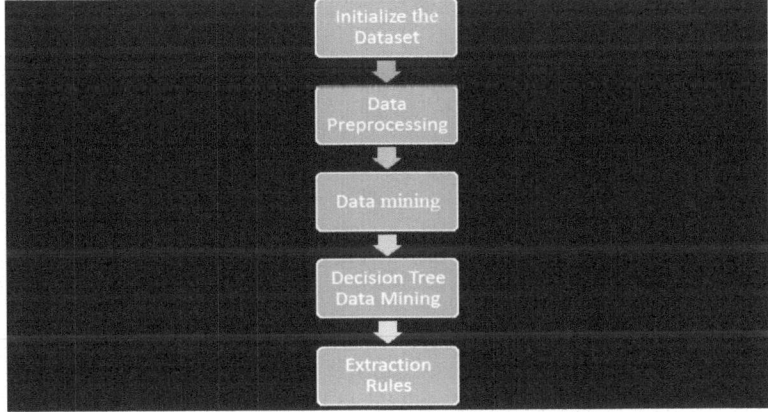

Fig. 2. This figure illustrates the workflow of decision tree algorithm

at the leaf nodes, and then steps through the sample's attributes in accordance with the branches.

Decision trees are constructed using quality criteria based on information theory. Several calls to library functions throughout the calculation greatly lengthen the calculation time since the information retrieval rate formula uses many logarithmic operations. It substantially simplifies the computation of the method by using the Taylor and McLaughlin mathematical formulas to calculate the algorithm's information gain rate. This approach is offered as a fix for the issue. The new formula's analysis reveals that the categorical information entropy is constant regardless of how the state attribute's information acquisition rate value is determined. The classification accuracy in this proposal by leaving out each component to employ the algorithm's formulas.

3.2 ANN Algorithm

ANNs are computational and structural models that imitate biological brain networks. Employed most frequently when estimating or approximating functions. A neural network's main layers are input, hidden, and output. In reality, the network has one neuron in the output layer and one neuron for each characteristic in the input layer. Two output neurons are present in an issue of multiple classification. The buried layers' thickness and number of neurons are carefully regulated. The output of the neuron is produced by comparing the total input received by the neuron to its threshold after processing the activation function. There are just two outcomes: 0 and 1. The neuron will not be activated by a value of 0. The neuron is activated with a value of 1. It typically selects the sigmoid function as the activation function since the step function is erratic and unreliable. The function's value varies from 0 to 1, as you can see from the function diagram. In other words, 0 to 1 is the range of function values. Input values that vary across a greater range during the interval can be compressed using the sigmoid function, commonly known as the squashing function. A neural network is created by merging several neuron units at specific levels and the structure of ANN is explained in Fig. 3.

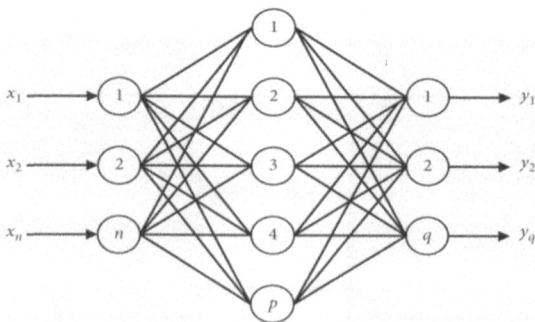

Fig. 3. This figures depicts the working of perceptron

Every arbitrary nonlinear function can be approximated using ANN's 3-layer perceptron and nonlinear optimization capabilities. But, ANNs also require development in other areas.

- There have been numerous training sessions, and the convergence rate is just slowly increasing. ANN algorithms need to be trained extensively in order to perform well in basic and common settings. Longer training sessions are necessary in more difficult conditions. When we use gradient descent to reduce the loss function, the result is invariably a jagged image and poorer algorithm performance. It can appropriately raise the impulse term, improve the error function, and adaptively alter the learning rate in order to more effectively address this issue and speed up the convergence of the ANN network.
- There is no guarantee of the global optimum, and it is simple to settle for the local optimum. As a result of the gradient descent strategy used by the backpropagation method, the weight space is a parabola with numerous minimum points. As a result, there is no single solution that works for all training methods.

The study of ANN improvement is currently quite active. Some of the most well-liked optimisation strategies include Conjugate Gradient Algorithm, Newton Algorithm, Levenberg-Marquardt Algorithm, Additional Impulse Algorithm, Variable Algorithm, Adaptive Learning Rate Method, RPROP Method, and others. Of the aforementioned algorithms, the LM algorithm has the highest durability and the quickest convergence speed.

4   Results and Discussion

The suggested joint optimisation model's enhanced decision network and enhanced ANN are depicted in Figs. 4 and 5. The findings of a comparative experiment based on a single upgraded model, which was conducted to verify the effectiveness of this joint model. With the lowest error index and maximum accuracy index, the joint optimisation model may deliver the best performance, as illustrated in Fig. 6. The accuracy and dependability of the joint optimization model are confirmed by its performance, which is superior to

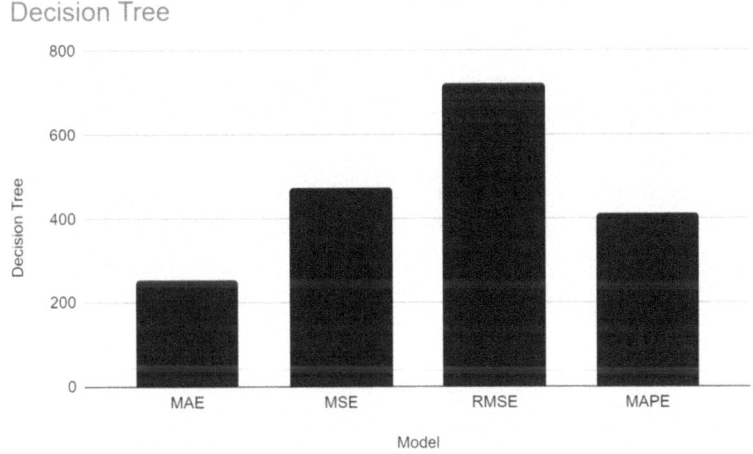

Fig. 4. Analysis of Enhanced Decision Tree

any single improved model. It is also clear that while ANN's performance is greater than the IDT model's, it still falls short of the joint optimization model described in Fig. 7.

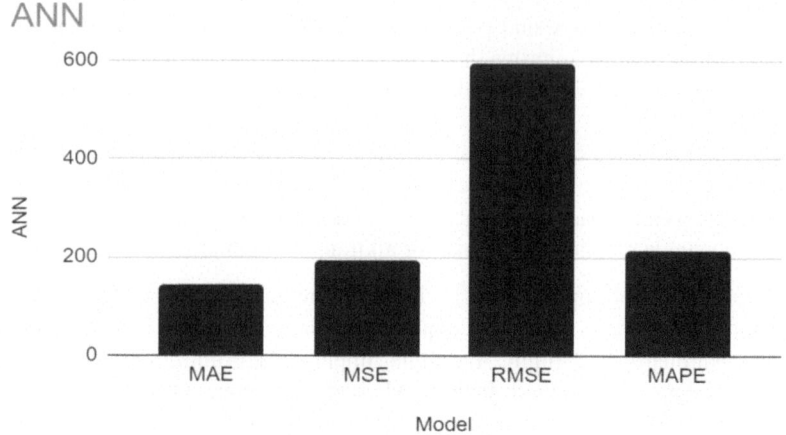

Fig. 5. Analysis of Enhanced ANN

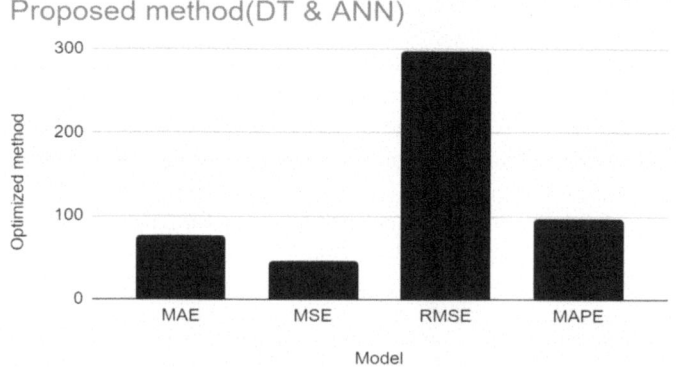

Fig. 6. This bar chart depicts the Analysis of Optimized Model

Additionally, it enhances the initial ANN and decision tree using the corresponding technique. A comparative experiment was also conducted to confirm the efficacy of this improvement measure, and the results are displayed in Fig. 6.

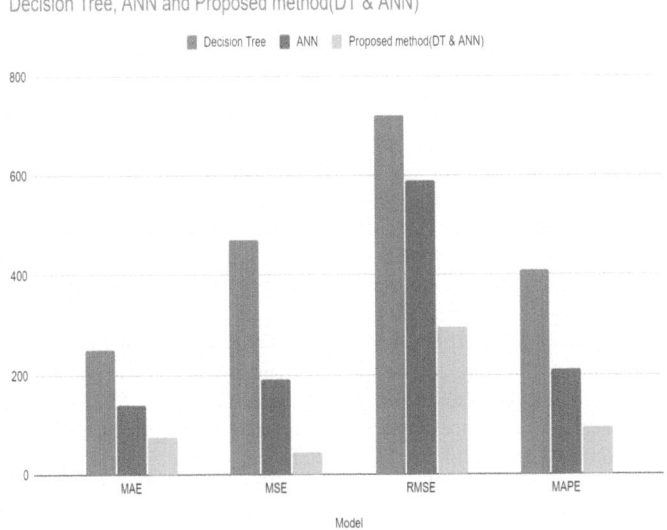

Fig. 7. This graph shows Comparison Analysis of DT, ANN and Proposed Model

5 Conclusion

This proposal aims to enhance the effectiveness of mental health intelligence assessment by integrating an optimized decision tree algorithm with an optimized artificial neural network. In conclusion, the integration of an optimized decision tree algorithm with an optimized artificial neural network holds great promise in enhancing mental health intelligence assessment. The proposed approach can empower consumers by providing valuable insights into their mental well-being, assisting them in resolving psychological issues, and fostering self-esteem.

The presented proposal marks the initial phase of an extensive journey towards creating a robust and reliable mental health intelligence assessment system. Future enhancements encompass expanding the dataset to improve accuracy and generalizability by ensuring diverse representation. Further research is needed for optimizing the decision tree algorithm and artificial neural network through hyperparameter tuning and exploring advanced techniques. Real-time monitoring and support should be considered by incorporating wearable devices or mobile applications for continuous mental state assessment and timely interventions. Addressing ethical concerns regarding data privacy and consent is critical, emphasizing strong data protection measures. A user-centered design approach involving in-depth user studies and feedback will lead to a more intuitive system. Collaboration with mental health professionals is essential for validation and alignment with established mental health practices. Lastly, conducting a comprehensive long-term impact evaluation will measure the system's effectiveness in promoting well-being and addressing psychological issues over time.

References

1. Grossiord, C., Buckley, T.N., Cernusak, L.A., et al.: Plant responses to rising vapor pressure deficit. New Phytol. **226**(6), 1550–1566 (2020)
2. Ma, C., Yang, Z., Xia, R., et al.: Rising water pressure from global crop production-A 26-yr multiscale analysis. Resour. Conserv. Recycl. **172**, 105665 (2021)
3. Schmitt, M.T., Neufeld, S.D., Mackay, C.M.L., Steenbergen, O.D.: The perils of explaining climate inaction in terms of psychological barriers. J. Soc. Issues **76**(1), 123–135 (2020)
4. Spinelli, M., Lionetti, F., Pastore, M., Fasolo, M.: Parents' stress and Children's psychological problems in families facing the COVID-19 outbreak in Italy. Front. Psychol. **11**, 1713 (2020)
5. Li, W., Yang, Y., Liu, Z.H., et al.: Progression of mental health services during the COVID-19 outbreak in China. Int. J. Biol. Sci. **16**(10), 1732–1738 (2020)
6. Kotera, Y., Laethem, M.V., Ohshima, R.: Cross-cultural comparison of mental health between Japanese and Dutch workers: relationships with mental health shame, self-compassion, work engagement and motivation. Cross Cult. Strat. Manag. **27**(3), 511–530 (2020)
7. Pei, J., Zhong, K., Li, J.: ECNN: evaluating a cluster-neural network model for city innovation capability. Neural Comput. Appl. **34**, 12331 (2021)
8. Xiang, Y.-T., Zhao, Y.-J., Liu, Z.-H., et al.: The COVID-19 outbreak and psychiatric hospitals in China: managing challenges through mental health service reform. Int. J. Biol. Sci. **16**(10), 1741–1744 (2020)
9. Holingue, C., Kalb, L.G., Klein, A., Beasley, J.B.: Experiences with the mental health service system of family caregivers of individuals with an intellectual/developmental disability referred to START. Intellect. Dev. Disabil. **58**(5), 379–392 (2020)
10. Ravikumar, S., Kannan, E.: Analysis on mental stress of professionals and pregnant women using machine learning techniques. Int. J. Image Graph **23**, 2350038 (2023). https://doi.org/10.1142/S0219467823500389
11. Wand, T., Buchanan, S.H., Derrick, K., Harris, M.: Are current mental health assessment formats consistent with contemporary thinking and practice? Int. J. Ment. Health Nurs. **29**(2), 171–176 (2020)
12. Jaeyalakshmi, M., Vijay, K., Jayashree, K., Priya Vijay, A.: Cloud based healthcare data storage system using encryption algorithm. In: Sugumaran, D., Pal, S., Le, D.-N., Jhanjhi, N.Z. (eds.) Recent Trends in Computational Intelligence and Its Application, pp. 486–491. CRC Press, London (2023)
13. Pocobello, R., Sehity, T., Negrogno, L., Minervini, C., Guida, M., Venerito, C.: Comparison of a co-produced mental health service to traditional services: a co-produced mixed-methods cross-sectional study. Int. J. Ment. Health Nurs. **29**(3), 460–475 (2020)
14. Haque, A.: Mental health concepts in Southeast Asia: diagnostic considerations and treatment implications. Psychol. Health Med. **15**(2), 127–134 (2010)
15. Latha, G.C.P., Sridhar, S., Prithi, S., Anitha, T.: Cardio-vascular disease classification using stacked segmentation model and convolutional neural networks. J. Cardiovasc. Dis. Res. **11**(4), 26–31 (2020)
16. Furst, M.A., Bagheri, N., Salvador-Carulla, L.: An ecosystems approach to mental health services research. BJPsych Int. **18**(1), 23–25 (2021)
17. Srinivasan, S.P., Shanthi, D.S.: A seed yield estimation modelling using classification and regression trees (CART) in the biofuel supply chain. J. Biomed. Imaging Bioeng. **1**(1), 8–12 (2017)
18. Guo, C., Tomson, G., Keller, C., Soderqvist, F.: Prevalence and correlates of positive mental health in Chinese adolescents. BMC Public Health **18**(1), 1–11 (2018)
19. Baek, J.-W., Chung, K.: Multi-context mining-based graph neural network for predicting emerging health risks. IEEE Access **11**, 15153–15163 (2023). https://doi.org/10.1109/ACCESS.2023.3243722

20. Priya, V., Sathya Sofia, A.: Review on malware classification and malware detection using transfer learning approach. In: 2023 5th International Conference on Smart Systems (2023)

21. Areán, P.A., Ly, K.H., Andersson, G.: Mobile technology for mental health assessment. Dialogues Clin. Neurosci. **18**(2), 163 (2016)

22. Halim, Z., Khan, G., Shah, B., Naseer, R., Anwar, S., Shah, A.: On the utility of parents' historical data to investigate the causes of autism spectrum disorder: a data mining-based framework. IRBM **44**(4), 100780 (2023)

23. Tan, C., Lin, J.: A new QoE-based prediction model for evaluating virtual education systems with COVID-19 side effects using data mining. Soft. Comput. **27**(3), 1699–1713 (2023)

24. Malathi, S., et al.: Prediction of cardiovascular disease using deep learning algorithms to prevent COVID 19. J. Exp. Theor. Artif. Intell. **35**, 791 (2021). https://doi.org/10.1080/0952813X.2021.1966842

25. Sairam, U., Voruganti, S.: Mental health prediction using deep learning. Int. J. Res. Appl. Sci. Eng. Technol. (IJRASET) **10**, 782 (2022)

Topological Navigation of Path Planning Using a Hybrid Architecture in Wheeled Mobile Robot

Vengatesan Arumugam$^{(\boxtimes)}$ and Vasudevan Algumalai

Department of ME, Saveetha Institute of Medical and Technical Science, Chennai, India
venkatesana9006.sse@saveetha.com

Abstract. In this research article, a hybrid architecture for topological path planning for wheeled mobile robots was proposed. Topological representations became more popular for path planning in wheeled mobile robots. The topological navigation of an ultrasonic sensor in sensing the vehicle number and landmarks identification (L1, L2, L3, L4, L5, and L6) was explored. Landmarks were represented as binary numbers (1001, 1100, 1010, 0110, 0011, 1110), and an additional landmark was added (L7 in 0111) at the entry position point. The artificial landmarks facilitated quick sensing of landmarks' identification numbers and vehicles as well. The shortest distance was determined using Dijkstra's algorithm from the start to the goal position point. The distinctive locations for (N, S, E, W) movements were observed from the left and right movements of the vehicles. The total area covered by landmarks was 1080 × 900 m, and the ultrasonic distance sensor could detect a range of 2 to 80 cm. The mobile robot moved in square, rectangle, linear, and triangular patterns, with an average distance of 19.7 mm and an angle error of 0.55°. The landmarks recognition range was 99%, and the identification range was 99.5%. The localization distance ranged between 0.25 and 0.31 s, while the localization estimation took 0.002 s.

Keywords: Path planning · Topological navigation · Hybrid architecture · Wheeled mobile robot · Dijkstra's algorithm · Artificial landmarks · Vehicle land parking area (L1 · L2 · L3 · L4 · L5 · L6 and L7)

1 Introduction

Wheeled mobile robots are becoming more prevalent in a variety of industries, such as manufacturing, logistics, and service applications. These robots are able to perform tasks, interact with their environment, and navigate autonomously in changing contexts [1]. Path planning, which entails choosing the best route from a starting point to a desired destination while avoiding obstacles and considering other factors, is a fundamental component of mobile robot navigation. Grid-based or geometric techniques, such as those used by A* or Dijkstra's algorithm, are frequently used in conventional path-planning algorithms [2]. To get around these restrictions, researchers have created topological navigation techniques that make use of the environment's natural structure to direct the path-planning process of the robot [3]. The robot can travel effectively by employing

R. Geetha et al. (Eds.): AAIMB 2023, CCIS 2202, pp. 32–44, 2025.
https://doi.org/10.1007/978-3-031-73065-8_3

this graph representation and concentrating on high-level decisions rather than precise geometry calculations [4]. Topological navigation visualizes the environment as a graph, with nodes standing in for different geographic features and edges for crossing over or connecting these features [5]. The benefits of both geometric and topological approaches are combined in a hybrid design for topological navigation [6]. It combines the global topological representation of the environment with the local geometric data collected from sensors like ultrasonic distance sensors [7]. With this combination, the robot can benefit from the topological representation's superior decision-making ability while still having a fine-grained view of its immediate surroundings [8].

The Path planning navigation of wheeled mobile robots in two concepts are graph search techniques and potential field planning. This design of the robot's path in a fully observed environment is known as global path planning [9]. Planning robots' routes the modelling of impediments in the task environment and the creation of search algorithms for finding the intended path are two challenges presented by roving in an unknown environment [10].

Earlier attempts to overcome these issues involved learning a navigational learning system that could memorize the courses, followed by a search strategy to determine the shortest path [11]. One of the major challenges with mobile robots is path planning. Navigation makes up the low-level velocity control mechanism [12]. In production environments, hospital settings, and various robot-based service environments, mobile robots have emerged as a crucial component. Different wheel-based driving techniques have been created to increase productivity in settings where mobile robots are used [13]. In this work, a quadtree representation of free space was searched using Dijkstra's algorithm. But the necessary search can be computationally expensive. In recent work, a method for improving search efficiency is described that modifies the A* algorithm [14]. The free space for a robot is represented by path planning as a network of connected collision-free paths. The following path is planned using the roadmap, a collection of collision-free paths [15]. The mobile robot must gather information about its surroundings in order to maneuver. When it comes to velocity, the robot's speed is dependent on the distance between obstacles found to be from where it is at the moment. This strategy is used to guide the robot as it travels to its destination spot securely [16]. Both manual placement and random placement are options for such items. The path planning button initiates the path planning approach automatically and displays the trajectory of the robot [17]. Many of the behaviors that an interacting robot would exhibit are human impersonations [18]. Algorithms for sensing and recognizing the surroundings can be used to implement certain behavior [19]. The behavior includes object following, tactile reflection, path planning, obstacle avoidance, visual contact, speech detection, and object following [20].

1.1 Problem Identification

In the research article, a hybrid architecture was utilized in wheeled mobile robots for topological navigation path planning, employing Dijkstra's shortest path algorithm or A* search. The minimum distance point was reached at 21 cm using this algorithm. The artificial landmarks were used to quickly reach the start of the goal position point. The mobile robot sensed the ultrasonic distance sensor within 10 cm for recognizing vehicles. The identification range was 99.5%, and the landmarks' recognition range was 99%.

2 Proposed Methodology

The flow diagram for topological navigation in path planning was depicted in Fig. 1. The path planning was carried out using Dijkstra's shortest path algorithm with a wheeled mobile robot utilized for topological navigation. The research article identified the topological navigation of WMR (Wheeled Mobile Robot) in landmarks 1, 2, 3, 4, 5,6, and 7. The subsequent vehicle moved from its identifying position point to determine the landmarks' start and goal position points. Similarly, the positions of the remaining landmarks were also identified.

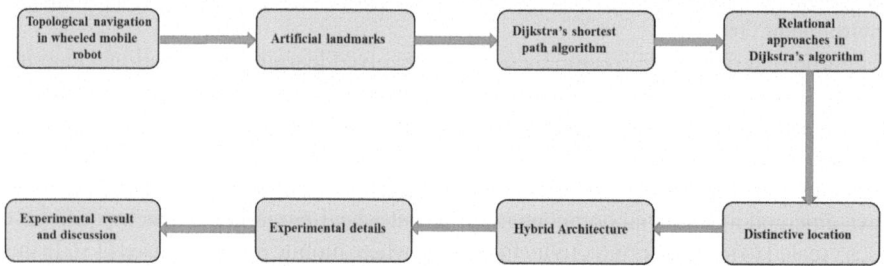

Fig. 1. Flow diagram for topological navigation in path planning.

2.1 Topological Navigation Path Planning for Wheeled Mobile Robot

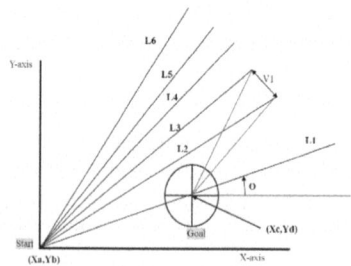

Fig. 2. Start and goal point reaching in topological navigation.

The start position point was denoted by the coordinates (X_a, Y_b), and it was connected to six landmarks: L1, L2, L3, L4, L5, and L6. The landmarks were used as reference points to reach the goal position (X_c, Y_d) with a specific coordinate angle θ. Denoted the direction and position of the robot coordinates and the goal position of reference as $g = (X_a, Y_b, \theta)^T$. The start position point coordinate Pa = (Xa, Ya, a) and the target position point coordinate Pb = (X_b, Y_b, b) can both be used in Dijkstra's shortest path algorithm. The vectors Pr = {Pa... Pb} are used to indicate the output techniques of the shortest path connection between the start and target nodes. Figure 2 displayed the starting and goal point-reaching process in topological navigation.

Path Following: The objective is followed from the starting point to the ending position. Thus, the goal position point (Xc, Yd) and start position point (Xa, Yb) coordinates. Determine the magnitude and orientation of the robot's resultant forces in the desired direction [21].

$$\theta = tan^{-1}\left(\frac{(X_c - X_a)}{(Y_d - Y_b)}\right) \tag{1}$$

if $X_c \neq X_a$

$$g = \left(\frac{(X_c - X_a)^2}{(Y_d - Y_b)^2}\right)^{1/2} \tag{2}$$

The robot's starting coordinate position point (start) and ending coordinate position point (goal), which are made up of sub-goals recognized by artificial landmarks, were both attained. Attractive force: Subgoal, potentially repulsive: obstacles [21].

$$R_{o(rep)} = C\left[\frac{EP_o}{d^2}\right]\hat{u} \tag{3}$$

where,

$R_{o(rep)}$ = Gross potential derivative from the degree and path ϴ.
C = Constant of proportion
E = Related to the robot
p_o = Related to the obstacle
d = I X_o–X_b I
û = Unit vector in the path of the obstacle [21].

$$\hat{u} = \left(\frac{(X_0 - X_b)}{|X_0 - X_b|}\right) \tag{4}$$

$$R_{uir} = B_i \times \overline{\upsilon} \tag{5}$$

$$\overline{\upsilon} = \left(\frac{(X_0 - X_s)}{|X_0 - X_s|}\right) \tag{6}$$

Total forces,

$$R_{net} = \sum R_{o(rep)} + R_{atr} \tag{7}$$

Resultant forces at angle θ

$$\theta = tan^{-1}\left(\frac{(R_y)}{R_x}\right) \tag{8}$$

where,

  X_o = Vector position of the robot
  X_b = Vector position of the obstacle
  Xs = Coordinate sub-goal
  R_{atr} = Attractive force

2.2 Artificial Landmarks for Topological Navigation

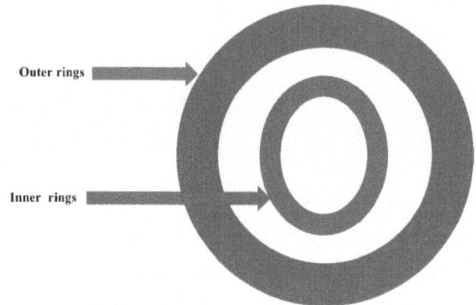

Fig. 3. Artificial landmarks for outer and inner rings.

Artificial landmarks are places or things that are deliberately placed in a space to help topological navigation for a variety of systems, such as robots, autonomous vehicles, or other agents. These landmarks act as quick-to-find, simple-to-identify reference points that can be utilized for localization and orientation. Figure 3 Artificial landmarks for outer and inner rings, hence set of concentric circular rings outer rings are blue and inner rings are red.

2.3 Dijkstra's Shortest Path Algorithm Using Path Planning

Figure 4 shows as Dijkstra's shortest path algorithm was utilized. The algorithm started at position point S and reached the goal position point T. The path taken was S-C-B-F-T, with a minimum shortest distance of 21. The remaining points were assigned high values, specifically 32, 32, 28, 32, and 39. Tables 1 and 2 illustrate the shortest paths of all the nodes leading to the goal position.

2.4 Relational Approaches for Dijkstra's Algorithm

In Fig. 5(a), the relational network representations of a vehicle land parking layout were presented. The dimensions of the vehicle land parking areas, namely L1, L3, L4, L6, and L7 were measured as 170 × 150 m, while the adjacent parking areas, L2, and L5, had dimensions of 200 × 150 m. Figure 5(b) illustrated the entry and exit points of the landmarks. The distances and times associated with each landmark were as.
  follows:

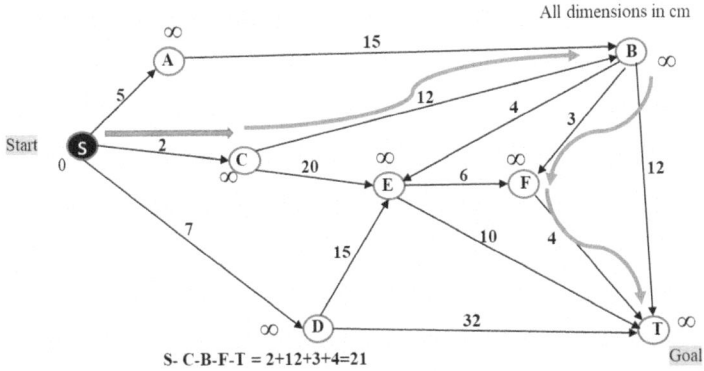

Fig. 4. Dijkstra's shortest path algorithm.

Table 1. Shortest path of all the nodes in a network.

S No	Visited	Unvisited	Weights
1	{S}	{A, B, C, D, E, F, T}	{0, ∞, ∞, ∞, ∞, ∞, ∞, ∞}
2	{S, A, C, D}	{B, E, F, T}	{0, 5, 2, 7, ∞, ∞, ∞, ∞}
3	{S, A, C, D, B}	{F, T}	{0, 5, 2, 7, 14, ∞, ∞, ∞}
4	{S, A, C, D, B, E}	{T}	{0, 5, 2, 7, 14, 22, 28, ∞}
5	{S, A, C, D, B, E, F, T}	-	{0, 5, 2, 7, 14, 22, 28, 32}

Table 2. Shortest path in goal position.

S No	Path Travelling point	Total distance (cm)
1	S-A-B-T	$5 + 15 + 12 = 32$
2	S-C-B-F-T	$2 + 12 + 3 + 4 = 21$
3	S-C-E-F-T	$2 + 20 + 6 + 4 = 32$
4	S-C-B-E-T	$2 + 12 + 4 + 10 = 28$
5	S-D-E-T	$7 + 15 + 10 = 32$
6	S-D-T	$7 + 32 = 39$

The seventh landmark (L7) had a distance of 1 m and a time of 20 s. (entry). Sixth landmark (L6) had a distance of 5 m and a time of 50 s. Fifth landmark (L5) had a distance of 10 m and a time of 90 s. Fourth landmark (L4) had a distance of 15 m and a time of 140 s. First landmark (L1) had a distance of 20 m and a time of 200 s. Second landmark (L2) had a distance of 25 m and a time of 270 s. Third landmark (L3) had a distance of 30 m and a time of 350 s (Exit).

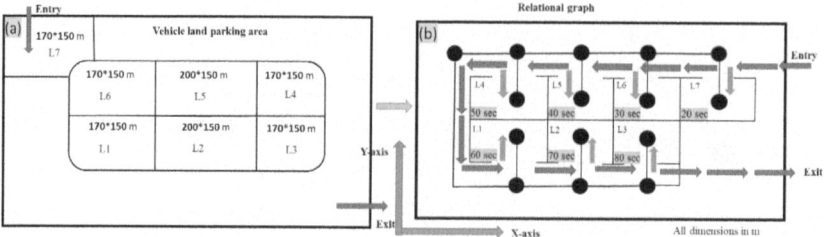

Fig. 5. (a). Vehicle land parking area. (b). Relational network representations of a Vehicle land parking layout.

Relational approaches depict the universe as a node and edge system in the form of a graph. Nodes stand for entry points, markers, or objectives. Edges show that there is a navigable path connecting two nodes and that there is a spatial relationship between them. Edges may also have additional information associated with them, like the distance to an approximate position (N, S, E, W), the type of terrain, or the navigational techniques required. Utilizing conventional layout approaches, a single origin shortest route algorithm, proposed by Dijkstra, paths can be calculated between two points. The edges of the relational graph, which they used to describe the world, reflected the direction and separation between the nodes.

2.5 Distinctive Location for Landmarks

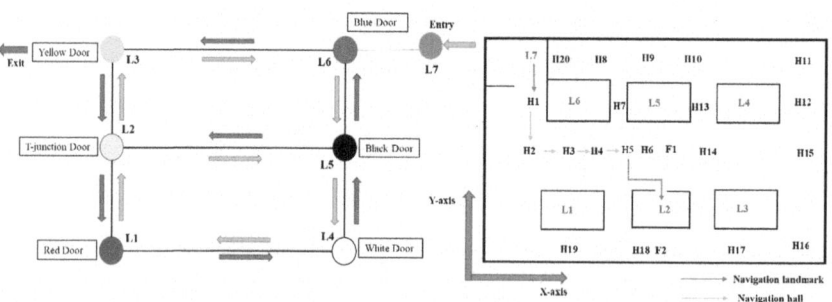

Fig. 6. Local control strategies for landmarks position point. Navigation processes for L7 to L2.

In Fig. 6 shown the local control strategies for landmarks' position points were presented. The identification of landmarks' position points (L1, L2, L3, L4, L5, L6, and L7) was accomplished using different colours, including blue, black, white, red, T-junction, yellow door fixed, and blue colour arrow marks indicating vehicle exits. Additionally, orange colour arrows denoted vehicle entry points from landmarks. As a result, the mobile robot could function to detect the vehicle number through sensor readings. The different colours used in the landmarks' position points allowed for easy vehicle identification, and binary codes were also available for each landmark (L1, L2, L3, L4, L5, sand L7), Hence red colour indicated to coordinate point. As shown in Fig. 6 Navigation process for L7 to L2. The calculates the route given in the first line based on the output. The second line indicates that the direction of travel is from L7 to H1. Since the user did not specify where the door was, the Task Manager chooses to open the door and begin by searching for the door. The door can be located to the south by looking through the database.

Step by step Processes [22]:

L7 ▼ L2

L7 - H1 - H2 - H3 –H4- H5 - L2

Moving beginning L7 to H1, going SOUTH

Navigation for a door

Ultra-rapidly in the direction of SOUTH

MOTOR MOVING AHEAD IS ACTIVE

Door found; initialization ended

ACTIVE MOVE THROUGH DOOR MOTOR

passed through the door, ending Nominal Behaviour

Moving from H1 to H2, forward to SOUTH

In navigating hall behavior

turning forward the: SOUTH

turned to face in the direction of the hall - Initialization finished

ACTIVE EAST HALL FOLLOWS THE MOTOR

Found hall - Nominal Behavior terminated

Moving from H2 to H5, forward to EAST

When navigating a hallway,

turn to the: EAST

Turned to face in the direction of the hall - Initialization terminated

vision looking for door relative: 90 (right side)

HALL FOLLOWS MOTOR ACTIVE

Found door (vision) - Terminated Nominal Behaviour

Moving from H5 to L2, going SOUTH

As it relates to door behavior

In search of a door in the direction of the: SOUTH

Left-side wall (on the right ground true)

WALL FOLLOW MOTOR ACTIVE

Found door - Initialization terminated

DYNAMIC TRAVEL OVER DOORS MOTION

passed through the door, ending nominal behaviour Target accomplished!

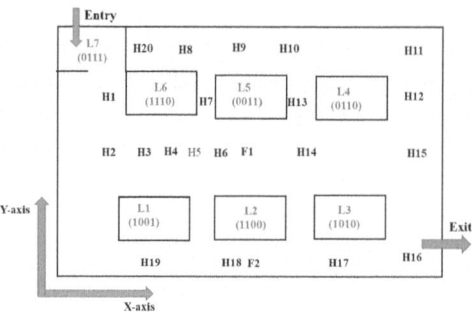

Fig. 7. Metrics map of a landmark layout.

2.6 Topological Navigation Path Planning for Using a Hybrid Architecture

The entryway was represented in orange colour, located adjacent to the identification point of landmark L7 (0111). Similarly, the remaining landmarks were represented as follows: L1 (1001), L2 (1100), L3 (1010), L4 (0110), L5 (0011), and L6 (1110). In Fig. 7, a metrics map of a landmark layout was presented. The input map consisted of three node types: binary number (B), landmarks number (L), and Entryway (E). As there were only four possible directions for edges between nodes, they were designated as north (N), south (S), east (E), and west (W), with N randomly assigned on the map.

3 Experimental Details

The landmarks in the parking area were set up for the experimental environment. Markers were positioned inside the L1, L2, L3, L4, L5, L6, and L7 areas. Each landmark was printed with a diameter of 250 m, as the dimensions of the area had been previously discussed. This resulted in a total distance of 1080 × 900 m for all the landmarks. The positioning and orientation of the landmarks were carefully measured.

Fig. 8. (a), (b) Wheeled mobile robot for path planning navigation in side view and top view.

The wheeled mobile robot used in topological navigation for path planning approaches can be used in Dijkstra's algorithm. Physical robotics components like wheels, sensors, motors, etc. are used in mobile robot functions together with artificial intelligence combinations. Industrial, food, and shipping sectors all use mobile robots on wheels. Figure 8(a), (b) Wheeled mobile robot for path planning navigation in side view and top view. The specification of the wheeled mobile robot given below Table 3.

Table 3. Specification of the wheeled mobile robot.

S No	Name of the parts	Quantity	Materials	Dimensions (mm)
1	Arduino Uno	01	Micro (Pcb)Controller	L*W*H = 75*54*12
2	Chassis	01	Acrylic	L*W*H = 215*150*3
3	Bo motor	02	Plastic Gear Motors	L*W*H = 65*23*27
4	wheels	02	Plastic Rubber	$\Phi = 65$
5	Caster wheel	01	Steel with zinc plated	L*W*H = 49*32*21
6	Battery holder	01	Plastic	L*W*H = 62 × 15 × 25
7	Lithium ion battery 3.7 voltage	02	Nickel & Cobalt	L*W*H = 4 × 1 × 1 cm
8	L298N Dc motor drive	01	Copper	L*W*H = 44 × 44 × 28
9	Ultrasonic distance sensor	02	Piezoelectric materials	L*W*H = 45 × 20 × 15

The ultrasonic distance sensor measures distance ranging from 2 to 80 cm, and all the landmarks are identified using binary numbers. The vehicle parked at landmark L1 with the mobile robot using the sensor to sense the vehicle's number. The mobile robot utilized Dijkstra's shortest path algorithm for topological navigation.

3.1 Experimental Result and Discussion

In the above setting, the experiments were conducted to showcase the landmark recognition of typical scenarios, as depicted in Fig. 9. The figure illustrates how various environmental elements such as lights, alarms, moving robots, electrical boxes, and landmark identification were photographed. The effective recognition of landmarks was made possible by the planned landmark design and ambient conditions. The vehicle's identification number and the subsequent binary number fixed in landmarks L1 to L7 were detected by the mobile robot. The number of landmarks spotted in relation to all landmarks was measured by the sensors' detection rate. Table 3 demonstrates that the mobile robot sensor detected the vehicle number and parking space. The robot moved in square, rectangle, line, and triangle patterns. The average distance was 19.7 mm, while the angle inaccuracy was 0.55°. The identification position point in Table 4 shows that

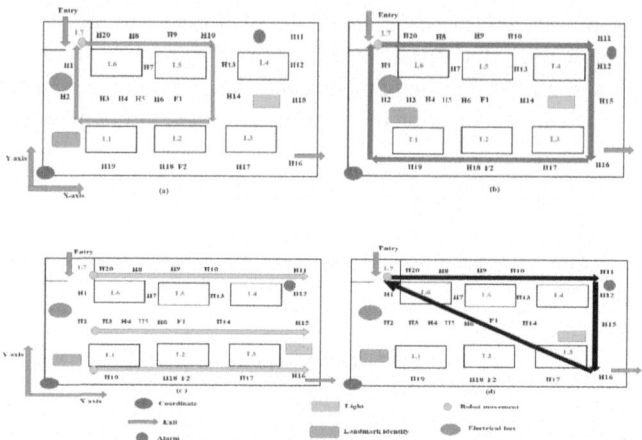

Fig. 9. (a), (b), (c) and (d) Shows the Square, Rectangle, Linear, and Triangle for Landmark arrangements and moving of robot in experiment.

Table 4. Landmark accuracy.

S No	Experimental path		Square	Rectangle	Linear	Triangle
1	Distance error	Average(mm)	18.6	15.4	20.4	24.5
	Distance error	STD (mm)	7.8	8.6	8.2	9.2
2	Angle error	Angle (θ)	0.7	0.4	0.6	0.5
	Angle error	STD (θ)	0.5	0.2	0.3	0.2

Table 5. Landmark identification position point [23].

S No	Identification point	Result %	Experimental result %
1	Landmark recognition range	98%	99%
2	Landmark Identification range	99%	99.5%

the landmarks' recognition range was 98% and increased by 99% in the experimental result, and the landmarks identification range was 99% and increased by 99.5%. The local running condition process for the picture input localization distance between 0.25 and 0.31 s. Full image detection for the car was about 0.30 s, and localization estimation was about 0.002 s (Table 5).

4 Conclusion and Future Scope

The topological navigation of path planning of the wheeled mobile robot was approached using a hybrid architecture. Landmarks (L1, L2, L3, L4, L5, L6, and L7) were mentioned with binary numbers (1001, 1100, 1010, 0110, 0011, 1110, and 0111) in vehicle parking area identification. Dijkstra's shortest path algorithm was used, and the minimum distance reached was 21 cm. The start-to-goal position point was determined using this algorithm. The artificial landmarks for topological navigation were quick to find the landmarks' location and orientation. The relation network vehicle land parking layout dimensions were determined. The position point moved from vehicles (N, S, E, W). The location of landmarks position points identified different colours in the door position point. The landmarks of navigation experienced a step-by-step process in L7 and L2, Hence the sensor navigation could recognize a range of 2 to 80 cm. The landmarks were arranged in square, rectangle, linear, and triangle patterns, with an average distance of 19.7 mm and an angle accuracy of 0.55°. The landmarks recognition range was 99% in the experimental phase, and the landmarks recognition range increased by 1% while the identification range increased by (99% to 99.5%) 0.5%. Full image detection of vehicles neighboring 0.30 s, and localization estimation neighboring 0.002 s.

The future scope was to explore two or more topological navigation methods in path planning and to utilize multiple robots with different algorithms. The landmarks area would be expanded on the ground floor position points, and the vehicle area would be increased by up to seventh floors. The range of ground allowing for landmarks parking identification the dimension area 3000 × 2500 m, from the ground floor to the seventh-floor area.

References

1. Staicu, S.: Mobile wheeled robots. In: Dynamics of Parallel Robots, pp. 277–308. Springer, Cham (2018). https://doi.org/10.1007/978-3-319-99522-9_11
2. Arthanari, S., Vinod B.: Sampling-based algorithms for path planning of a mobile robot. In: Authorea (n.d.). https://doi.org/10.22541/au.158456509.94127297
3. Achat, S., Marzat, J., Moras, J.: Path planning incorporating semantic information for autonomous robot navigation. In: Proceedings of the 19th International Conference on Informatics in Control, Automation, and Robotics (2022). https://doi.org/10.5220/0011134300003271
4. Reily, B., Zhang, H.: Team assignment for heterogeneous multi-robot sensor coverage through graph representation learning. In: (2021) IEEE International Conference on Robotics and Automation (2021). https://doi.org/10.1109/icra48506.2021.9561082
5. Haddad, M., Bouguessa, M.: Topo detect: framework for topological features detection in graph embeddings. Software Impacts **10**, 100139 (2021). https://doi.org/10.1016/j.simpa.2021.100139
6. Topological Spaces. In: Geometric and Topological Inference, pp. 3–9 (n.d.). https://doi.org/10.1017/9781108297806.002
7. Aliew, F.: An approach for precise distance measuring using ultrasonic sensors. In: IECMA (2022). https://doi.org/10.3390/iecma2022-12901
8. Reguii, I., Hassani, I., Rekik, C.: Mobile robot navigation using planning algorithm and sliding mode control in a cluttered environment. J. Robot. Control JRC **3**(2), 166–175 (2022). https://doi.org/10.18196/jrc.v3i2.13765

9. Patnaik, S.: Path planning. In: Robot Cognition and Navigation, pp. 39–58. Springer, Berlin (2006). https://doi.org/10.1007/978-3-540-68916-4_3

10. Jesuthas, N.J.A., Somaskandan, S.: Path-finding and planning in a 3D environment an analysis using bidirectional versions of Dijkstra's: Weighted A*, and Greedy Best First Search Algorithms. In: 2nd Asian Conference on Innovation in Technology (ASIANCON) (2022). https://doi.org/10.1109/asiancon55314.2022.9909251

11. Multi-Agent Assisted Shortest Path Planning using Monte Carlo Tree Search (2023). https://doi.org/10.2514/6.2023-2655.vid

12. Wang, J., Herath, D.: How to Move? Control, navigation and path planning for mobile robots. In: Foundations of Robotics, pp. 205–238. Springer, Berlin (2022). https://doi.org/10.1007/978-981-19-1983-1_8

13. Kim, C., Suh, J., Han, J.-H.: Development of a hybrid path planning algorithm and a bio-inspired control for an Omni-wheel mobile robot. Sensors **20**(15), 4258 (2020). https://doi.org/10.3390/s20154258

14. Binbakir, T.: Improving Bees Algorithm Using Gradual Search Space Reduction (2021). https://doi.org/10.21203/rs.3.rs-790818/v1

15. Kumar, S., Sikander, A.: A modified probabilistic roadmap algorithm for efficient mobile robot path planning. Eng. Optim. **55**, 1–19 (2022). https://doi.org/10.1080/0305215x.2022.2104840

16. Acosta, D., Fariña, B., Toledo, J., Acosta, L.: Improving mobile robot maneuver performance using fractional-order controller. Sensors **23**(6), 3191 (2023). https://doi.org/10.3390/s23063191

17. Ghafil, H.N., Jármai, K.: Path and trajectory planning. In: Optimization for Robot Modelling with MATLAB, pp. 123–155 (2020). https://doi.org/10.1007/978-3-030-40410-9_6

18. Booth, S.: Aligning robot behaviors with human intents by exposing learned behaviors and resolving misspecifications. In: Companion of the 2023 ACM/IEEE International Conference on Human-Robot Interaction (2023). https://doi.org/10.1145/3568294.3579971

19. Cain, S.: The Necessity of Certain Behaviors, pp. 127–142. University of Pittsburgh Press, Pittsburgh (2011). https://doi.org/10.2307/j.ctt5hjnr3.11

20. Wu, Z., Yin, D., Xiao, J.: Research on path planning strategy based on dynamic object obstacle avoidance. In: 4th International Conference on Information Science, Electrical, and Automation Engineering (ISEAE 2022) (2022). https://doi.org/10.1117/12.2639535

21. Esparza, D., Savage, J.: Topological mobile robot navigation using artificial landmarks. In: Latin American Robotics Symposium and Competition (2013). https://doi.org/10.1109/lars.2013.54

22. Murphy, R.R.: Introduction to AI Robotics. MIT Press, Cambridge (2019)

23. Zhong, X., Zhou, Y., Liu, H.: Design and recognition of artificial landmarks for reliable indoor self-localization of mobile robots. Int. J. Adv. Rob. Syst. **14**(1), 172988141769348 (2017). https://doi.org/10.1177/1729881417693489

Abnormal Behaviour Detection in Surveillance Videos

R. Vijayakumar[1]([✉]), D. Sorna Shanthi[1], B. Bhuvaneswaran[1], and M. Pragadeesh[2]

[1] Department of CSE, Rajalakshmi Engineering College, Mevalurkuppam, India
r.vijayakumar91@gmail.com
[2] Department of IT, Rajalakshmi Engineering College, Mevalurkuppam, India

Abstract. People from all walks of life have always been concerned about public safety issues. There has been a meteoric rise in the number of security cameras used for monitoring both public and private settings in recent years. The key to avoiding public safety hazards is now in spotting anomalous human behaviour on film, thanks to advances in video detection technologies. The ability to recognise anomalous human behaviour is crucial, especially in social settings like those found in student groups. Existing algorithms for detecting anomalous human behaviour tend to focus on outdoor activity detection, and their performance indoors is subpar at best. The majority of a student's time is spent inside, and most modern classrooms have some form of surveillance technology. This investigation focuses on identifying anomalous actions taken by indoor humans, and it does so by employing a novel abnormal behaviour detection framework. In this project, we are detecting the human abnormal activity using tensorflow pose estimation model by estimating the location of key joints of the person. This model can estimate not only one person but also many persons in the videos. This method has better detection performance.

Keywords: Video surveillance · abnormal Behavior · CNN · video based detection

1 Introduction

There has been a dramatic growth in the use of surveillance cameras to keep an eye on both public and private settings. The majority of the time, automated analytic methods used today can pinpoint precisely predefined ideas of anomalous behaviour, such an intruder in a restricted area. Alternatively, operators either continuously record or monitor the video feeds. The ideal analytic programme for many visual surveillance applications, however, would automatically interpret the image and sound an alarm if anything suspect was detected. The project's overarching goal is to identify instances of human-caused environmental abnormality.

R. Geetha et al. (Eds.): AAIMB 2023, CCIS 2202, pp. 45–55, 2025.
https://doi.org/10.1007/978-3-031-73065-8_4

2 Literature Survey

[1] Most fall detection systems rely on wearable or otherwise complicated equipment based on wearable sensors, environmental sensors, and computer vision. They can, however, restrict the aged person's normal activities. This study provides a conceptual verification module that uses Wi-Fi signals to recognise human fall behaviour, based on the indoor propagation theory of wireless signals. The network's feature extraction capability can be enhanced because [2] the CBAM method incorporates both channel and spatial attention processes. Finally, the convolutional neural network is used to extract the features that will be used to detect and report pilots' anomalous driving behaviour. The results demonstrate the viability and applicability of the deep learning algorithm based on the enhanced YOLOv4 method in tracking pilots' anomalous driving behaviour during the flight's manoeuvring phase. According to the data, the calling phase has a mAP of 87.35%, an accuracy of 75.76%, and a recall of 87.35%, which is much better than the original YOLOv4 algorithm's findings.

[3] Human posture is comprehended by detection, extraction, and recognition procedures on visual data, allowing for early detection of problematic behaviour. -Many sophisticated practical projects are introduced after an overview of relevant work in the field of aberrant human behaviour analysis. They suggest a gesture recognition system that might be used to operate common household gadgets. To reduce customer discomfort, we made use of the accelerometers and gyroscopes found in most modern smart watches. For feature analysis, feature learning, and feature representation in sensor signals, 1D-CNN-biLSTM networks have been proposed [4]. In order to reduce the number of frames that must be calculated, it is possible to effectively remove a significant number of unnecessary frames from the long video by detecting the motion amplitude of the long video. Using the written question's response as a guide, the trained semantic model then matches the two to determine which video segment best answers the question [5].

The proposed CapsNet architecture's efficacy can be measured by playing around with the dynamic routing between capsule layers. The proposed SensCapsNet outperforms the CNN and LSTM baseline approaches on two testing datasets, with an F1-score of 77.7% and 70.5% for 1 route, respectively [6]. Human motion features are extracted from RGB video frames using transfer learning to improve detection accuracy. The VGGNet-19 model is used to pre-train a convolutional neural network (CNN) that is then used to extract descriptive features. After that, a Binary Support Vector Machine (BSVM) model is constructed with the feature vector as the input. The effectiveness of the proposed framework is determined by its accuracy, area under the curve, and equal error rate. In experiments [7], the UMN dataset performed at an accuracy of 97.44% and an area under the curve (AUC) of 0.9795, whereas the UCSD-PED1 dataset performed at an accuracy of 86.69% and an AUC of 0.7987. The proposal was made due to the fact that elderly people living alone have significant care needs. The rising expense of nursing care and the resulting shortage of health care professionals have contributed to a growth in the demand for assisted living in the home in recent years. Therefore, the use of information and communication technology for the purpose of detecting abnormal behaviour in the context of home healthcare is a growing field of research [8].

To evaluate the performance of the proposed CapsNet architecture, one can experiment with the dynamic routing between capsule levels. The proposed SensCapsNet achieves an F1-score of 77.7% and an F1-score of 70.5% for 1 route over the CNN and LSTM baseline techniques on two testing datasets, respectively [9]. To enhance detection accuracy, RGB video frame human motion features are extracted using transfer learning. In order to extract descriptive features, a convolutional neural network (CNN) is pre-trained using the VGGNet-19 model. As a next step, we feed the feature vector into a model built using Binary Support Vector Machines (BSVMs). Accuracy, area under the curve, and equal error rate are measures of the suggested framework's efficacy. Experiments [10] showed that the UMN dataset achieved 97.44% accuracy and 0.9795 AUC, while the UCSD-PED1 dataset achieved 86.69% accuracy and 0.7987 AUC. Because older persons who live alone have such high care requirements, this idea was proposed. The demand for in-home care has increased in recent years due to the rising cost of nursing care and the resulting scarcity of health care providers. As a result, the study of how to identify abnormal behaviours using information and communication technologies in the context of home healthcare is expanding [11].

[12] They presented a system to identify unusual activity in online examinations. The convenience, adaptability, and friendliness of online examinations have contributed to their rise in popularity. One of the biggest problems with online exams is keeping an eye out for the examiner's strange behaviour as the test progresses. Traditional monitoring programmes have a blind spot when it comes to spotting out-of-character actions by testers. [13] For the purpose of keeping tabs on psychiatric patients, they suggested a technique for detecting anomalous human behaviour. The spatial and temporal characteristics of human actions define typical behaviour. Human behaviour is unpredictable and complex, making the detection of anomalous behaviour challenging. It shifts in pace and form at different times.

They created a skeleton-based aberrant behaviour detection and a secure partitioned CNN model (SP-CNN) [14] to extract human skeleton keypoints and to achieve safely collaborative computing by deploying separate CNN model layers on the cloud and the IoT device. Since the sensitive video data is not being transmitted, the danger of privacy disclosure is objectively lowered as a result of the numerous CNN layers processing the data outputted by the IoT device. We also develop an encryption method that makes advantage of channel state information (CSI) to further protect the confidentiality of the data. [15] They suggested using shoes with pressure sensors to track people's movements.This inspired the development of pressure-sensing shoes. Embedded in the shoe is a measuring module and a pressure-sensing insole. Walking, running, standing, sitting, going up and down stairs, and cycling were all studied to see if their respective foot pressure distributions might be used to categorise user activities.

3 System Architecture

The system architecture in Fig. 1, deals with the flow from data collection to the final result. Firstly, the training phase is videos converted into frames and feed into the pose estimation model. This model gets the skeleton form of a person from the image. Using the skeleton form, the feature is extraction from the skeleton body and then the features feed to the fully connected layer with tanh activation function. The output of the fully connected layer enters into the output layer with the softmax function and it classify the abnormality. Now the live video input enters into the model by converting into frames and process through the model and detects the abnormality from the videos.

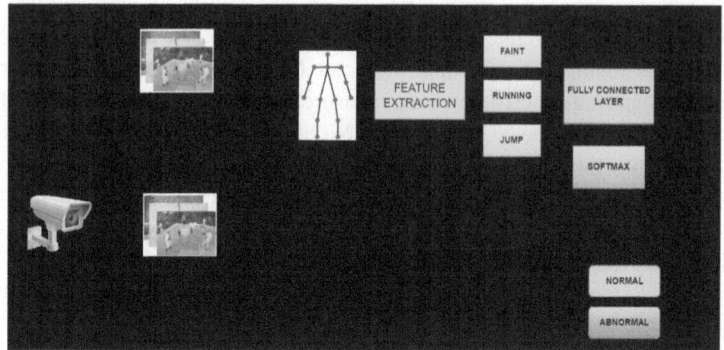

Fig. 1. System Architecture

4   System Implementation

4.1   Video Capturing

In this module the input data or initial level data are collected. Where the input data should be in the form of video with a format of mp4, mpeg, mov, avi and so on. These video images are taken as input, which up on processed by the detection system provides results for abnormal occurance of activities. This kind of video capturing can be performed by the model where it initially splits video of larger size into smaller one and each smaller videos are further converted into frames and these frames are then upon processing helps in capturing of object presence in the particular frame.

4.2   Open Pose Skeleton

OpenPose is the first real-time multi-person system can identify each of a subject's 135 key-points in a single shot, including the body, hands, faces, and feet. Scientists from Carnegie Mellon University offered the suggestion. They have made the plugin available as a Unity add-on, including Python source code and a C++ implementation. The OpenPose repository is the place to get these items.

First, we run the image through a basic convolutional neural network (CNN) to get feature mappings of the input. The researchers employed the initial 10 nodes of the VGG-19 network for their work. Part confidence maps and part affinity field are produced using a multi-stage convolutional neural network (CNN) pipeline after the feature map has been processed (Fig. 2).

When the ground-truth example features unusual or inverted stances, open stance has trouble making an accurate prediction. Because of the overlapping PAFs, greedy multi-person parsing fails in extremely crowded photos when people are overlapping, and as a result, the technique tends to blend annotations from different persons while missing others.

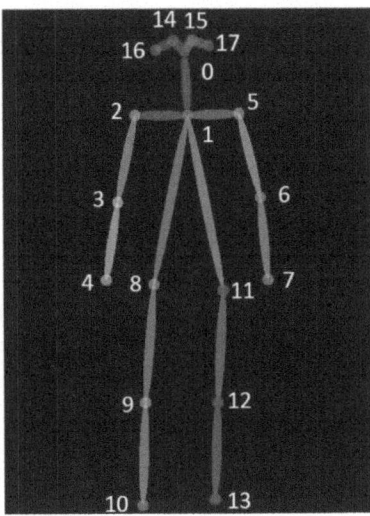

Fig. 2. Openpose Skeletal points

4.3 Model Explanation

Combining ideas from both picture localization and classification, object detection is a challenging problem. A list of the image's object classes is generated by the algorithms, together with bounding boxes that are aligned with the image's axes and provide information about the objects' relative positions and sizes.

4.3.1 Falling

In the literature, there are three main sorts of falls distinguished by whether the actor is standing, sitting, or lying down. Possible outcomes for each scenario are as follows:

1. The character is initially seen going across the scene, but subsequently collapses and is discovered lying face-down.
2. A person who remains motionless while lying flat on their back may later be seen to tumble backward, forward or to the side before coming to a stop on the ground

4.3.2 Violence

When it comes to violence, the focus of this research is on isolated incidents involving only two people, such as a brawl, a kick, or a punch. Fights typically follow patterns such as striking, kicking, jabbing, object hitting, and clinching.

4.3.3 Joint Detection of Falling, Loitering, and Violence

The system may confuse one abnormal activity for another if their characteristics are similar enough, which is why combination detection can be challenging.

4.4 Alerting

After the training and testing of the model, we further move to the last module. In which if the model finds something abnormal or if the activity doesn't fit to the environment, the model is then alert the user with the alert box which intimates about the abnormal behaviour of a human both in real-time and in recorded video files.

5 Implementation and Result

The home page of the detection system is shown in the Fig. 3. The two major features of this detection system are the Real-time and Video based detection. User can select any one of the listed categories to proceed further. PyQt5 have been used in order to make the detection system more interactive to the user.

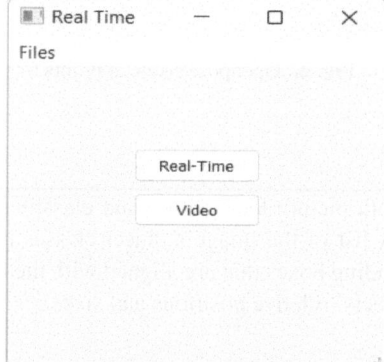

Fig. 3. Main menu

In order to make the detection system uses the webcam or other camera utility devices, permissions have to be provided before undergoing further procedures. To gain the access of the camera from the windows or any other operating system, user should allow the detection system to gain access to the camera only while running the GUI. This permission can be granted by using another user interface as shown in the Fig. 4. where the user must specify the type of camera the user wishes to use and get permission for using that particular camera device.

If the user chooses Real-time in the main menu as shown in the Fig. 5, a pop up message appears to grant camera access as shown in Fig. 5. Once the access is permitted, the webcam begins to run in the background. If the webcam or the surveillance device detects human presence inside the frame it starts recording that particular time interval. With the help of deep learning model frame-wise joints are generated for every human action. At the left, a side panel appears which contains information such as,

1) Object captured and its count
2) Action detected
3) Probability of the action performed

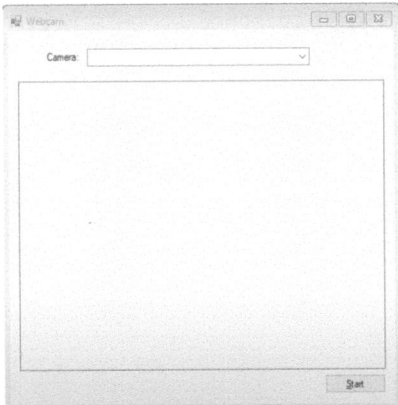

Fig. 4. Camera access

4) Behaviour detection.

At initial level of detection class and probability of the class can be detection. Whenever the detection system detects the human action inside the camera frame, It conducts the deep analysis to track the keyframe movements trace the frame based skeletal lines, based on that the model is trained in such a way to produce a list of actions which looks more similar to the action that is undergone in real time as shown in Fig. 5.

Fig. 5. Output of Real-time Category

The Real-time activity can be performed by using may devices which is mounted with a camera facility. Here the Real-time human activity has been detected as shown in the Fig. 6.

The another option user left to use is Video-based activity detection. This option mainly used in the place of investigation of videos which occurred in the past. If the

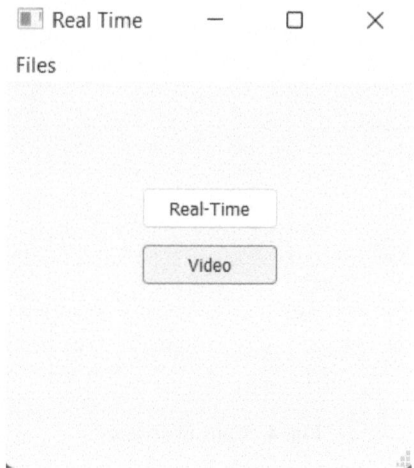

Fig. 6. Video-based Category

security or the surveillance operator gets suspicious upon a certain incident this option plays a vital role. Where it helps in detection of abnormal behaviour in videos by converting it into frames and check for abnormality in each frame. By clicking the Video option in the main menu as shown in the Fig. 6, a file explorer box appears. Where the user could select any video of their choice.

If the user selects Video option a File Explorer dialogue box opens. Where the user is allowed to select the video file they need to check for any abnormal activity. The selected video file is then displayed. For every time interval in the video the generated images get compared with the sample images in the datasets that is already available. Based on the action undergone by the human in the particular frame the model detects the what kind of activity is being performed and alerts if there is any abnormal behaviour.

As shown in Fig. 7, the user selects the video of a man walking in a mp4 format. Where this mp4 file gets converted into avi file which contains short videos mainly used for storing large amount off video files. As the detection system uses datasets which contains large amount of actions and for each action it contains hundreds of avi sample datasets. So in order to make the prediction more accurate these avi files then converted into frames and these frames helps the model to detect the action of the human and classify the behaviour accordingly.

User can also check the performance and accuracy of the detection system by providing different inputs in both realtime and video formats. In Fig. 8 the user selected video which does different actions to check the compatibility.

While processing the video captured both on realtime and recorded video using the deep network model, whenever the model detects any abnormal behaviour it lets the user know about the abnormality by means of alert box which helps the security know that there is an emergency and necessary actions could be taken about the situation. The system alerts the user with the help of alert message and a warning sign as shown in the Fig. 9.

Fig. 7. Output of Video-based detection

Fig. 8. Output of Video-based detection

Once the detection process is done the process can be terminated by clicking upon the close button on the top right corner. By clicking on the close button, an confirmation dialogue box appears as shown in Fig. 10. It is a confirmation box. By clicking OK the user could terminate the detection process and close the program.

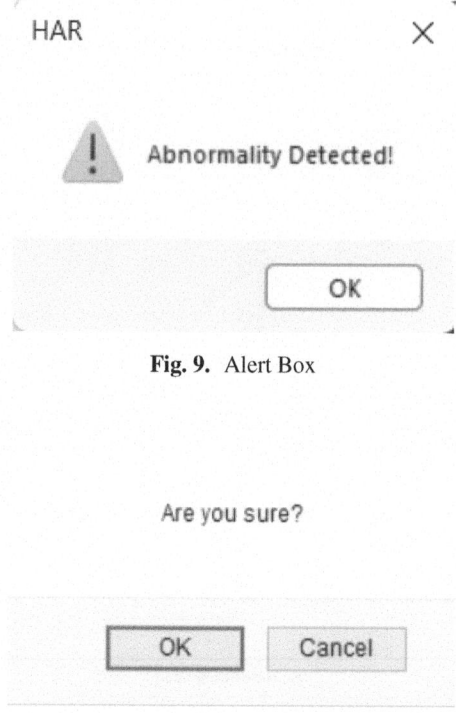

Fig. 9. Alert Box

Are you sure?

Fig. 10. Confirm Box

6 Conclusion

There are several HAR (Human Activity Recognition) systems present nowadays. But this paper deals with some innovative way of training the model in such a way in order to get accurate output. This was basically designed for the purpose of CCTV operators, securities and coordinators to ease their work. This detection system is designed with developing a model by training it with video clips or images as a input. Security and surveillance department get interested towards our proposed system as it efficiently replaces their work.

As far now the detection which are recommended is only on school and college premises. In future, we can implement it for all type of environment like roadsides, hospitals, supermarkets, ATM booths etc. We can also focus on increasing the efficiency and accuracy of Machine Learning model.

References

1. Cai, Q., Deng, Y., Li, H., et al.: Survey on humanaction: recognition based on deep learning. Comput. Sci. **47**(4), 85–93 (2022)
2. Bo, H.: Deep Learning on Video Based Human Action Recognition. Shanghai Jiao Tong University (2018)

3. Wang, H., Klaser, A., Schmid, C., et al.: Action recognition by dense trajectories. In: Proceedings of IEEE Conference on Computer Vision and Pattern Recognition, pp. 3169–3176 (2011)
4. Yang, L.: A Research of Behavior Recognition Algorithms Based on Multi-Features Fusion. University of Electronic Science and Technology of China (2013)
5. Han, X.X.: Study of Human Behavior: Recognition Based on Multi-feature Fusion. Nanjing University of Posts and Telecommunications, (2019)
6. Hu, Y., Yang, J.: Tracking algorithm based on block color histogram and mean shift. J. Syst. Simul. **21**(10), 2936–2939 (2009)
7. Wu, H., Yuan, Y., Wei, L., et al.: On entropy, similarity measure and cross-entropy of single-valued neutrosophic sets and their application in multi-attribute decision making. Soft. Comput. **22**, 7367–7376 (2018)
8. Fang, H.S., Xie, S., Tai, Y.W., et al.: RMPE: Regional multi-person pose estimation. In: IEEE International Conference on Computer Vision (ICCV) (2017)
9. Lin, W., Mi, Y., Wu, J., et al.: Action recognition with coarse-to-fine deep feature integration and asynchronous fusion. In: Thirty Second AAAI Conference on Artificial Intelligence (2018)
10. Simonyan, K., Zisserman, A.: Two-stream convolutional networks for action recognition in videos. In: Advances in Neural Information Processing Systems, pp. 568–576 (2014)
11. Ma, M.: Study on Human Pose Estimatio: Tracking and Human Action Recognition in Videos. Shandong University (2017)
12. Tran, D., Bourdev, L., Fergus, R., et al.: Learning spatiotemporal features with 3D convolutional networks. In: Proceedings of the IEEE International Conference on Computer Vision, pp. 4489–4497 (2015)
13. Bochkovskiy, A., Wang, C.Y., Liao, H.Y.M.: YOLOv4: Optimal Speed and Accuracy of Object Detection. In: IEEE Conference on Computer Vision and Pattern Recognition (2020)
14. Wu, Z., Wang, X., Jiang, Y., et al.: Modeling spatialtemporal clues in a hybrid deep learning framework for video classification. In: IEEE Transactions on Multimedia, pp. 461–470 (2017)
15. Donahue, J., Hendricks, L.A., Rohrbach, M., et al.: Long-term recurrent convolutional networks for visual recognition and description. IEEE Trans. Pattern Anal. Mach. Intell. **39**(4), 677–691 (2017)
16. Ravikumar, S., Kirubasri, G.: Video surveillance for intruder detection and tracking. In: Recent Trends in Computational Intelligence and Its Application: Proceedings of the 1st International Conference on Recent Trends in Information Technology and its Application (ICRTITA, 22), p. 409. CRC Press (2023)
17. Jaeyalakshmi, M., Vijay, K., Jayashree, K., Priya Vijay: A cloud based healthcare data storage system using encryption algorithm. In: Recent Trends in Computational Intelligence and Its Application, pp. 486–491. CRC Press (2023)
18. Carmel Mary Belinda, M.J., Antonykumar, K., Ravikumar, S., Kulkarni, Y.R.: A Collaborative Data Publishing Model with Privacy Preservation Using Group-Based Classification and Anonymity. In: Machine Learning Paradigm for Internet of Things Applications (2021)

ISAApp – Image Based Smart Attendance Application

Aritra Dutta$^{(\boxtimes)}$, G. Suseela, G. Niranjana, Pushpita Boral, Pranav Gupta, and Subha Bal Pal

Department of Networking and Communications, School of Computing, Faculty of Engineering and Technology, SRM Institute of Science and Technology, Kattankulathur, Tamil Nadu 603203, India
ad9382@srmist.edu.in

Abstract. This research paper introduces a Python-based implementation of a facial recognition system utilizing face recognition and Open CV libraries. The system has diverse applications, including security, surveillance, social media, and entertainment. By employing a pre-trained neural network, the system enables users to select an image file and accurately detect and recognize faces within it. Additionally, it includes functionality to encode faces from a specified folder, enabling comparison against the faces in the target image. By employing a blend of face detection, feature extraction, and machine learning algorithms, the system effectively identifies, and labels faces in images. Its effectiveness illustrates the potential of facial recognition technology for numerous applications, while also underscoring the ethical and privacy concerns associated with such technology. Addressing these issues and developing more precise and dependable systems would require further research. The primary objective of the system is to offer an improved solution for classroom attendance, ensuring the avoidance of false attendance records and significantly reducing the time required for attendance taking, through the application of machine learning solutions.

Keywords: Digital Image Processing · Digital Attendance · Face Recognition · Support Vector Machine (SVM) · Face Net · Image Registration · Object Detection · Image Segmentation · Pattern Recognition Methods · Convolutional Neural Networks (CNN)

1 Introduction

Attendance management holds paramount importance in educational institutions as it enables them to ensure the safety and security of students by tracking their presence in classrooms. However, traditional attendance systems, including roll-call, sign-in sheets, and RFID systems, have proven to be inefficient and susceptible to manipulation, leading to inaccurate attendance records.

This paper proposes a novel solution to address the shortcomings of existing attendance systems in educational institutions.

R. Geetha et al. (Eds.): AAIMB 2023, CCIS 2202, pp. 56–67, 2025.
https://doi.org/10.1007/978-3-031-73065-8_5

1. Time-Consuming Process

The conventional roll-call system requires teachers to individually call out each student's roll number and verify their presence, making it a time-consuming process. Additionally, ensuring accuracy and preventing proxy attendance demand considerable effort from teachers. The proposed face recognition-based attendance system offers a streamlined approach. With automated face detection and recognition, the system swiftly identifies and records students' attendance without the need for manual verification, saving valuable classroom time.

2. Manual Process

Both roll-call and sign-in sheets necessitate physical attendance recording, which is labor-intensive and prone to errors. Relying on manual methods increases the likelihood of mistakes and poses administrative challenges. The proposed face recognition system eliminates the need for manual data entry, as it automatically captures students' attendance based on facial features. This digitization not only improves accuracy but also frees up administrative resources for more productive tasks.

3. False Attendance

Conventional sign-in sheets and RFID systems are vulnerable to false attendance, as individuals can easily provide fraudulent signatures or use someone else's RFID card to mark attendance on their behalf. These practices undermine the integrity of attendance records and create accountability issues. In contrast, the proposed face recognition system offers a secure and reliable solution. By uniquely identifying each student based on facial patterns, it ensures that only the actual individuals present in the classroom are recorded accurately.

The face recognition-based attendance system leverages the power of artificial intelligence and machine learning algorithms to perform robust and efficient attendance tracking.

By implementing this technology, educational institutions can improve the accuracy and reliability of their attendance records, ensuring a safer and more secure learning environment for students. Furthermore, this solution presents an opportunity to optimize administrative processes, allowing educators to focus on core teaching responsibilities. However, it is vital to address ethical and privacy concerns associated with facial recognition technology to ensure its responsible and ethical implementation.

By striking a balance between technological advancements and ethical considerations, the proposed face recognition-based attendance system holds immense promise in revolutionizing attendance management in educational settings.

2   Related Work

Attendance Management has been a crucial task in many domains, including education, healthcare, and workplace management. Various approaches have been proposed and evaluated to manage attendance, including traditional methods such as pen and paper and modern methods such as RFID and biometric-based systems. In recent years, image-based attendance systems have gained significant attention due to their ability to provide real-time attendance management, reduce manual effort, and improve accuracy. In this section, we review previous works on attendance systems and image-based attendance systems.

2.1 Pen-and-Paper Attendance Systems

Pen-and-paper attendance systems are one of the traditional methods used to record attendance in various institutions, including educational settings, workplaces, and events. These systems involve manually taking attendance by using a paper sheet or a register where each student or participant signs their name or marks their presence. While pen-and-paper systems are straightforward to implement and do not require significant financial investment, they come with several technical limitations and drawbacks:

1. Potential for Errors: Manual data entry can lead to errors in pen-and-paper attendance systems, such as misspelt names, illegible handwriting, and accidental omissions, impacting the accuracy of attendance records.
2. Lack of Real-Time Monitoring: Pen-and-paper systems lack continuous real-time monitoring of attendance, as data is recorded only at specific moments, hindering immediate actions or emergency responses based on attendance data.
3. Time-Consuming Data Management: Managing attendance records manually in larger institutions can be time-consuming and cumbersome, involving collecting paper sheets, collating data, and entering it into a database, leading to inefficiencies and delays.
4. Lack of Data Insights: Pen-and-paper systems do not offer data insights or analytics since attendance data is not stored digitally, limiting the ability to analyze patterns, track trends, or make informed decisions based on attendance data.
5. Difficulty in Data Sharing and Accessibility: Sharing attendance data with stakeholders becomes challenging as it is stored physically, requiring manual effort for distribution, hindering efficient communication and collaboration among relevant parties.

2.2 RFID-Based Attendance Systems

Radio-frequency identification (RFID) technology has been widely used in attendance tracking systems in various settings, including educational institutions and workplaces. Several studies have explored the use of RFID-based attendance systems and their advantages over traditional paper-based systems. For instance, [1] developed an RFID-based student attendance tracking system that achieved a high level of accuracy and real-time monitoring. [2] proposed an RFID-based attendance system for schools that was found to be effective in reducing the time and effort required for attendance management. [3] designed and implemented an RFID-based student attendance management system that offered several advantages, such as automatic record keeping and ease of data management.

Despite the advantages of RFID-based attendance systems, they also have limitations that must be considered before implementation.

1. Risk of Loss or Damage: RFID tags or cards used in attendance systems can be lost or damaged, leading to inaccuracies in attendance records. If a student or participant loses their RFID tag or card, their attendance may not be recorded properly, affecting the overall accuracy of the system.
2. Higher Implementation Cost: Implementing RFID-based attendance systems can require a higher initial investment compared to traditional paper-based systems. The cost includes purchasing RFID tags or cards, RFID readers, and the necessary infrastructure to support the technology.
3. Privacy Concerns: RFID-based attendance systems involve collecting and storing personal data, such as the identity of students or employees. Concerns about privacy arise as this data can potentially be misused or accessed without authorization. Proper measures must be put in place to safeguard sensitive information and comply with data protection regulations.
4. Data Security: Ensuring the security of the data transmitted and stored in RFID-based attendance systems is critical. Unauthorized access to the system or data breaches could compromise the integrity of attendance records and raise serious security issues.
5. Scalability: The system's scalability should be considered, especially in large institutions or organizations with a growing number of attendees [15]. The system should be able to handle increased attendance data without compromising performance.

2.3 Biometric-Based Attendance Systems

Biometric-based attendance systems are another type of electronic attendance tracking system that uses biometric data, such as fingerprints or facial recognition, to identify students or employees and record attendance.

Several studies have explored the use of biometric-based attendance systems and their advantages and limitations. For example, [4] developed a biometric-based attendance system for a university that offered high accuracy and real-time monitoring. [5] implemented a biometric-based attendance system in a secondary school and found that it was effective in reducing absenteeism and improving attendance tracking. [6, 12] designed and evaluated a biometric-based attendance system for a workplace and found that it was reliable and easy to use.

Despite the advantages of biometric-based attendance systems, they also have limitations that must be considered.

1. Higher Implementation Cost: Biometric-based attendance systems often involve the use of specialized hardware and software for biometric data collection and processing. The cost of acquiring and installing biometric sensors, such as fingerprint scanners or facial recognition cameras, along with the development of sophisticated algorithms, can be higher compared to other electronic attendance tracking systems [14].
2. Maintenance and Upkeep Expenses: Biometric systems require regular maintenance and calibration to ensure accurate and reliable performance. Technical expertise and ongoing support may be needed, adding to the overall implementation cost.
3. Privacy Concerns: Biometric data, such as fingerprints or facial features, is highly sensitive and unique to individuals. Collecting and storing this data raise privacy concerns [11], as it could be misused if not adequately protected. There is a need

to establish robust data protection protocols and ensure compliance with privacy regulations.

4. Technical Limitations: Biometric systems may face challenges in certain conditions, such as poor lighting affecting facial recognition accuracy or fingerprint readability issues due to dirty or wet fingers. These technical limitations could result in occasional errors or false rejections.

5. Scalability: As the number of users increases, biometric systems need to handle a growing database of biometric templates. Ensuring scalability and efficient data handling is essential in large-scale implementations.

To address these concerns, some studies have proposed using privacy-enhancing technologies, such as encryption and secure data storage, to protect biometric data. For example, [7] proposed a biometric-based attendance system that uses secure data storage and encryption to protect sensitive data.

Overall, biometric-based attendance systems offer several advantages over traditional paper-based systems and other electronic attendance tracking systems, such as accuracy and real-time monitoring.

3   Methodologies

The technique which we have deployed for attendance monitoring is face recognition. The main sub-divisions being image collection, face recognition, attendance registration. The system is designed to eradicate false and erroneous attendance which is often recorded using manual methods. It is useful in saving time, providing efficient and correct data, reduce manual labor and ultimately improve the entire process of attendance monitoring.

3.1   Image Collection

The first step in this digital attendance system is to provide the data for the model to process, which are the images of the classroom. The lecturer captures three to four images of the entire classroom, so that all the students are covered. These images are then fed into the system using the user-friendly interface of the application.

1. **Image Capture Procedure**: To ensure adequate coverage of the classroom, the lecturer is responsible for capturing three to four images from different angles and viewpoints. This step is essential to account for variations in lighting conditions and student positions. The use of multiple images enhances the robustness of the system and reduces the chances of missing any students during attendance tracking.

2. **Data Preprocessing and Quality Assurance**: Before the images are used for attendance tracking, data preprocessing techniques may be applied to enhance image quality and remove any artifacts that might affect recognition accuracy. Quality assurance checks may also be performed to ensure that the images meet the required standards for accurate facial recognition.

3. **Documentation and Metadata**: To maintain transparency and traceability, the system documents each image's metadata, such as the date and time of capture, the

lecturer responsible, and any relevant contextual information. This documentation assists in tracking the source of the images and serves as a reference for future analysis.

3.2   Face Recognition

In the proposed attendance monitoring system, the next critical step after collecting group pictures of the classroom is face detection and recognition. To achieve this, We have utilized image processing libraries, such as OpenCV and face recognition, which are readily available and integrated into Python. Additionally, the Tkinter library is employed to facilitate the loading of images into the AI model and to display the encoded information.

1. **Training the Model**: We have trained the AI model using image processing libraries like OpenCV and face recognition. We have harnessed the capabilities of OpenCV to process the images and prepare them for face detection and recognition tasks.
2. **Face Detection using OpenCV**: The face recognition library, an essential component of the system, is responsible for detecting all the faces present in the group images. By leveraging the functionality of this library, the system identifies and locates the facial features of individuals, including the eyes, nose, mouth, and chin. This process is crucial as it serves as the foundation for subsequent steps of facial recognition.
3. **Utilizing Facial Features for Attendance Monitoring**: [10] Tracing and manipulating facial features play a pivotal role in various applications, including the proposed attendance monitoring system. The extracted facial features are instrumental in identifying and recognizing individuals accurately, ensuring reliable attendance tracking. By using the face recognition library, the system gains insights into the unique characteristics of each person's face, enabling precise identification.
4. **Image Compression for Energy Efficiency**: [8] Considering the potential storage and processing requirements of handling numerous images, We have employed energy-efficient image compression techniques. Image compression helps optimize the storage space and reduces computational load, contributing to the system's overall efficiency.

3.3   Attendance Registration

The pictures are processed and the faces which are detected are matched with the existing database of the institution. Once the faces are successfully matched, their details are extracted and the name and timestamp are read, which is then converted and stored in the form of a.csv file. This automates the entire process of attendance marking and reduces the unnecessary hassle of roll call and other manual methods.

These methods are not only prone to error, but also hamper the performance of students in a classroom because often they are more attentive towards their attendance registration rather than the content being delivered in the class by the lecturer. With the use of our application, student attendance can be easily registered and monitored by the mentor and can be viewed anytime they wish to. The students can also keep a check on their attendance percentage by simply logging into our user-friendly portal.

4 Implementation

The system we developed aims to enhance accessibility and reduce costs. Its primary goal is to enable teachers to concentrate more on delivering lectures rather than spending time on attendance-taking procedures, thereby providing them with ample teaching time.

To initiate the attendance process, a teacher captures a photograph of the entire classroom using the cameras installed beforehand. After verifying the image, they upload it into the system, which automatically marks each student's attendance (Fig. 1).

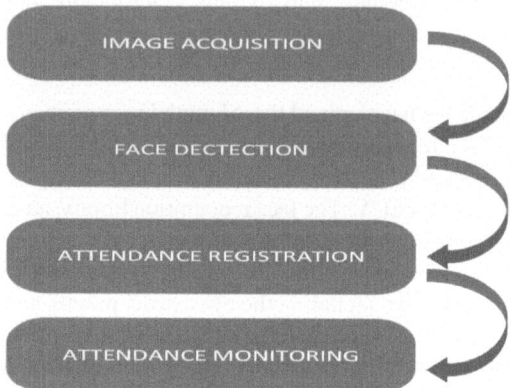

Fig. 1. Process flow of Implementation

Image Acquisition

The process of image acquisition is a crucial step in the attendance tracking system, as it lays the foundation for accurate facial recognition and identification. In this section, we elaborate on the image acquisition process, emphasizing the authentication, flexibility, and optimization aspects.

1. **Authentication for Access**: To maintain the security and integrity of the attendance tracking system, the lecturer must undergo an authentication process to access the image acquisition feature.

   This authentication ensures that only authorized personnel, such as teachers or instructors, can capture images for attendance tracking. It prevents unauthorized access to sensitive data and helps maintain the system's confidentiality.
2. **Sources of Image Acquisition**: The system offers two primary sources for image acquisition: the pre-installed cameras within the classroom and mobile devices. The pre-installed cameras, strategically positioned within the classroom, provide a convenient and non-intrusive means of capturing images without disrupting the learning environment. On the other hand, the flexibility of capturing images using mobile devices allows lecturers to take attendance in various settings, including temporary or remote classrooms.
3. **Image Quality and Visibility**: To ensure precise facial recognition, it is imperative that the images acquired by the lecturer offer sufficient quality and visibility of all

students' facial features. Clear and well-lit images facilitate the accurate detection of facial landmarks, aiding the subsequent recognition process. The lecturer is encouraged to capture images from optimal angles, minimizing obstructions and shadows that may hinder feature detection.

Face Detection

After obtaining the group images of the classroom, the subsequent critical step involves the detection and recognition of faces. To achieve this, we have utilized trained models by leveraging Python's built-in image processing libraries, including OpenCV and face recognition. Additionally, the tkinter library plays a key role in facilitating image loading and encoded information display.

OpenCV, a popular open-source software library for computer vision and machine learning, serves as a foundational framework for a wide range of computer vision applications. By leveraging OpenCV, we can effectively identify and manipulate facial features in the acquired images, accurately determining the precise locations and contours of essential facial components such as eyes, nose, mouth, chin, and more. This capability is particularly crucial for the successful operation of our attendance monitoring system.

For the extraction of distinctive facial features and subsequent recognition, we have employed the face recognition library. This library enables the precise detection and analysis of all faces present in the given images. Additionally, OpenCV is employed to seamlessly display the target images (Fig. 2).

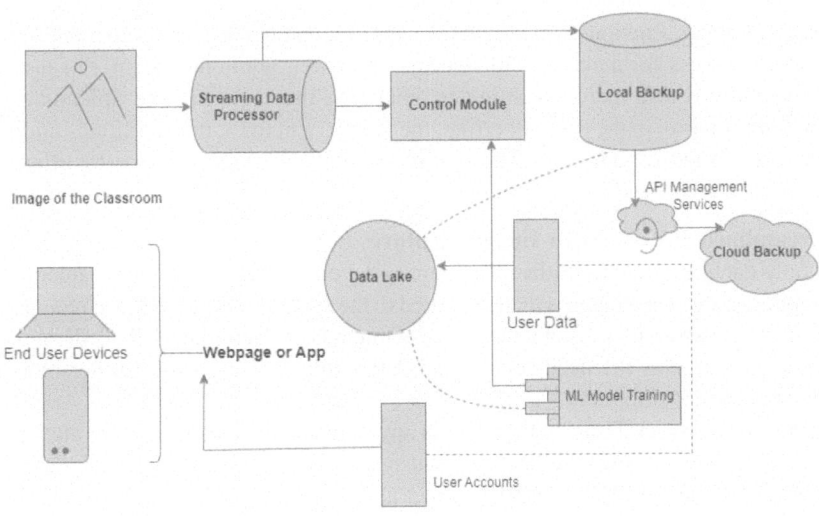

Fig. 2. System Architecture

Attendance Monitoring

The attendance monitoring phase is a critical step in the proposed system, where the algorithm detects and records the attendance of each student based on the acquired images. The process involves facial recognition, identification, and data storage, providing convenient access to attendance records for both students and lecturers through the system's web application or mobile application.

1. **Attendance Recording and Storage**: Once the students are successfully recognized, their attendance is recorded in real-time. The system creates a digital record for each student, marking their presence in the class. The attendance records are then stored securely within the system's database, ensuring data integrity and privacy.
2. **Accessible Web and Mobile Applications**: To provide seamless access to attendance records, the system offers both a web application and a mobile application. Through the web application, lecturers can easily view and manage attendance records for their respective classes. On the other hand, students can access their individual attendance records through the mobile application, allowing them to keep track of their attendance progress.
3. **Real-Time Updates**: Attendance records are updated in real-time as students are recognized during class sessions. This real-time updating ensures that attendance data remains accurate and up-to-date, providing an accurate representation of students' attendance statuses.

5 Evaluation

Time Consumption
The implementation of this process has resulted in a substantial time reduction for attendance-taking. Previously, it would take approximately ten minutes to complete the attendance process for a class of 50 students. However, with our system, this task has been streamlined to a mere two minutes, involving the capturing and uploading of a photo. Furthermore, in the event of errors, the recovery time for our solution is minimal, taking just a little over a minute. These improvements highlight the superior efficiency of our system compared to conventional methods such as roll calls and sign-in sheets.

1. **Streamlined Process with Image Capture**: In the proposed system, the lecturer simply captures a photo of the classroom using either the pre-installed cameras or a mobile device. This image is then uploaded to the system, where the facial recognition algorithms swiftly identify and record the attendance of each student. This streamlined process eliminates the need for manual data entry or paperwork, further reducing overall time consumption.
2. **Minimal Recovery Time**: In the event of any errors or discrepancies in the attendance data, the recovery time with the proposed system is minimal. The system's real-time updates and the ability to quickly correct and reprocess data ensure that any errors are rectified promptly. The recovery time, as reported in the research, takes just a little over a minute, further underscoring the system's efficiency in handling potential issues (Fig. 3).

Model Accuracy
In our testing, we found that the model achieved an accuracy level between 75% and 85%.

Fig. 3. Time consumption

Remarkably, we identified a maximum of 6 false positive detections in a classroom with a total of 50 students. This highlights the superior accuracy of our model in comparison to other currently available models in the market. It is noteworthy that by incorporating multiple photos of the classroom from various angles, we can enhance the accuracy to 90%. However, it is crucial to consider that this improvement comes at the expense of increased overall time required for the attendance process.

1. **Superior Accuracy Compared to Market Models**: One significant finding from the testing is that the proposed model outperforms other currently available models in the market. With only a maximum of 6 false positive detections in a classroom of 50 students, the system demonstrated its ability to accurately recognize and identify students' faces. The low number of false positives indicates that the model is effectively distinguishing between different individuals, minimizing the risk of misidentifications.
2. **Enhancing Accuracy with Multiple Photos**: We observed that the accuracy of the model can be further improved by incorporating multiple photos of the classroom from various angles. By utilizing different viewpoints, lighting conditions, and student positions, the model gains more robustness in handling potential variations in appearance. This enhancement can lead to an accuracy level of 90%, significantly raising the system's reliability.
3. **Trade-Off with Time Consumption**: However, it is crucial to consider the trade-off between accuracy improvement and time consumption. As mentioned, capturing and processing multiple photos take more time, which could impact the overall efficiency of the attendance process. Therefore, striking a balance between accuracy and time consumption becomes an essential consideration for system implementation (Fig. 4).

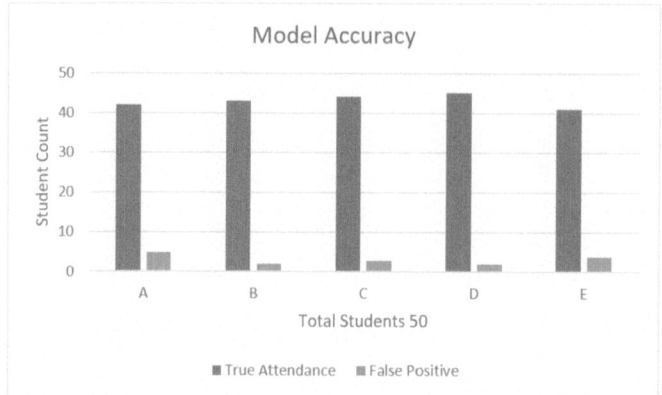

Fig. 4. Data Visualization showing Model Accuracy and false positive

6   Conclusion

In conclusion, the research paper has successfully showcased the capabilities of facial recognition-based attendance systems, leveraging artificial intelligence and machine learning algorithms to enhance attendance recording across different environments. The implementation effectively tackles privacy and security concerns through the use of face recognition and OpenCV libraries, and its performance has been rigorously evaluated through experimental testing. The system's ability to identify students and continually improve its facial pattern recognition with new data has proven to be a valuable solution to the attendance problem addressed in the paper.

Nevertheless, recognizing the challenges that accompany any novel technology, it is crucial to acknowledge the existing limitations and obstacles. Future research endeavors should prioritize enhancing the accuracy and reliability of facial recognition-based attendance systems while also addressing potential issues related to bias and misuse.

Looking ahead, several areas hold promise for future enhancements:

1. Algorithm Optimization: There is a clear goal to further refine the algorithm, prioritizing faster processing to increase the system's efficiency and responsiveness.
2. Diverse Scenario Recognition: Expanding the capabilities of the model to identify individuals in various scenarios beyond the classroom setting will broaden its applicability and utility.
3. Offline Functionality: Working towards enabling the system to function offline without connectivity by storing essential model data and results within devices for a limited duration will provide seamless attendance recording even in connectivity-challenged environments. By pursuing these avenues of improvement, facial recognition-based attendance systems have the potential to revolutionize attendance recording practices and enhance overall efficiency across diverse settings. It is imperative to continue refining and responsibly deploying such systems, ensuring they align with ethical standards and respect individual privacy and rights.

References

1. Aljuaid, H., Altuwaijri, S.: RFID-based student attendance tracking system. IEEE Access **7**, 5262–5270 (2019)
2. Basha, M.A., Ramana, B.V.: RFID based attendance system for schools. Int. J. Pure Appl. Math. **119**(12), 2791–2795 (2018)
3. Wang, W., Li, C., Li, L., Li, Y., Li, H.: Design and implementation of RFID-based student attendance management system. J. Ambient. Intell. Humaniz. Comput. **10**(2), 829–836 (2019)
4. Karim, M.A., Rahman, M.T.: Development of biometric attendance system for a university. J. Eng. Appl. Sci. **13**(1), 174–182 (2018)
5. Khan, F.M., Ahmad, I., Haque, M.A., Alam, M.: Biometric-based attendance system using deep learning for secondary schools. J. Ambient. Intell. Humaniz. Comput. **12**(1), 625–635 (2021)
6. Sharma, M., Shukla, P., Mittal, S.: Biometric attendance system for workplace. J. Ambient. Intell. Humaniz. Comput. **12**(4), 4377–4384 (2021)
7. Shrivastava, P., Bhargava, S.A.: Secure biometric attendance system using face recognition. J. Ambient. Intell. Humaniz. Comput. **12**(5), 5775–6578 (2021)
8. Suseela, G., Phamila, Y.A.V., Niranjana, G., Ramana, K., Singh, S., Yoon, B.: Low energy interleaved chaotic secure image coding scheme for visual sensor networks using pascal's triangle transform. IEEE Access **9**, 134576–134592 (2021). https://doi.org/10.1109/ACCESS.2021.3116111
9. Reddy, A.K., Kumar, A.V.: Real-time student attendance monitoring using bluetooth low energy (BLE) beacon technology. Int. J. Distrib. Sens. Netw. **16**(1), 1550147720902481 (2020)
10. Kim, J., Kim, J., Lee, K., Lee, S.: A smart classroom system using facial recognition for attendance management. J. Supercomput. **75**(5), 2657–2675 (2019)
11. Patel, S., Raval, H., Shah, S.: IoT-based smart attendance system using fingerprint recognition. Int. J. Adv. Trends Comput. Sci. Eng. **9**(2), 46–52 (2020)
12. Yu, Q., Liu, H.: An intelligent attendance monitoring system using facial recognition in smart cities. Clust. Comput. **24**(1), 1605–1616 (2021)
13. Thakur, V., Kumari, A., Dey, N.: Biometric-based attendance system using machine learning techniques. In: Proceedings of the Third International Conference on Smart Systems and Inventive Technology, pp. 1095–1102 (2020)
14. Singh, A.K., Patel, R.B., Kumar, A.: Development and evaluation of a fingerprint-based attendance system for educational institutions. Int. J. Intell. Syst. Appl. **13**(2), 14–24 (2021)
15. Sardana, G., Sharma, M.: An efficient RFID-based attendance monitoring system for corporate offices. In: 2020 2nd International Conference on Trends in Electronics and Informatics (ICOEI), pp. 1252–1257. IEEE (2020)

A Self-learning Ai-Based Information Leak Protection System

M. Jaeyalakshmi, P. Rohit Gangadhar$^{(\boxtimes)}$, M. Srivatsan, and M. Bhavani

Department of Computer Science and Engineering, Rajalakshmi Engineering College, Mevalurkuppam, India
rohitgangadhar.p.2019.cse@rajalakshmi.edu.in

Abstract. In order to prevent sensitive information leakage, sometimes called data leakage or loss to unauthorized recipients, organizations use information security systems to prevent it. In fact, data can leak through a variety of channels. While it is impossible to completely prevent it from happening, there are certain things one can do to decrease the likelihood of it happening. All financial institutions manage client information for commercial purposes. Humans used to search these digital documents for sensitive information, which was labor-intensive and expensive. In order to analyze material using cutting-edge data mining, statistics, and machine learning methods across multiple data dimensions, a smart and trustworthy system is required. According to the research, AI-based information leak prevention solutions that require no configuration can be created by utilizing LSTM to categorize document pictures according to the existence of NPI and PII semantic signatures. This system is made to be actively utilized as a tag SD pictures when they are storing data alarm systems. It may also serve as a real-time checkpoint for data loss brought on by documents that are being utilized or transported. The proposed model offers a protection mechanism against information loss by utilizing the most sophisticated LSTM-based binary classifier in artificial intelligence.

Keywords: Artificial Intelligence · LSTM(Long Short-Term Memory) · Machine Learning · Data mining · Digital Document · Deep Learning

1 Introduction

Title insurance is indemnity insurance, and unlike other insurances, it covers the past. This means that title insurance shields the insured from losses and claims resulting from a flaw in the title to the property that existed even before the policy's start date and that the insured was unaware of. So, to put it simply, title insurance is protection against any loss brought on by a flaw in the property's title [1]. Before providing insurance to a buyer or mortgagor of a property, title insurance companies conduct title searches on public records. Title insurers search public records after a real estate contract is signed and an escrow account is opened to look for any title issues. An examination of very old land records may be necessary as part of the research [2]. Interestingly, over a third of all title

R. Geetha et al. (Eds.): AAIMB 2023, CCIS 2202, pp. 68–78, 2025.
https://doi.org/10.1007/978-3-031-73065-8_6

searches indicate a title issue that needs to be resolved before the deal closes.To establish the legal title to a piece of property, to claim it, and to learn if it is the subject of any claims, one conducts a title search, which is an investigation of public records. or not. 1 The title may contain typographical errors and unresolved build code issues that give it a "dirty" appearance. Title insurance protects borrowers and homebuyers from financial loss or property damage caused by liens, obstructions, or flaws in the ownership of the property. Claims against title that are frequently present include tax arrears, mortgage liens, home equity lines of credit (HELOCs), land titles, and competing wills [7, 13–16].

Sensitive data requests can be handled and processed independently by AI, preventing data from getting into the wrong hands. Additionally, it eliminates the possibility of human error for industries looking to add a second layer of security to their data. For a variety of purposes, AI systems can efficiently analyze and use data, but they are unable to discern the meaning of the data. AI-based data protection and security technologies can dramatically increase a company's ability to protect the privacy of its sensitive data. The administrative burden on compliance administrators may be lessened by technologies like natural language processing (NLP), and technologies like accurate and efficient identity management may be made possible by technology like intelligent facial recognition [4]. All forms of personal data may be tracked and protected by AI systems, which can also assist a business make sure that its security measures adhere to all relevant standards and regulations on private rights. AI is establishing itself as a potent technology friend for companies in terms of data security and privacy.

One of the primary focuses of security teams is to lessen the possibility of and damage caused by data breaches as they defend an organization's digital assets and infrastructure against assaults. The term "information leakage" refers to the unauthorised disclosure of private, proprietary, or otherwise sensitive information. Leaks of sensitive data can occur when hackers gain access to an unprotected server or application, or when an employee is duped by a social engineering assault into divulging confidential information or giving away login credentials for a protected network.

In the hands of cybercriminals, a data breach can lead to fraud, stolen identities, money, and even access to otherwise secure networks. Information that has been illegally obtained can even be sold in underground markets on the deep and dark web.

The safeguarding of sensitive information is a significantly bigger operational challenge for enterprises (IS). The term "SI" in the context of this article refers to information that, while not directly identifying a specific person, is nonetheless related to that person and transmits confidential information or information that, if made public, could put that person in risk. The SI contains data from NPI, PII, and PI [5]. Network infrastructure cannot match the speed at which artificial intelligence (AI) systems can recognize, reroute, and process requests for private data [7]. AI tools can also handle complex data requests in a more time- and money-efficient way.

2 Related Work

The ability to recognise objects on the page is necessary for comprehension of the document. Granularity may affect how well an object performs. The intrinsic document image hierarchies taken into account in this study include block-level area object recognition. The simultaneous object classification, bounding box recognition, and page object masking are all done using the terminal network [1]. The Area-Based Convolutional Neural Network (R-CNN) technique is the model used for this network. To identify page objects in document images, the R-CNN mask is modified in this project. The convolutional neural network (CNN) framework comprises feature pyramid networks (FPNs), area recommendation networks (RPNs), and region feature extraction of interest (ROI). The system supports bounding box regression, mask prediction, and label classification terms. Our strategy broadens the default Mask R-CNN framework for document image processing. It is capable of both logic comprehension and layout analysis [2].

This study uses an R-CNN mask-based network to identify hierarchical page objects in document pictures and to produce end-to-end results that include object classification, bounding box, and page object hiding. Along with block-level area object detection, different levels of granularity will be taken into account. A latex-based synthesis generation technique was used to increase the size of the training dataset [3]. Three outputs are generated by this method: a bounding box, a mask, and a classifier. Documents that need to be protected include identification cards, licences, passports, and delivery receipts since they include personal data. Privacy is applicable to a wide range of real things from a present security standpoint, which is restricted to online services.

The creator of this concept suggests an offline document security method that can safely protect identification cards and significant papers shared online that contain private information. The theft of messages and packages containing personal data, such as names, addresses, phone numbers, and other details, and using that data for criminal activity can also be avoided [4]. Two-dimensional barcodes and a way of safeguarding offline documents based on the user's private key are suggested by the article's author. Webcams, smart glasses, and smartphones are a few examples of client devices that can be used with the suggested methodology. In order to protect user privacy, the server hides offline documents [5]. The client then securely restores the hidden offline documents.

The server specifically creates and distributes offline documents that have a QR code over private information that has been encrypted with the user's secret key and stored in a two-dimensional barcode [7]. In contrast to a server with cloaking capabilities, a client can only reveal private information to the rightful owner by decrypting offline documents using the user's private key and hiding the resulting data from the QR code once user authentication is finished. Future work will focus on precisely extracting unidentifiable regions from the client device to enhance the precision of anonymous offline document retrieval and lessen the effect of noise. The author then intends to experiment with different offline document formats by integrating better offline document recovery functionality into smart glasses or smartphones [8].

Data masking is a potent technique that can be used to protect document images from tampering or illegal tampering, unlike standard black and white patterns like barcodes and rapid response codes. Strong. The author of this paper suggests a dependable water-marking method to safeguard genuine documents using ordinary adversarial networks (GAN). The input material is first given the right shape using geometry matching [9]. Using the aforementioned networks, the generated document is then extracted from the input document and used as a template for data detection and masking in the following step.

After that, it will use an algorithm to hide a secret piece of information within the document, creating a watermarked document with minimally altered content under normal observation. In addition, It will also present a technique that, by comparing the pixel values between the generated and watermarked documents, finds hidden data from the watermarked document. It will use pseudo random numbers to encode the secret information before hiding it to increase security. Finally, It will show that, when compared to cutting-edge methods, our approach provides high data detection precision and competitive performance [10].

Network-based document clustering groups documents based on the significance and strength of their connections. This technique can be used with a variety of information categories that describe the significance of documents and how they relate to one another [11]. The author has set up a probability network graph, developed a generalized model of probability, and used computational methods to cluster the extensive literature in this study. The most crucial phase of the clustering process is calculating the similarity between documents. In this study, the author ranks the documents externally to determine their quality before comparing the documents' latent vectors to determine how similar they are to one another. Been looked into. The suggested approach outperforms those currently in use. Additionally, a network supporting multi-label clustering has been found, and a productive computational method for the probabilistic general model has been suggested [12].

3 Existing System

3.1 User-Based Document Management Mechanism in Cloud

The development of cloud computing benefits greatly from the document. By using an electronic document, the user can receive and exchange information. It has a wealth of information and many different representations. But security is also put in jeopardy. First, suggest a cutting-edge user-based document secure management system that uses re-encryption to address the secure requirement for cloud-based documents. In order to combine document creation encryption with access control, the re-encrypted key will be generated in accordance with the access control requirements [20].

3.2 Automatic Authenticity Verification of Printed Security Documents

Specifically designed security documents were examined in the current test. All of these documents—bank checks, various tickets, such as lottery tickets and airline tickets—as well as legal documents, certificates, notes, and postage stamps—are of the same type from a security perspective. Criminals are increasingly attempting to produce forged copies of these documents. This study set out to develop a standardized methodology for quickly determining if these sensitive materials were legal. The idea of authenticity against duplication was discovered in the feature space after the suggested method successfully extracted security features from the document picture. The bank check is taken into account as routine throughout testing. A support vector machine is employed to validate these criteria [4, 18, 19].

3.3 Machine Authentication of Security Documents

This technique extracts security objects computationally from the document images, allowing the concept of authenticity or duplication to be determined in the object space. Bank checks are the standard for current experience. Neural networks as well as support vector machines are both used to validate these tests.

3.4 Document Encryption Through Asymmetric RSACryptography

The best method for data security is cryptography with asymmetric keys. The RSA (Rivest-Shamir-Adleman) algorithm is the most popular choice for asymmetric cryptography. The most common type of encrypted document that is attached to an email. The file contains the file types.docx,.pptx,.xlsx,.pdf, and.jpg, as well as.mp4. Public and private keys can be sent privately by sending digital documents created during encryption. The private key created during encryption is used to decrypt the file at the other end of a digital communication.

3.5 Printed Document Authentication Using Watermarking Technique

The use of modern, sophisticated electronic tools like scanners and computers can make forgeries of official printed documents very simple. There is an urgent need to find solutions to the threat of the counterfeiting of such documents because the forged documents are typically impossible for human eyes to detect. Using the watermarking technique, it is possible to embed data in printed documents that is used to verify their authenticity. Since the embedded information may be invisible to the human eye, forgery is more difficult for the attackers [21–24]. The watermarked document's embedded watermark is removed in order to verify the document's owner. The printed document may, however, experience printing and scanning (PS) distortion during the verification process, thus it is essential to address the noise, undesired rotation, and any other printing and scanning-related degradation. To solve this problem, this system employs a watermarking method [25].

4   Proposed System

The suggested model, which uses a binary classifier and is based on contemporary LSTM techniques, offers a protection mechanism against information loss as shown in Fig. 1.

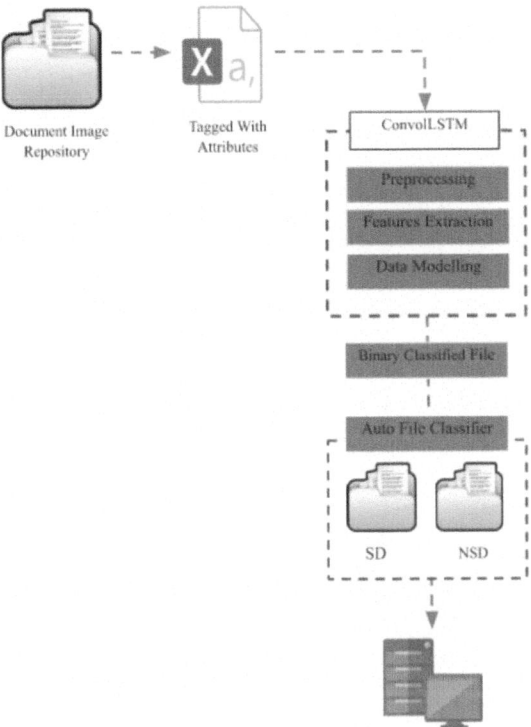

Fig. 1. System Architecture

Use a binary classification model to establish decision boundaries and identify the combined n-gram features of documents. LSTM uses and extracts tagged data to categorize document images as SD (Secure Document) or NSD. Figure 2 illustrates the system workflow.

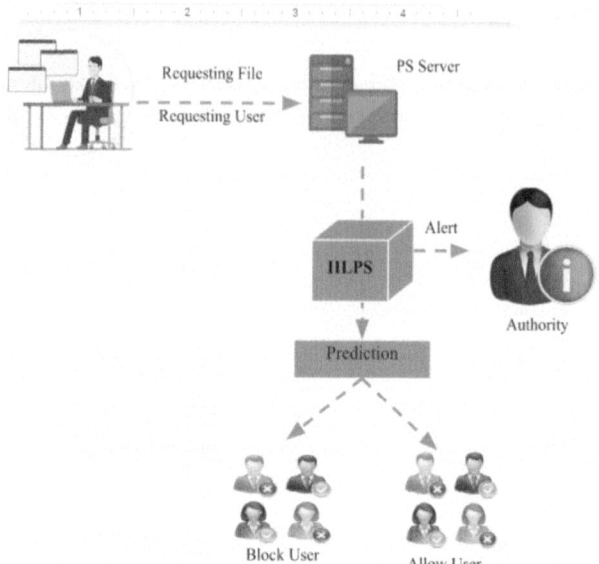

Fig. 2. System Workflow

4.1 Data and Database

Imagine a situation where a business needs to keep track of the names of hundreds of employees so that each can be uniquely identified. The business then compiles data on each of these employees. When I refer to "data," I really mean that the business gathers various pieces of information about a particular audience. Therefore, this object could be anything, like a laptop, mouse, or a real-world object like a person.

- **Database Management System**

  A database management system (DBMS) is computer software that exchanges data via communications between databases, users, and other applications. The database allows for the modification, retrieval, and deletion of data. Any type is acceptable, including but not limited to strings, integers, and pictures.

- **Structured Query Language (SQL)**

  The fundamental tool for managing and accessing a relational database is SQL. It can access subsets of data, edit databases, add, update, or delete rows of data, among other things, by utilizing SQL. Following are the various SQL subsets:

- **DDL** (Data Definition Language) – It enables a variety of database operations, including the creation, alteration, and deletion of objects.
- **DML** (Data Manipulation Language) – It permits Its will to access and modify data. The database can be used to retrieve, update, delete, and insert data.
- **DCL** (Data Control Language) – It enables control over who has access to the database. Granting or revoking access permissions is an example.

- **TCL** (Transaction Control Language) – It permits Its decision-making process to handle database transactions. Example: Set Transaction, Save Point, Commit, Rollback, etc.
- **MySQL**

The fundamentals and complexities of MySQL are covered in the MySQL lesson. Our MySQL training is useful for both novices and experts. The MySQL relational database management system is built on the widely used language, Structured Query Language, for updating and accessing database records. MySQL is free and open source software and is released under the GNU license. This is supported by Oracle Company. For managing databases and manipulating data, MySQL offers a variety of SQL queries. In addition to creating and deleting tables, these queries also allow for the creation, deletion, update, and selection of records. Additionally, MySQL interview questions are offered to aid in your understanding of MySQL databases.

- **Apache Web Server**

The web server, along with PHP, MySQL, JavaScript, and CSS, is the dynamic web's fifth hero. This refers to the Apache web server in the context of this book. The web server performs a lot more work in the background than we've discussed, even though we've only covered a small portion of what it does during the HTTP server/client transaction. For instance, Apache supports a wider variety of file types than HTML; It is capable of processing a wide range of data, including Flash files, images, MP3 audio files, and RSS feeds. Web clients are required to notify servers of any HTML elements they encounter. It's not necessary for them to be static files like GIFs. Yes, PHP can also generate files for you, including images and other file kinds, either instantly or in advance of a later serving. Usually, it calls runtime modules to accomplish this or uses precompiled Apache or PHP modules.

A separate collection of modules that Apache provides is. The security-related modules, in addition to the PHP module, are the most crucial for your requirements as a web developer. Thanks to important innovations, he will figure out how to use some of these modules to improve his power. It will be covered in later chapters of the book. Although MySQL, Apache, and PHP are the three most popular engines in their respective categories, it is still unknown whether open source was involved in their creation. The fact that they are open source, however, merits emphasising because it means that teams of developers working together in the community construct the features they wish to modify.

- **Web Framework**

Programmers can build web applications using the Online Application Framework, also known as the Web Framework, which is a set of tools and modules that frees them from low-level issues like protocols, management, threading, etc. Integrated Developing Environments (IDEs), which offer many additional capabilities including in-editor debugging and programme testing, as well as function explanations and much more, are far superior to specialized programme editors in terms of programming productivity.

5   Result and Discussion

All corporate and public sectors worldwide now regularly store and manage digital documents in the digital age. A digital package, also known as a heterogeneous document stream, is created by bulk scanning physical documents and storing them in a digital archive. To simplify robotic automation (RPA), document flows must be automatically segmented into a consistent and independent subset of multi-page documents by defining appropriate document boundaries. The confusion matrix and all measures obtained from it, including overall accuracy, F1 score, recall, and accuracy, are used to test the model. Matching to the range from 6 to 9 on the model for determining whether a document contains secured content. If a clean document was projected to be Secured rather than a Secured document being predicted to be Not Secured, the firm would not be as negatively impacted by a data leak.

The cost of validation directly relates to accuracy. The expense of the system would ultimately increase if a false detection needed to be confirmed by a human partner. F1-Score evaluates how well the two measurements mesh. The Recall and Precision scores of the proposed MLP model are 0.917 and 0.921, respectively, outperforming all baseline classifiers employing the composite feature space.

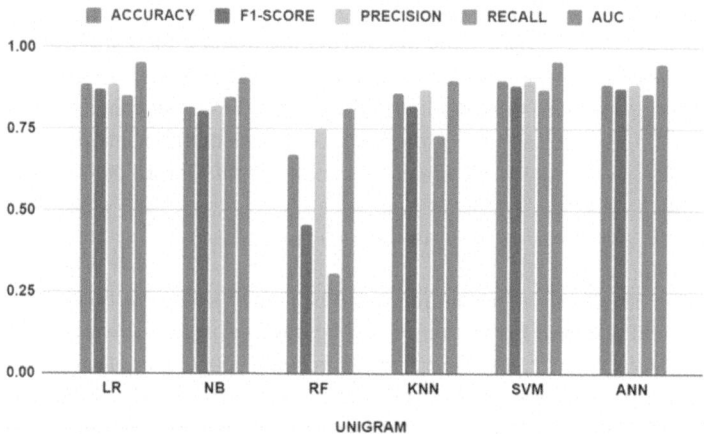

Fig. 3. Comparison of the baseline model's accuracy, F1-Score, precision, recall, and AUC metrics

The model generates an F1 score of 0.9420, which is much higher than the F1 scores of other baseline classifiers, such as the SVM classifier, which has marginally higher accuracy and AUC values. A graph displaying the accuracy readings of several models is kept on file to enable for visual comparison. The proposed global feature-based fusion model's results are shown in Fig. 3 to be the most trustworthy, hence it has been given consideration for further validation. Thereafter, no more steps are taken to validate more models.

6   Conclusion

The purpose of an enterprise IT Automated Document Management System (ADMS), where business operations are automated using RPA, is to extract information from documents. Digital technical content that requires little user input. It is suggested that LSTM be used to create a self-learning, intelligent system for preventing information leaks that categorises document pictures as SD or NSD based on the existence of NPI and PII semantic signatures. This is done without the need for any formal configuration. Clear obstruction The technique is anticipated to be actively deployed as a data storage idle SD image early warning system. As opposed to the technique that is advised, other advanced methods. The test employed sample data from the organization's digital document repository, and the indicators of predictive accuracy were noticeably better and within the bounds specified by the organization's information security monitoring team as to what was considered to be acceptable. The system architectures and methods that have been described are not just applicable to digital documents from the field of information technology.

References

1. Ahmad, M.S., Bamnote, G.: Data leakage detection and data prevention using algorithm. Int. J. Comput. Sci. Appl. **6**(2), 394–399 (2013)
2. Al-Fedaghi, S.: A conceptual foundation for data loss prevention. Int. J. Digit. Content Technol. Appl. **5**(3), 293–303 (2011)
3. Alhindi, H., Traore, I., Woungang, I.: Preventing data leak through semantic analysis Internet Things, (2019). https://www.sciencedirect.com/science/article/abs/pii/S25426605 1930126X?via%3Dihub
4. Hauer, B.: Data and information leakage prevention within the scope of information security. IEEE Access **3**, 2554–2565 (2015)
5. Raman, P., Kayacık, H.G., Somayaji, A.: Understanding data leak prevention. In: Proceeding 6th Annual Symposium Information Assurance (ASIA), p. 27 (2011)
6. Alhindi, H.: A framework for data loss prevention using document semantic signature. Ph.D. dissertation, Department Electronic Computer Engineer University Victoria, Victoria, BC, Canada, (2019)
7. Guha, A., Samanta, D.: Hybrid approach to document anomaly detection: An application to facilitate RPA in title insurance. Int. J. Autom. Comput. **18**(1), 55–72 (2021)
8. Lu, Y., Huang, X., Ma, Y., Ma, M.: A weighted context graph model for fast data leak detection. In: Proceeding IEEE International Conference Communications (ICC), pp. 1–6, (2018)
9. Alneyadi, S., Sithirasenan, E., Muthukkumarasamy, V.: A survey on data leakage prevention systems. J. Netw. Comput. Appl. **62**, 137–152 (2016)
10. Buczak, A.L., Guven, E.: A survey of data mining and machine learning methods for cyber security intrusion detection. IEEE Commun. Surv. Tutor. **18**(2), 1153–1176 (2015)
11. Rajdeep, C., et al.: A novel machine learning based feature selection for motor imagery ecg signal classification in internet of medical things environment. Future Gener. Comput. Syst. **98**, 419–434 (2019)
12. Cheng, L., Liu, F., Yao, D.: Enterprise data breach: causes, challenges, prevention, and future directions. Wiley Int. Rev. Data Min. Knowl. Disc. **7**(5), e1211 (2017)
13. Malik, D., Leung, K.: Information leakage & data loss prevention. IT Assurance & Governance Information, (2009)

14. Farràs, O., Ribes-González, J.: Provably secure public-key encryption with conjunctive and subset keyword search. Int. J. Inf. Secur. **18**(5), 533–548 (2019). https://doi.org/10.1007/s10 207-018-00426-7

15. Samuel, P., Jayashree, K., Babu, R., Vijay, K.: Artificial Intelligence, machine learning, and IoT architecture to support smart governance. In: Kavita Saini, A., Mummoorthy, R.C., Gowri Ganesh, N.S. (eds.) AI, IoT, and Blockchain Breakthroughs in E-Governance, pp. 95–113. IGI Global, Pennsylvania (2023). https://doi.org/10.4018/978-1-6684-7697-0.ch007

16. Abhijit, G., Debabrata, S.: Real-time application of document classification based on machine learning. In: International Conference on Information, Communication and Computing Technology, pp 366–379. Springer, (2019)

17. Kumar, K.A., Ravikumar, S., David, S.A.: Compression and decompression of encrypted image using wavelet transform. J. Comput. Theor. Nanosci. **15**(11–12), 3528–3532 (2023). https://doi.org/10.1166/jctn.2018.7656

18. Barbara, H.: Data leakage prevention. In: Proceedings of the 16th International Conference on Enterprise Information Systems-Vol. 2 pp 361–367. SCITEPRESS-Science and Technology Publications Lda (2014)

19. Prithi, S., Sumathi, S.: A survey on recent DFA compression techniques for deep packet inspection in network intrusion detection system. J. Electr. Eng. **17**(3), 14 (2017)

20. Jaeyalakshmi, M., Vijay, K., Jayashree, K., Priya Vijay: A cloud based healthcare data storage system using encryption algorithm. In: Recent Trends in Computational Intelligence and Its Application. CRC Press: pp. 486–491 (2023)

21. Karamani,B.: Improving data loss prevention using classification. In: International Conference on Emerging Internetworking, Data & Web Technologies. Springer: Cham, pp 183–189 (2018)

22. Katz, G., Elovici, Y., Shapira, B.: Coban: a context based model for data leakage prevention. Inf. Sci. **262**, 137–158 (2014)

23. Kaur, K., Gupta, I., Singh, A.K.: Data leakage prevention: e-mail protection via gateway. J. Phys. **933**, 012013 (2017)

24. Anusha, S., Elakkiya, N., Vijayakumar, R.: Separable reversible data hiding in encrypted image using dual data embedding with histogram shifting. (2020)

25. Janger, E.J., Schwartz, P.M.: The gramm-leach-bliley act, information privacy, and the limits of default rules. Minn. L. Rev. **86**, 1219 (2001)

Deep Learning

Enhancing Abnormal Object Detection in Camera-Based Systems Through Computer Vision and Deep Learning Techniques

K. Veena[✉], NagaHemanth Murari Allagadda, A. Sai Simha Reddy, A. Deepa, M. Selvi, and P. Kathambari

Departmeent of CSE, Sathyabama Institute of Science and Technology, Chennai, India
veena.cse@sathyabama.ac.in, nagamurari03@gmail.com

Abstract. With the proliferation of cameras in our households for security purposes, the task of evaluating anomalies occurring within their field of view has become increasingly complex. An anomalous event refers to an occurrence deviating from the norm, exhibiting unusual behavior that happens infrequently. To bolster public safety, surveillance cameras are deployed in high-traffic areas, hospitals, banks, and shopping malls. However, challenges arise due to camera placement limitations and the insufficient number of cameras compared to human monitors. The detection of abnormal events such as crimes, illegal activities, and traffic accidents is a crucial objective in video surveillance. In response, our proposed system harnesses the power of computer vision and deep learning to accurately detect anomalous events in real time. By leveraging advanced algorithms, our solution offers improved precision in identifying and flagging unusual activities, thus enhancing overall security and enabling prompt response measures.

Keywords: abnormal things · Computer vision · camera · detection · anomaly detection · video analysis · event class

1 Introduction

Society is developing unexpectedly, which corresponds to a large and dense population growing at a speedy charge. It is not possible to expect whilst a terrorist attack, a car twist of fate, a fight, or any of the myriad different forms of social unrest will arise. The need for a dramatic boom in social safety is pressing and pressing. The range of surveillance cameras established in public places and public transportation systems is also growing swiftly every year, and cameras are in nearly each aspect of human beings every day's lives. Perhaps the camera could be similar to human eyes and could give human beings the capability to "see" what's occurring around them in order that they do not get harm. This could be accomplished for safety reasons. Computers can mimic the human brain in lots of methods, including the potential to make alternatives. For computational and investigative purposes, the pc extracts the video information from the photographic surveillance. This allows the computer to interpret the content of the image inside the

R. Geetha et al. (Eds.): AAIMB 2023, CCIS 2202, pp. 81–96, 2025.
https://doi.org/10.1007/978-3-031-73065-8_7

commentary scene, which in turn lets in it to come across deviant conduct, discover, warn and cause an alarm. Escalators, that are a kind of transportation gadget, are widely used in department stores, buildings, instructional institutions and diverse different public places to facilitate the float of human beings. The purpose is that escalators help human beings get from one level to some other quicker. Accidents that have an effect on the protection of manual exertions are much more likely no longer handiest in widespread, however mainly amongst college students who represent a special demographic group. This occurs due to the fact college students still have an immature feel of autonomy, rather than our bodies which have now not but reached their complete ability. In addition, via checking the presence of students on the escalator, the yawning of the escalator may be averted, which saves strength and enters lifestyles. In addition, by way of calculating the passenger flow within the escalator, first-rate manipulation may be obtained. As college students come to be greater awareness of the dangers they gift, it's far even more essential to preserve constant vigilance at the escalator to save you well-timed safe falls.

2 Literature Survey

The book "Image Processing, Analysis, and Machine Vision" by Sonka, M., Hlavac, V., and Boyle, R. provides a comprehensive overview of various methods and techniques in image processing and computer vision in the book Sonka [2014].

Joana Alexandra Bozdog, Todea Daniel-Nikusor, Marcel Antal; Claudia Antal, Tudor Cioara, Ionut Angelus, Ioan Salomi, "Human Behavior and Anomaly Detection Using Machine Learning and Wearable Sensors", 17th IEEE International Conference on Intelligent Computing and Data Processing (ICCP), 2022 [1]. This article targets to pick out capacity anomalies in human conduct by using detecting and reading them using a set of non-invasive wearable sensors. Joana Alexandra Bozdog proposes a allotted experimental net machine that uses sensor statistics and device getting to know algorithms to music human conduct and detect anomalies. Many characteristic configurations and feature selection strategies, in addition to guide labeling, have been used to monitor gaining knowledge of. When abnormalities are observed within the behavior of the aged, the caregiver is notified. Finally, the paper demonstrates the implementation and capability of the machine the usage of the Fitbit smart bracelet and integration with the Fitbit Cloud. The effects had been received the usage of a publicly available interest dataset and a selection of engine shape detection algorithms and mastering capabilities that promise an accuracy of 87% and it's with an F1 score of 0.9.

Huifang Qian, Xuan Zhou, Mengmeng Zheng, "Anomalous Behavior Detection and Recognition Based on Multilayer Residual Network", 4th IEEE 2020 [2].

Huifang Qian proposes Now residual community structure to stumble on and perceive abnormal conduct of human beings in video. Multiple residual network modules consist of human frame function detection and popularity. Based at the above, this newsletter proposes Network Residual Detection (d-Res) as a multi-scale goal detection approach to achieve the detection fee and impact of the human frame. This is used to extract spatial functions of anomalous behavior, even as the translation-residue reputation network (r-Res) extracts deep capabilities of images for green category of anomalous conduct. Experiments with the UTI dataset to assess the overall performance of the proposed algorithm. The consequences display that the proposed method can successfully come across and understand anomalous conduct in real scenes.

Weihu Zhang, Chang Liu, "Research on Abnormal Human Behavior Detection Based on Deep Learning", International Conference on Reality and Intelligence (ICVRIS), 2011 [3]. To higher estimate the conduct of anomalous people in a complex scene, a Gaussian mixture version changed into used to define the clear contours of moving targets on the device and perform Gaussian filtering on them to cast off noise results at the scene. The important vicinity of human movement in the video is extracted by way of calculating the center point of the pixel and drawing a field based totally on that boundary. The Farneback set of rules obtains spatiotemporal optical flow information. To achieve the result, the weighted merge approach plays two methods at the output of the softmax network. The consequences could display that the conduct of the class reached a completely correct 91.2%, and the conduct of the peculiar recognition changed into 92%.

In the paper Redmon et al. [2016] the YOLO (You Only Look Once) method introduced in the paper is a unified approach for real-time object detection. It performs object detection by dividing the input image into a grid and predicting multiple bounding boxes and class probabilities in a single pass through a convolutional neural network. YOLO achieves a balance between accuracy and speed, outperforming other state-of-the-art object detection methods and making it suitable for applications such as video surveillance and autonomous driving.

In the paper titled "A hierarchical approach for abnormal event detection in crowded scenes" by Huang, Y et al. [2017], the authors proposed a hierarchical framework for abnormal event detection in crowded scenes. The method involves three stages: spatial-temporal feature extraction, event classification using hierarchical clustering, and anomaly detection based on a fusion of scores. One potential disadvantage of this paper is that it may have limitations in handling complex and dynamic scenes where multiple abnormalities occur simultaneously or overlap with each other. Additionally, the hierarchical clustering approach may be sensitive to parameter settings, potentially impacting the performance and generalizability of the method. It's worth considering these factors while applying the proposed approach in real-world scenarios.

In the paper Hassanpour, R et al. [2018] the authors proposed a method for abnormal event detection by combining multimodal sensor data and a spatio-temporal deep autoencoder. However, one potential disadvantage of this paper is that it may rely heavily on the availability and synchronization of multimodal sensor data, which could be challenging and costly to obtain in practical scenarios.

In the paper Sabokrou et al. [2018] the authors proposed a method for fast anomaly detection in crowded scenes using a fully convolutional neural network (FCN). One potential disadvantage of this paper is that the method's performance may be sensitive to the quality and diversity of the normal training samples, as it heavily relies on learning the normal scene representation.

The method proposed in the paper by Hasan et al. [2016], focused on learning temporal regularity for anomaly detection in video sequences. However, a potential disadvantage is that the method may struggle to accurately detect anomalies in complex or dynamic scenes with diverse temporal patterns, which can lead to false negatives. Furthermore, the effectiveness of the approach may heavily depend on the availability and quality of annotated training data, limiting its applicability in scenarios with limited labeled data.

In the paper Sultani, W et al. [2018] authors proposed a method for real-world anomaly detection in surveillance videos using a two-stream spatio-temporal autoencoder. However, one potential disadvantage of this paper is that the method's performance may be affected by variations in lighting conditions, camera viewpoints, and other environmental factors, which can impact the accuracy of anomaly detection.

In the paper Chen, J et al. [2020] the authors proposed a method for abnormal event detection in crowded scenes by enhancing foreground extraction and utilizing a motion descriptor. However, a potential disadvantage of this paper is that the method's effectiveness may be affected by complex scenes with occlusions or high crowd density, which can hinder accurate foreground extraction and potentially lead to false positives or missed detections.

In the paper Li, Z et al., the authors proposed a method for detecting anomalous events in videos by learning deep representations of appearance and motion. However, a potential disadvantage of this method is that it may heavily rely on the availability of labeled training data, which can be time-consuming and costly to acquire, limiting its applicability in scenarios with limited labeled data. Additionally, the approach's performance may be affected by variations in lighting conditions, camera viewpoints, or environmental factors, which can impact the accuracy of anomaly detection in real-world settings.

In the paper by Hu, Q., et al. [2019] one potential disadvantage of the method is that it may require a large amount of labeled training data for each object class, which can be time-consuming and challenging to obtain, hindering its applicability in scenarios with limited labeled data.

In the paper Li, W et al. [2014] a potential disadvantage of the method is that it may face challenges in accurately localizing anomalies in highly crowded scenes with complex interactions and occlusions, leading to potential inaccuracies in anomaly detection and localization.

In the paper Zhang, X et al., a potential disadvantage of the method is that it may suffer from limited generalizability to new or unseen anomaly types, as the deep convolutional neural network features might be biased towards the anomalies present in the training data, limiting its effectiveness in detecting novel or rare anomalies.

Long Wen et al. [2023], the authors provide a survey of methods used for oriented object detection in remote sensing images. As a comprehensive survey, the paper does not focus on a specific method but rather reviews and analyzes various approaches employed in the field of oriented object detection in remote sensing imagery. Qinghua Guo et al. [2023] the authors proposed an enhanced camera-based method for individual pig detection and tracking in smart pig farms.

3 Research Methodology

3.1 Active Problems with the Current System

Today's gadget is designed with a statistical version to discover human conduct and assigns an approximate distribution that minimizes the performance of the prediction model.The modern-day system most effective detects the strength of detection that is already available in real CCTV anomaly detection systems.

The modern gadget performs poorly in terms of download and execution times and has low precision. However, pursuing a degree and a career don't necessarily serve as means of severing ties (Fig. 1).

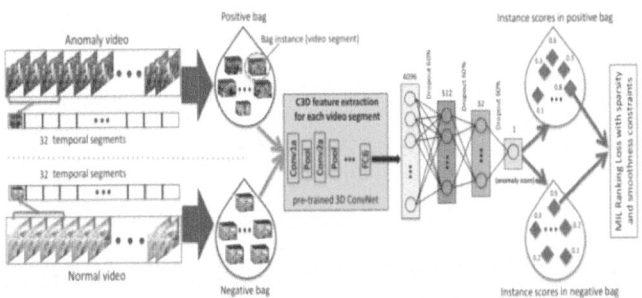

Fig. 1. Architecture diagram

Feasibility Study

It is a facility plan to improve the popularity of the server at this level and the proposed business notion with the maximum not unusual plan and some fee estimate. In the analysis of the system, it's far vital to perform a feasibility look at of the proposed gadget. Feasibility research require a few expertise of the primary device requirements.

Three key considerations involved in the feasibility analysis are

- Economical feasibility.
- Technical feasibility.
- Operational feasibility.

Description of Proposed System Selected Methodology or Process Model Algorithm

As in everyday life, many cameras had been hooked up in our houses for protection purposes due to technological advances. How to evaluate anomalies that seem within the eyes, mainly in frequent conditions, can be intricate. An event that is deflected to the nearest area due to a few uncommon conduct that takes place occasionally is known as an anomalous occasion. To improve public safety, surveillance cameras are utilized in crowded locations, in hospitals, banks, and buying malls. As a result, there may be a hassle with the location of safety cameras and the insufficient quantity of cameras in comparison to human video display units. Detection of odd activities along with crimes, unlawful activities, site visitors injuries, and many others. Video surveillance is an crucial venture. Atypical activities generally tend to occur a good deal much less regularly than normal sports. A precedence requirement is the development of shrewd laptop imaginative and prescient algorithms for automatic video anomaly detection. The cause of a real-time anomaly detection gadget is to detect any hobby that deviates from everyday conduct patterns. Once an anomaly has been detected, similarly category algorithms may be assigned to one of the particular tasks.

The important capabilities and contributions of this paintings are as follows:

- Autoencoders make it smooth to extract representative functions rather than manually created ones.
- Hierarchical learning can be performed thru a couple of hidden layers of the encoder.
- Spatial and temporal encoders are used to resolve the hassle of anomaly detection, which is called the outlier detection hassle thru spatiotemporal order.

Description: In this venture we are able to perceive anomalies in video using a spatiotemporal version and LSTM.

Input: User will provide video or animation as input.

Output: Determine if there are any anomalies in the video or now not.

Module 1: Data Acquisition and Data Preprocessing

Data acquisition

To enforce deep learning, a method referred to as records mining is used to collect information from various assets. Data have to be stored in a way suitable for the supposed cause.

We acquired each of the 10 tables the usage of the Avenue UCSD Ped1 dataset received from Kaggle. The Avenue dataset consists of 21 clips for assessment and 16 for training. Films have been filmed for a complete of 30,652 (15,328 teaching, 15,324 take a look at) shots at the CUHK Avenue campus.

The schooling version supplied in Table 1 makes use of only the usual images and take a look at films with anomalous and normal results, every video is much less than a minute. Going down the steps and taking the subway is the usual component. They include human beings taking walks inside the opposite course, walking, backpacks, and other unusual activities.

Data Preprocessing

Data preprocessing is using records mining to extract new information in a beneficial and efficient form. Data preprocessing transforms statistics into a layout that mining, machine getting to know, and different facts processing operations can perform more quickly and

Table 1. Training set

Data set	Avenue	UCSD Ped 1	UCSD Ped 2
Training	16	34	16
Testing	21	36	12

successfully. To make sure reliable consequences, the methods are commonly carried out at the very starting of system getting to know and AI improvement pipelines.

The method of this records is very simple. Each version calls for a positive form of input, so pre-processing is important, the uncooked photograph facts must be transformed to deliver it consistent with valid inputs to the model. In this level the following steps are executed;

1. The length of each extracted member to 227×277.
2. All pixel values are scaled from 0 to at least one so that all enter image frames are on the equal level.
3. Normalization is carried out by subtracting each body from the common of the worldwide image, which"

$$\text{Global Mean Image} = \mathbf{X/Y}$$

- Where X = sum of all pixel values in every body of the dataset, Y = wide variety of pixels.
- The picture files are transformed to gray scale for dimensionality discount.
- All tables are procedurally normalized to have a unit variance and an average of zero.

Change the Color Channel

We will convert the object's RGB channel to GreyScale. A grayscale image is an photograph wherein the price of every pixel is a separate pattern that incorporates only light depth information. Grayscale most effective produces grayscale images, a kind of monochrome black and white or grey. Black has the lowest, and white has the best.

Some human beings want to convert their RGB images to grayscale for various reasons.

Module 2: Building Model

We will create a spatiotemporal model with an LSTM version the usage of Convolution and LSTM layers to construct our model.

Spatiotemporal Analysis and Models

Due to the creation and use of recent computational algorithms that permit the analysis of spatiotemporal databases, the analysis of spatiotemporal information is a brand new region of research.

Because spatiotemporal patterns appear when time and space data are accrued and feature at the least one spatial and one temporal character, we used this method. A spatiotemporal phenomenon that occurs at a selected time t and a selected location x is described by the event in a spatiotemporal facts set.

When reading spatiotemporal information, each temporal and spatial correlations should be taken into account. The records evaluation system is tons more complex whilst both the temporal and spatial dimensions of the facts are considered;

1. Continuous and discrete modifications in spatial and non-spatial houses of spatial items.
2. The mixed impact of spatio-temporal locations on every other.

LSTM Models
Long Term Short Memory Networks or LSTMs are used in deep studying. Many recurrent neural networks (RNNs) can study lengthy-term dependencies, specifically in duties concerning collection prediction. In addition to character information points, which include snap shots, the LSTM has feedback, which allows it to method a whole collection of records. Among different matters, translation and speech popularity system is used. A precise version of RNN called LSTM indicates awesome performance in a extensive range of duties. Because of this it additionally helps in anomaly detection.

Convolutional layers specific functions from enter photograph frames from videos. Using least-squares enter data, convolution operations analyze photo properties and maintain spatial relationships between elements. In the context of image processing, the convolution operation is the enter point of the picture phase and the local filter.

Defining Loss and Optimizer
We will define the loss feature as MSE (Mean Square Error) and the optimizer as SGD (Stochastic Gradient Descent).

MSE Loss Function
The mean squared error (MSE) is the maximum regularly used regression loss function. The loss is the average of the discovered squared variations among the actual and expected values.

$$MSE = \frac{1}{N} \sum_{i=1}^{N} y_i - \widehat{y}_i \tag{1}$$

SGD (Stochastic Gradient Descent)
A simple however powerful technique for becoming classifiers and linear regressors to convex loss functions such as (linear) guide vector machines and logistic regression is stochastic gradient descent (SGD). SGD has been gift within the discipline of engineering studies for a long time, but in the context of massive disciplines, it has simplest currently acquired a great deal attention. Large-scale and uncommon device getting to know troubles that frequently stand up in text classification and natural language processing are efficiently solved with SGD. Because of its superior performance and simplicity of use, we used this optimizer.

The actual slope is approximated by way of stochastic descent (or "on-line") gradient through the slope in a single model:

For n=1:N

$$w_i^n = w_i^{n-1} - n\frac{\partial E_n(w)}{\partial w_i} \tag{2}$$

$$w_i^n = w_i^{n-1} - n(y(x_n, w) - t_n)1\frac{\partial y(x_n, w)}{\partial w_i} \tag{3}$$

Loading CUDA GPU and Train

We have already divided the dataset into parts - schooling and checking out data. We run our training on a CUDA GPU and it's miles sturdy. The application can use the CPU and GPU concurrently way to NVIDIA's parallel computing platform called CUDA. We use this due to the fact NVIDIA is now the most popular GPU vendor for cloud computing and device getting to know. In addition, NVIDIA GPUs can be used with maximum Python GPU-prepared languages.

Creating Prediction

We will create a predictive characteristic to locate anomalies inside the video. The next step is to create a characteristic that could decide if inputs including pics, movies and time are anomalous or no longer. By using Python's Predict approach, we are able to are expecting the facts values of labels in a found out layout.

Module 3: Creating GUI

Tkinter will be used to build the GUI. Although there are alternative Python GUI frame-works, Tkinter is the only one that is part of the core library. Tkinter is really fortunate. As it is a go platform, the same same code runs on Windows, macOS, and Linux. Since Tkinter builds its graphics using local running device components, applications created with it resemble those that run on the platform they may be operating on. Tkinter is the name of Python's built-in GUI library.

Description of the Software to be Utilized and a Plan for Testing the Proposed Model/System

Anaconda is an open-source package deal supervisor for Python and R. It is a famous platform among information scientists for implementing Python and R. There are over three hundred libraries to be had for information science, so a strong distribution gadget for them should be taken into consideration. Some profession in the field. Anaconda makes it smooth to deploy and control applications. In addition, it has numerous equip-ment that can collect statistics thru synthetic intelligence and device learning algorithms. With Anaconda you may without difficulty set up, control, and proportion your Conda surroundings. In addition, while using Anaconda, you want to install any software which you want in a few clicks. Anaconda has many benefits, amongst which the subsequent are the most excellent: Anaconda is unfastened and open source. This approach that you may use it without any cash. In the information science region, Anaconda is the enterprise chief. It is also open-source, which has made it broadly famous. If you need to emerge as a data scientist, you have to realize the way to use Anaconda as opposed to Python, because each person recruits you to have this skill. This is essential records technology (Fig. 2).

Architecture/Overall Design of Proposed System

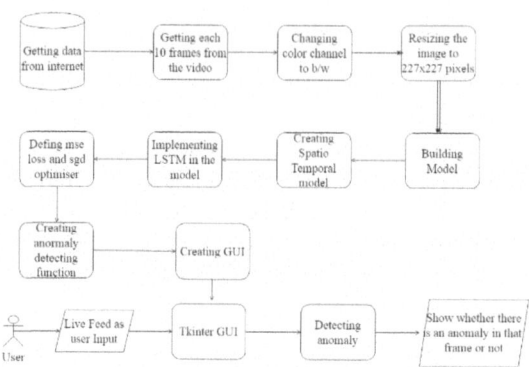

Fig. 2. System Architecture

4 Result

This project's main theme is to detect anomaly/abnormal activity detection and this was achieved successfully using the above methodology since humans cannot monitor all the cameras at all times so using these project we can categorize the situation, especially the abnormal situations like Arson, Assault, Burglary, Explosion, Fighting, Normal, Road Accidents, Robbery, Shooting, Shoplifting, Stealing, Vandalism and show us on the screen about the situation happening as a video and mention the particular frame of the above-mentioned situation., please refer below screenshot (Figs. 3, 4, 5, 6, 7 and 8).

Fig. 3. Accident happen

The below - mentioned graph represents the anomaly happened at a particular time, where the peak represents the anomaly score where the graph increases the anomaly score so that we can know that anomaly occurred at the particular frame.

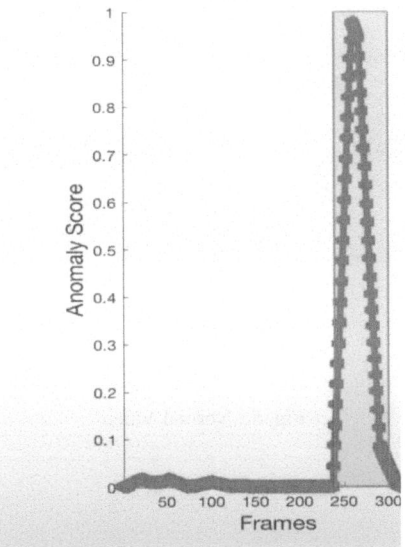

Fig. 4. Accuracy Score

Currently, we are giving input as an video and the model detects the abnormalities which are mentioned above and shows us the graph along with frame numbers. if we can implement this in CCTV cameras we can get an instant message or alert if any of the particular abnormal situations occurred so that we can easily get to know that abnormal situation has been detected.

Normal video

Fig. 5. Normal Video

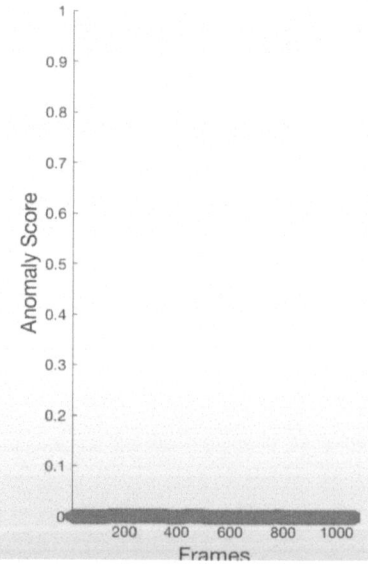

Fig. 6. Accuracy Score v/s Normal Video

This might be very useful when around crowded areas or any busy areas or at traffics, where abnormalities may occur frequently when a camera is recording the situation if there is an anomaly detected then among all cameras on the main screen in the security

Explosion

Fig. 7. Explosion

room the particular anomaly detected camera's recording is popped up on the main screen so that we can get to know that there is something wrong happened so that we can take necessary precautions or call to emergency Contacts like to call the police, fire station or to ambulance and let them know the place where that occurred so that lives can be saved without any delay.

The table above table tell us about the dataset which we used in this project and the table gives information about no.of videos length of dataset and its particular type of anomaly.

The report discussed the use of Spotting Anomalous Behavior detection using camera-based systems. The main aspect of the system's capacity to precisely identify the compilation of cases of diverse abnormal circumstances is criminal science specialists. An idea of the system and action module descriptions of recognition Outcomes of anomaly detection, which sought to identify unusual circumstances Techniques for evaluating the stances of prisoners were shown.

The additional task will involve the procurement of additional recordings that show the unusual circumstances and adoption of new techniques for identifying those behaviors.

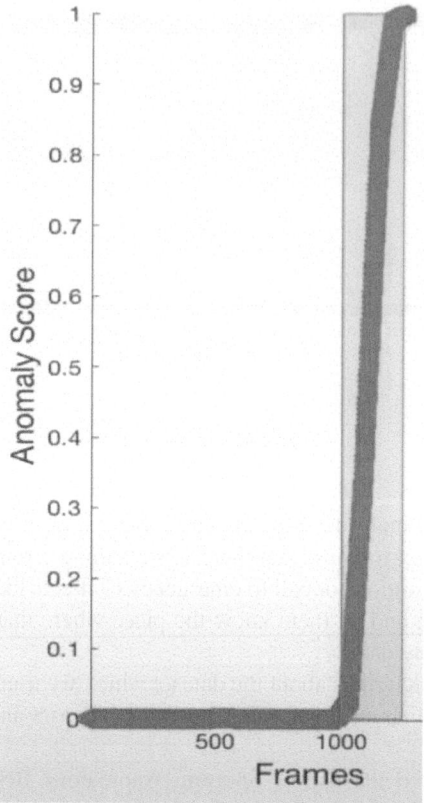

Fig. 8. Accuracy Score v/s Frames

Table 2. Dataset about no. of videos length of dataset and its particular type of anomaly

	# of videos	Average frames	Dataset length	Example anomalies
UCSD Ped1 [27]	70	201	5 min	Bikers, small carts, walking across walkways
UCSD Ped2 [27]	28	163	5 min	Bikers, small carts, walking across walkways
Subway Entrance [3]	1	121,749	1.5 hours	Wrong direction, No payment
Subwa Exit [3]	1	64,901	1.5 hours	Wrong direction, No payment
Avenue [28]	37	839	30 min	Run, throw, new object
UMN [2]	5	1290	5 min	Run
BOSS [1]	12	4052	27 min	Harass, Disease, Panic
Ours	**1900**	**7247**	**128 hours**	**Abuse, arrest, arson, assault, accident, burglary, fighting, robbery**

5   Conclusion

The proposed project offers a solution that allows for the detection of even the slightest abnormalities without the need for human monitoring. By leveraging advanced computer vision and deep learning techniques, our system can swiftly identify and respond to anomalous events, eliminating the delays typically associated with manual video monitoring. This capability is particularly crucial during critical moments, as prompt and decisive action is vital to regain control over unexpected situations.

By implementing this project, necessary actions can be taken swiftly, ensuring the safety and security of individuals in society. The automated detection of abnormalities minimizes the risk of human error and maximizes the efficiency of emergency response measures. With this technology at hand, precious moments during which abnormalities occur can be effectively managed, mitigating potential threats and safeguarding public well-being.

References

1. Ioana, A.B., Todea, D.-N., Marcel, A., Claudia, A., Tudor, C., Ionut, A., Ioan, S.: Human behavior and anomaly detection using machine learning and wearable sensors. In: IEEE 17th International Conference on Intelligent Computer Communication and Processing (ICCP) (2022)
2. Huifang, Q., Xuan, Z., Mengmeng, Z.: Detection and recognition of abnormal behavior based on multi-level residual network. In: IEEE 4th Advanced Information Technology, Electronic and Automation Control Conference (IAEAC) (2020)
3. Weihu Zhang, Chang Liu: Research on human abnormal behavior detection based on deep learning. In: International Conference on Virtual Reality and Intelligent Systems (ICVRIS) (2021)
4. Federico, A., Jiawei, Y., Syed, M.N.: Privacy-preserving online human behaviour anomaly detection based on body movements and objects positions. In: IEEE International Conference on Acoustics, Speech and Signal Processing (ICASSP) (2019)
5. Thomas, G., Dylan, S., Alexiei, D.: Detecting human abnormal behaviour through a video generated model. In: 11th International Symposium on Image and Signal Processing and Analysis (ISPA) (2019)
6. Rongyong, Z., Yan, W., Ping, J., Cuiling, L., Yunlong, M., Zhishu, Z.: Abnormal human behavior recognition based on image processing technology. In: IEEE 5th Advanced Information Technology. Electronic and Automation Control Conference (IAEAC) (2021)
7. Shih-Chung, H., Cheng-Hung, C., Chung-Lin, H., Por-Ren, T., Miao-Jian, L.: A video-based abnormal human behavior detection for psychiatric patient monitoring. International Workshop on Advanced Image Technology (IWAIT) (2018)
8. Redmon, J., Divvala, S., Girshick, R., Farhadi, A.: You only look once: unified, real-time object detection. In: Proceedings of the IEEE Conference on Computer Vision and Pattern Recognition (CVPR), pp 779–788 (2016)
9. Sonka, M., Hlavac, V., Boyle, R.: Image processing, analysis, and machine vision. Cengage learning (2014)
10. Huang, Y., Yang, L., Yang, M.H., Zhou, Y.: A hierarchical approach for abnormal event detection in crowded scenes. IEEE Trans. Image Process. **26**(2), 948–963 (2017)
11. Hassanpour, R., Mahoor, M.H.: Deep abnormal event detection using multimodal sensor fusion and spatio-temporal deep autoencoder. IEEE Trans. Image Process. **27**(7), 3382–3395 (2018)
12. Sabokrou, M., Fayyaz, M., Fathy, M., Klette, R.: Deep-anomaly: fully convolutional neural network for fast anomaly detection in crowded scenes. Comput. Visi. Image Understand. **172**, 88–97 (2018)
13. Hasan, M., Choi, J., Neumann, J., Roy-Chowdhury, A.K.: Learning temporal regularity in video sequences. IEEE Trans. Patt. Anal. Mach. Intell. **38**(10), 2026–2040 (2016)

14. Sultani, W., Chen, C., Shah, M: Real-world anomaly detection in surveillance videos. In: Proceedings of the IEEE Conference on Computer Vision and Pattern Recognition (CVPR), pp 6479–6488 (2018)
15. Chen, J., Chen, J., Chen, K.: Abnormal event detection in crowded scenes with improved foreground extraction and motion descriptor. Neurocomputing **383**, 81–92 (2020)
16. Li, Z., Zhang, X., Liu, Z., Yan, J.: Detecting anomalous events in videos by learning deep representations of appearance and motion. IEEE Trans. Image Proc. **27**(6), 2840–2852 (2018)
17. Hu, Q., Xu, Y., Lin, S., Liao, Y.: Object-oriented anomaly detection via deep anomaly cognition. Patt. Recogn. **88**, 536–545 (2019)
18. Li, W., Mahadevan, V., Vasconcelos, N.: Anomaly detection and localization in crowded scenes. IEEE Trans. Pattern Anal. Mach. Intell. **36**(1), 18–32 (2014)
19. Zhang, X., Patel, V.M.: Video anomaly detection using temporal aggregation of deep convolutional neural network features. In: Proceedings of the IEEE Conference on Computer Vision and Pattern Recognition (CVPR), pp 2921–2929 (2016)
20. Long Wen, Y., Cheng, Y.F., Li, X.: A comprehensive survey of oriented object detection in remote sensing images. Expert Syst. Appl. **224**, 119960 (2023). https://doi.org/10.1016/j.eswa.2023.119960
21. Guo, Q., et al.: Enhanced camera-based individual pig detection and tracking for smart pig farms. Comput. Elect. Agric. **211**, 108009 (2023). https://doi.org/10.1016/j.compag.2023.108009

Detection and Classification of Brain Tumor in Magnetic Resonance Images Using CNN

S. Nivedha$^{(\boxtimes)}$, A. Mani, and S. Muthukumarasamy

Department of CSE, S.A. Engineering College, Chennai, India
nivedha.s2805@gmail.com

Abstract. An abnormal growth of cells or tissues of mass amount growing inside the brain is called brain tumor, which is harmful to humans, as the tumor can be both cancerous and non-cancerous. Early diagnosis and accuracy of brain tumor helps the neurologists to initiate the treatment to their patients at early stages. To provide a detailed view and diagnosis of the brain tumor, MRI is commonly used. Deep learning is the process of teaching how the computer should perform and act in a way the human brain does. In deep learning, the neuron structure, which is this is inspired by a human brain is represented as the artificial neural network, as how the neurons play an important role in the human brain. The study of deep learning is applied in various fields as it gives a high level of accuracy. A Convolutional Neural Network algorithm is proposed to detect and classify the brain images in MRI scan efficiently. In the given MRI image, if the brain image has no tumor, then it is defined as normal, otherwise, if a tumor is predicted, then it is further classified into the tumor types such as Glioma, Meningioma and Pituitary tumor. The model result shows better accuracy of 98% in detecting the tumor from the MR image.

Keywords: Brain Tumor · Deep Learning · Magnetic Resonance Images (MRI) · Convolutional Neural Network (CNN)

1 Introduction

Brain tumor is a deadly disease, which grows inside a human brain, giving pressure and changes the function of the other brain tissues, which results in the tumor signs and symptoms [1]. The person who is suffering from the tumor needs an immediate treatment to prevent the risk coming from it. The tumor can be categorized into cancerous (malignant) and non-cancerous (benign), they are also called as primary brain tumors. The secondary brain tumor or metastatic are stated as where the tumor begins from the other part of the body and spread to the brain cells [2]. In general, the tumor needs an early diagnosis from a neurosurgeon or radiologists and the manual detection of tumor from magnetic resonance images (MRI) is very time consuming, as it is difficult to predict whether the tumor is present or not. MRI and Computed Tomography (CT) are mostly used for detecting the brain tumor. The computer- assisted model approaches are helpful in medical field, as it helps to make it easier for the surgeons to determine the case accurately.

© The Author(s), under exclusive license to Springer Nature Switzerland AG 2025
R. Geetha et al. (Eds.): AAIMB 2023, CCIS 2202, pp. 97–110, 2025.
https://doi.org/10.1007/978-3-031-73065-8_8

There are more than 150 types of brain tumor are recorded, but the two important types are primary and metastatic [9]. Metastatic tumors are seen in rare cases; mostly the people with lung cancer likely 40% develop this type of tumor. In a year, an estimation of 150,000 people suffers from this type of tumor. The tumors can be categorized into Glioma, Meningioma and Pituitary. Glioma is a type of brain or spinal cord tumor that grows from a glial cells brain cell that supports the nerve cells [12, 13].Glioma consists of 80% of a malignant tumor and 30% being all other types of brain tumors and central nervous system tumors [10, 11].Meningioma tumor is a non-cancerous and a primary central nervous system tumor which starts from the membranes around the brain or spinal cord. The tumors are graded as Grade I as low grade tumor, Grade II being the mid-grade tumor and Grade III being the high grade tumor (malignant). The Grade I tumor cells grow slowly while the Grade II tumor, after removing the tumor from brain, it has a high chance of regrowth and Grade III are malignant tumors because the tumors grows fast [14].

The Artificial Intelligence (AI) is a technique enabled to imitate the human behavior, whereas the machine learning is the process used to achieve artificial intelligence through different algorithms that are trained with datasets to achieve the desired results. Deep Learning is a subset of machine learning technique and that being a subset of artificial intelligence. Deep learning is the process of teaching how the computer should perform and acts in a way the human brain does. In deep learning, the structure is represented as the artificial neural network, as how the neurons play the important role in human brain. The structure of artificial neuron network is inspired by a human brain. The study of deep learning is applied in various fields as it gives a high level in accuracy.

A computer's or system's ability to extract meaningful information from digital photos, videos, and other visual inputs and take appropriate action or provide recommendations in response to that information is known as artificial intelligence (AI) in computer vision [15]. We can use a Deep Learning architecture that incorporates CNN (Convolution Neural Network), also known as NN (Neural Network), and other methods to detect brain cancers. An MRI scan is the best procedure for inspecting the brain and spinal cord and looking for malignancies. The images they produce are often more precise than CT scans.

Deep learning models are also known as deep neural networks, as neural network architecture is used in most of the deep learning methods. Mostly traditional neural network consist of 2 to 3 hidden layers while the deep neural network consists as many as 150. A Neural network also known as artificial neural network, where its structure is meant to be a representation of how a human brain operates, an interconnection between the nodes or neurons in a layered architecture and can be learned from the data. This is trained in order to learn to recognize the patterns, classify the objects and predicts the future events.

Typically neural networks are one of the machines learning approach, used mostly in modelling the non-linear relationships to recognize patterns and classifying the objects, processing sounds and images etc. The neural network architecture consists of input layers, one or more hidden layers and an output layer interconnected to each other as how human brain neurons are connected to process the data from one layer to another. The output of one layer is the input to the next processing layer. Each nodes or neurons present in the layer will be calculated according to the increase and decrease in the weight obtained at that time. To create and train a deep learning model and to classify object is to start the training from scratch, transfer approach and feature extraction.

A convolutional layer, a pooling layer, and a fully connected layer are the three layers that make up CNN's architecture. It belongs to the class of neural networks that analyses data using an architecture resembling a grid. The convolution layer is the fundamental part of CNN that is mostly in charge of computation. An input layer, an output layer, and multiple hidden layers make up convolutional networks. The layers of a convolutional network's neurons are organized in three dimensions (width, height, and depth), in contrast to a standard neural network. This allows the CNN to convert a three dimensional input volume into an output volume. Convolution, pooling, normalizing, and fully connected layers make up the hidden layers [16].

The deep learning algorithms are applied to train the model. Brain tumors are a cancerous one and the manual detection of tumor is time-consuming and difficult, as the constraint being predicting the tumor location accurately. The proposed methodology detects whether the MRI scan consists of a normal brain or tumor present. If the image consists of a tumor, it further classifies them into Glioma, Meningioma and Pituitary tumor. The acquired dataset is trained and tested to build the model that works efficiently.

2 Literature Review

2.1 Shah, H. A., Saeed, F., Yun, S., Park, J. H., Paul, a., & Kang, J. M. (2022). a Robust Approach for Brain Tumor Detection in Magnetic Resonance Images Using Fine Tuned Efficient Net. *IEEE Access*, *10*, 65426–65438.

Hasnain Ali Shah *et al.* [1] presented a new automated model which is an efficient finetuned Efficient Net B0 based transfer learning approach that is combined with the proposed recommended layers to help enhance the overall classification. The proposed model helped to identify the brain tumor from the MRI scans accurately with the overall performance accuracy with 98.87%. This model is compared with the other state-of-art method and determined that Efficient Net B0 model results shows better performance when compared to the other models.

2.2 Ottom, M. A., Rahman, H. A., & Dinov, I. D. (2022). Znet: Deep Learning Approach for 2D MRI Brain Tumor Segmentation. IEEE Journal of Translational Engineering in Health and Medicine.

Mohammed Ashraf Ottom et al. [3] proposed a novel 2D segmentation framework to classify the automated brain tumor from MRI scans. Using deep neural networks (DNN) and data augmentation techniques, this research proposes a novel framework for segmenting 2D brain tumors in MR images. The proposed strategy (Znet) is founded on the notion that data amplification, skip-connection, and encoder-decoder architectures transmit to thousands of fabricated instances the intrinsic affinities of a relatively limited number of expertly characterized tumors, such as low-grade glioma (LGG) patients in their hundreds to synthetic cases. In some cases of computer vision applications, the high pixel accuracy value is misleading, so the other evaluation metrics such as dice and Intersection over Union (IoU) are useful for auto segmenting the tumor from MRI scans. High values of the mean dice similarity coefficient were found in the experimental results (dice = 0.96 for model training and dice = 0.92 for the independent testing dataset). Other evaluation metrics, such as pixel accuracy (0.996), F1 score (0.81), and Matthews Correlation Coefficient (MCC) (0.81), were also generally high. The Znet model's capacity to locate and automatically segment brain tumors in MR images is demonstrated by the findings and visualization of the DNN-derived tumor masks in the testing dataset.

2.3 Ahmad, S., & Choudhury, P. K. (2022). on the Performance of Deep Transfer Learning Networks for Brain Tumor Detection Using MR Images. IEEE Access.

Several transfer learning-based deep learning techniques were examined, and their corresponding results were evaluated, by Saif Ahmad et al. [4] in order to choose the optimal CNN model for the detection of brain tumors from MR images. The deep learning framework is built using seven classical feature extractors, and the extracted features from each pre-trained model are categorized using five classical classifiers. In this study, the VGG-19-SVM model that performed the best displays the maximum accuracy, with a score of 99.39%.

2.4 Ismail, M., Prasanna, P., Bera, K., Statsevych, V., Hill, V., Singh, G.,...& Tiwari, P. (2022). Radiomic Deformation and Textural Heterogeneity (r-depth) Descriptor to Characterize Tumor Field Effect: Application to Survival Prediction in Glioblastoma. IEEE Transactions on Medical Imaging, 41(7), 1764–1777.

Radiomic-Deformation and Textural Heterogeneity (r-Depth), an integrated MRI-based descriptor, was introduced by Marwa Ismail et al. This descriptor includes measurements of the small changes in tissue deformations caused by mass effect throughout the surrounding normal parenchyma. For instance, the aggressive brain tumor glioblastoma (GBM) frequently results in brain herniation and poor outcomes due to the increase in intracranial pressure caused by the tumor load. The findings demonstrated the potential utility of the r-Depth descriptor as a complete and reliable MRI-based predictive predictor of disease aggressiveness and survival in solid tumors.

2.5 Hao, K., Lin, S., Qiao, J., & Tu, Y. (2021). a Generalized Pooling for Brain Tumor Segmentation. IEEE Access, 9, 159283–159290.

In order to combine maximum and average pooling, Kuankuan Hao et al. [6] initially presented a revolutionary generalized pooling (GP) approach with adaptive weights. This is the first effort to enhance models from the standpoint of pooling procedures for the segmentation of brain tumors. According to the experimental findings, generalized pooling outperforms conventional pooling techniques in its ability to segment brain tumors.

3   Proposed System Methodology

Recent studies have employed deep learning to increase the efficiency of computer-aided medical diagnostics in the study of brain cancer. They are crucial to the healthcare industry and serve as effective tools for treating many serious illnesses, such as skin cancer image analysis and the identification of brain diseases. For the categorization of brain tumors, DL techniques based on transfer learning and fine-tuning are favored and frequently utilized. The goal of this study is to carry out extensive testing with deep convolutional neural networks to automate the identification and categorization of brain tumors.

The proposed model was implemented on the open-access dataset obtained from kaggle which consists of 3264 images in it. The model was built in Python using the Keras and Tensor Flow frameworks. The performance of the model was trained on a computer system with specifications of Intel Core(TM) i3-6006U CPU at 2.00 GHz. Our system had a 64- bit operating system with 4GB memory and 930 GB HDD. In the first phase, we will select the dataset that are required. The selected dataset consists of four different labels such as no tumor, glioma, meningioma and pituitary tumor. The next step is to import the packages we need from keras and this paper used the Tensor Flow library to generate the image processing.

The Python-based Keras high-level neural network API can be used with Tensor.

Flow, CNTK, or Theano. It was created with the goal of promoting quick experimentation. In this case, the backend will be Tensor Flow. Architecture of convolutional neural networks (CNN) one of the most effective vision model architectures created to date is VGG16. The most notable aspect of VGG16 is that it consistently used the same padding and maxpool layer of 2x2 filters with a stride 2 and prioritized having convolution layers of 3x3 filters with a stride 1. The number 16 in VGG16 stands for the 16 layers with weights. This network is a huge network with about 138 million parameters.

The data consists of four labels such as no tumor, glioma, meningioma and pituitary classes. The images are assigned into 70% of training and 30% of testing validation data. Using the Image Generator, the training data is generated with the hyper parameters such as rescale $= 1./2555$ and shear range and zoom range being 0.2. Using Augmentations, an open source python library and an image augmentation technique, used for enhancing the image like random transformation rotation methods such as 90^0, 180^0, 270^0, horizontal and vertical flips [1]. Images measuring 64 x 64 x 3 were sent to the pre-trained neural network model, which automatically extracts the features, after data enhancement and augmentation. Using a flattened layer, we directed the feature sets from the Cong layer and transformed them into a 2D array. It is then delivered to the dense layer with 128 units after flattening. We combined a dense layer of 1 unit with sigmoid activation and rectified linear unit Relook as the activation function. The final classifier is the sigmoid. The logistic function's binary operation, which determines the threshold value, is performed. It assigns values of 0 and 1, with 0 denoting the absence of a tumor and 1 denoting a predicted tumor.

To measure the average difference between the expected and predicted values, we have used Cross Entropy (CE). The Adam optimizer was selected to achieve the least loss reduction during training. Adam was used because of its quick learning process, efficient usage and easy way of implementation. The batch size used is 32 and epochs being 25 is used as the stable duration for training the model.

Convolutional layers are produced by a collection of filters known as kernels and then applied to an input image. The convolutional layer produces a feature map, which is a representation of the input image with the filters applied. Deep learning uses convolutional layers of the pooling layer type. The spatial dimension of the input is decreased by pooling layers, which speeds processing and consumes less memory.

Pooling both expedites training and helps to reduce the number of parameters. The two main types of pooling are max pooling and average pooling. Max pooling uses the largest value, whereas average pooling uses the average value from each feature map. Pooling layers are frequently used after convolutional layers to reduce the size of the input before it is sent into a fully connected layer. The fully connected layer (CNN) is one of the most basic varieties of layers in a convolutional neural network. As the name implies, every neuron in a totally linked layer is completely coupled to every other neuron in the layer beneath it (Fig. 1). Fully linked layers are typically used in the final stages of a CNN when it is desirable to use the features found by the previous levels to make predictions [17].

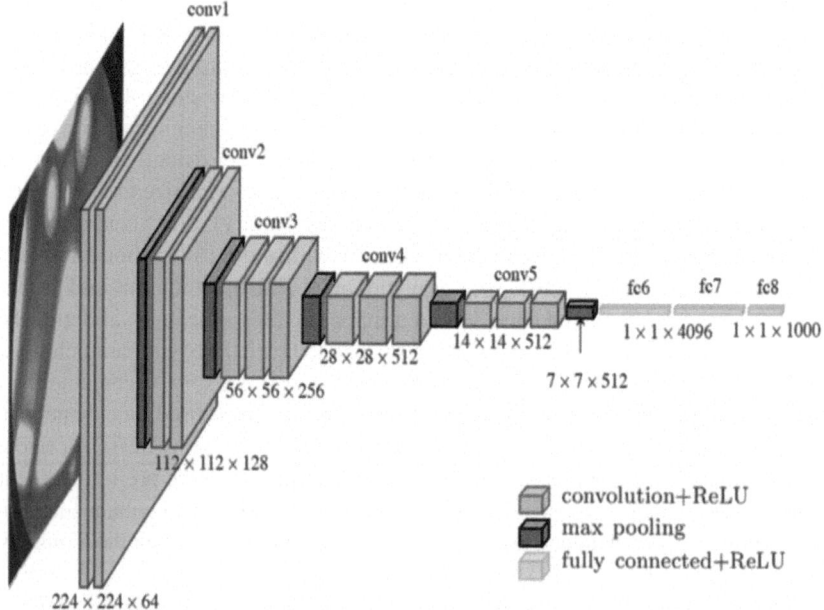

Fig. 1. CNN Architecture [40]

3.1 Brain MRI Dataset

The dataset is obtained from kaggle named brain tumor classification (MRI) [20] image dataset consisting of 3264 images with different categories (Figs. 2 and 3).

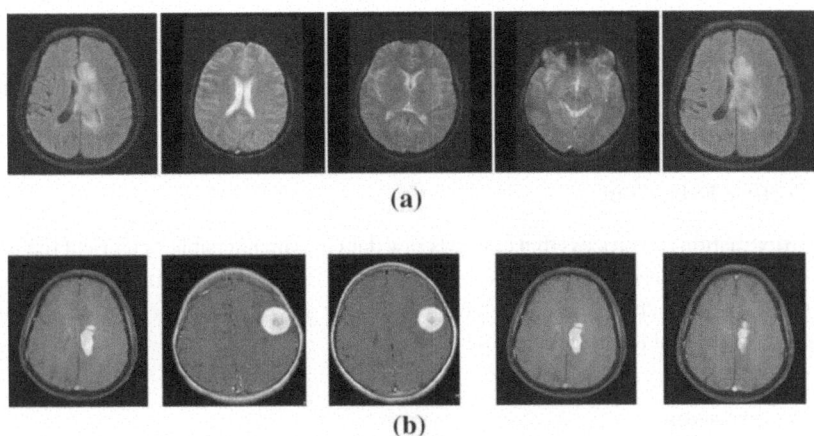

Fig. 2. Sample Brain Tumor MRI Datasets

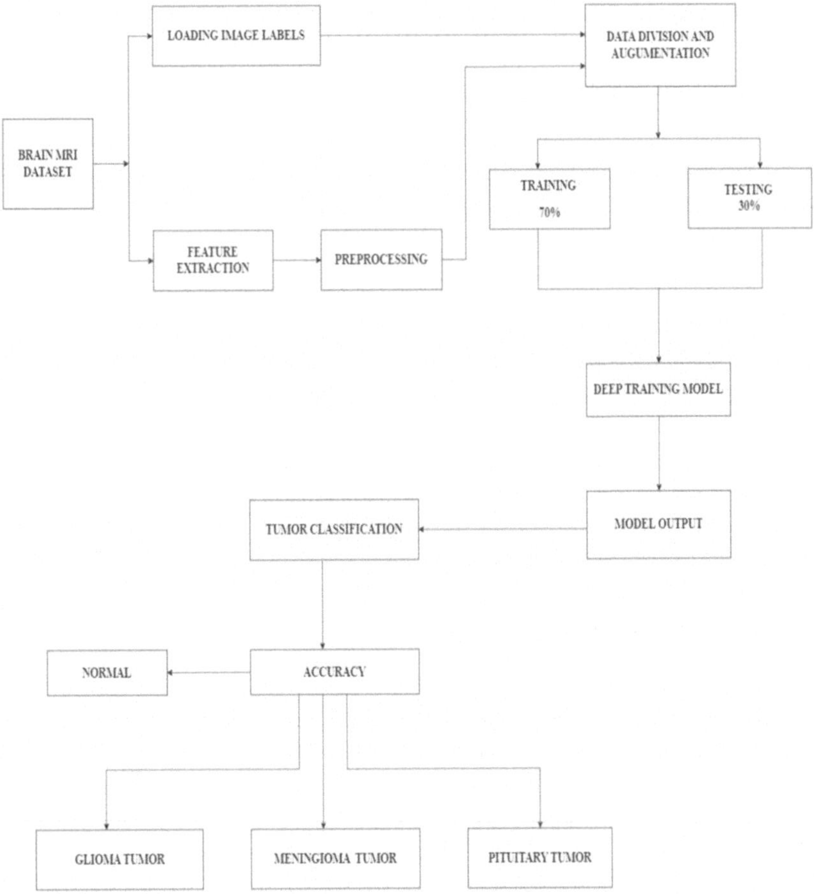

Fig. 3. Block Diagram for Detecting and Classifying the Brain Tumor from the MR Images

3.2 Loading Image Labels

In this step, we are grouping the images into subdirectories such as class1, class2 and class3. We label the MRI scan with tumor as 1 and 0 as no tumor.

3.3 Feature Extraction

Feature extraction is a process that converts raw data into manageable numerical features while preserving the original data set's information. Feature extraction is part of the dimensionality reduction process, which reduces the size and complexity of a starting collection of raw data into manageable chunks.

3.4 Pre-processing

This step is performed to classify the images into a standard format. Here, the images will be converted into grayscale with constant image pixel resolution as 64 x 64. This preprocessing helps to remove the noisy data and improves the quality of the image.

3.5 Data Division and Augmentation:

By adding significantly modified versions of already existing data or completely new synthetic data that is derived from existing data, data augmentations are used to increase the amount of data. It serves as a regularize and helps to reduce over fitting when a model is being trained.

3.6 Training and Testing:

The datasets are divided into two subsets. In order for the model to discover and learn patterns from our actual dataset, it is supplied with the initial subset, also known as the training data. In this way, it trains our model. The other subset is known as the testing data, and you may use it to evaluate the success and growth of your algorithms' training and to tweak or optimize them for better results. Training data typically outweighs testing data in size. This is due to the fact that the model needs as much information as possible to be able to recognize and learn valuable patterns.

3.7 Deep Training Model and Output:

After 70% training and 30% testing the model, the convolutional neural network model is built with various hyper parameters and optimization technique to detect the tumor. The model efficiency is proved when the input is given, the output should match the expected output. Generally, Python comes with Tkinter as a standard library module, which is a portable GUI (graphical user interface) development library. Tkinter provides a native look and feel for Python-coded GUIs on Windows, X-Windows, and Mac OS and provides an object-based interface to the free and open-source Tk library [18]. Graphical User Interfaces (GUIs) are made in Python using Tkinter, which is included in all popular Python distributions [19]. The sole framework in the Python standard library is this one.

The creation of GUI applications is rapid and easy because to the integration of Python and Tkinter. Tkinter makes it simpler to create a GUI application. By making use of Tkinter, we are building a GUI application for a user interface program to upload the MRI brain image to the application. The fileopenbox method helps to open the file manager immediately to load the image. Then by the applied algorithm, it predicts and detect whether the given image has tumor or not, finally the result is printed for the uploaded image (Fig. 4).

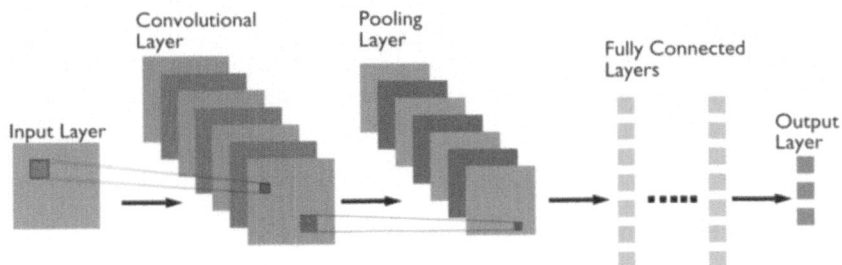

Fig. 4. Schematic representation of the proposed CNN layers architecture [40]

Convolutional neural networks also known as CNNs or ConvNets are a subset of deep neural networks used in machine learning and deep learning. In deep learning, this is used for pattern recognition in images and for other applications. CNN are also used for data analytics, computer vision, Natural Language Processing (NLP) and for various other fields. They are straightforward neural networks with at least one layer that uses convolutional matrix multiplication. ConvNets have demonstrated success in identifying objects, faces, and diseases. A convolutional neural network is made up of an input layer, an output layer, and multiple hidden layers [7]. The feed forward neural network, or CNN, is a well-liked method for categorizing and identifying images. The convolutional neural layers convolve the input before sending their output to the next layer. CNNs are created by modifying multilayer perceptron. The multilayer perceptions are completely linked networks because every neuron in the layer below is connected to every other neuron in the layer above. The "fully-connected" network denotes over fitting data. To build the CNN model, the following packages are needed 8:

Sequential: To initialize the neural network, this Sequential method is used. This arranges the layer to be in sequence.

Convolution2D: The number of filters that convolutional layers will learn from is the Conv2D parameter. It is an integer number that also establishes how many output filters will be used in the convolution. This import Conv2d from kerar.layers to perform convolutional operations. The Conv2D is used as we are working on images dataset, which is basically a 2 dimensional array. It consists of four arguments, namely, number of filter, kernel size of the filter shape, input shape of the image, and an activation function.

MaxPooling2D: A pooling procedure known as "max pooling" selects the largest element from the feature map area that the filter covers. Therefore, the feature map produced by the max-pooling layer would contain the standout features from the previous feature map. With this, the complexity of the model is decreased without sacrificing the model's performance or reducing the image's size.

Flatten: The flatten function is used to convert the pooled feature map into a single column and give it to the fully linked layer. The method of flattening, or turning a two-dimensional array of pooled picture pixels into a single vector, is used to turn all of the combined images into a continuous vector.

Fully Connected Layers: It is also referred to as hidden layers. Dense adds to the fully connected layer to the neural network, units to define the number of nodes that should be present in this hidden layer along with the rely activation function. The output layer consists of only one node, which is in binary classification with only one unit present, either 0 or 1, with sigmoid as an activation function.

4   Results and Discussions

The proposed convolutional neural network classifier model's training and validation outcomes are covered in this part. The model was developed using MR images from the Kaggle open access dataset. To increase the image quality and size, several pre-processing and data-augmentation approaches were applied. The suggested model was trained using the Keras API and used hyper parameters such the Adam optimizer and batch size. The suggested model was trained to effectively classify the tumor using 70% training data and 30% testing data, and it produced an accuracy result of roughly 98%.

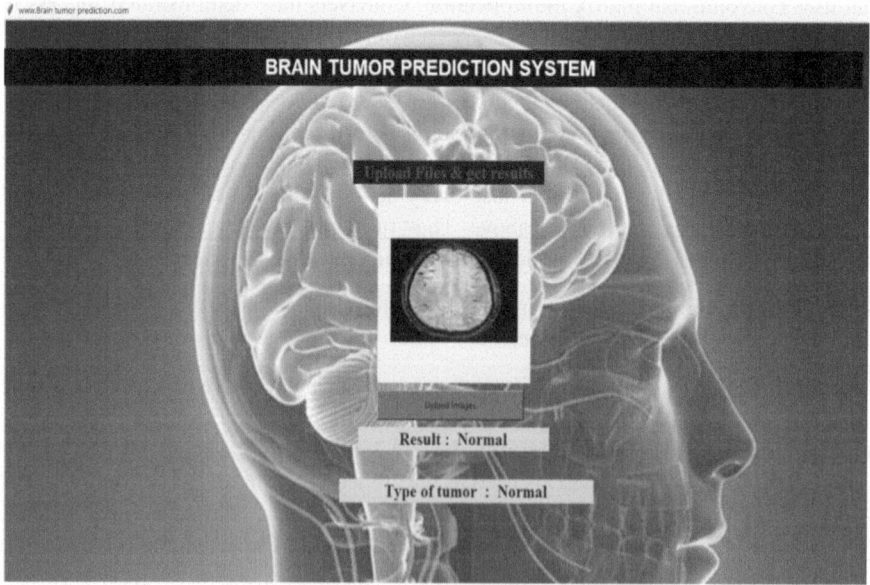

Fig. 5. Results of Normal Brain MRI Image

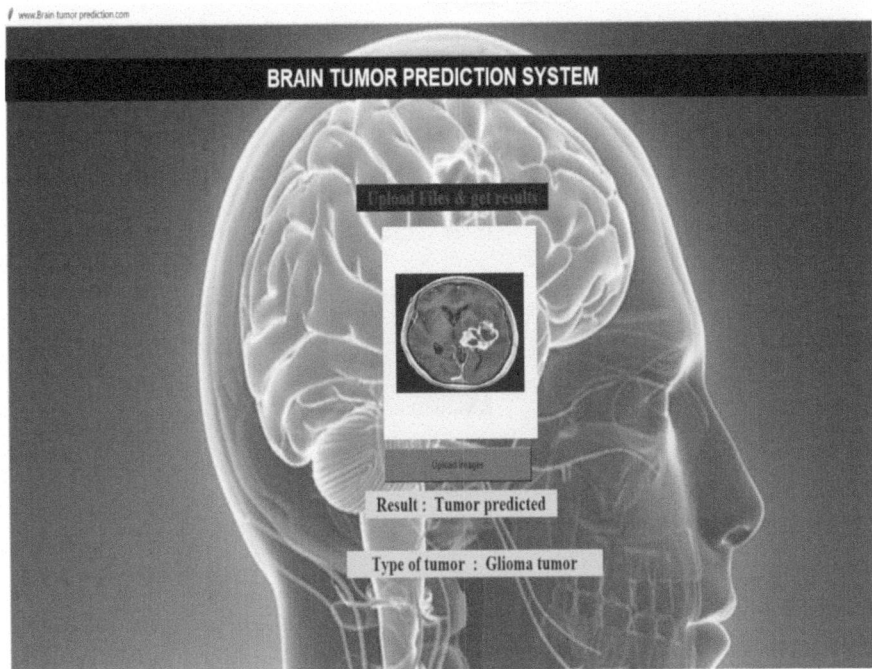

Fig. 6. Results of Glioma Tumor Brain Image

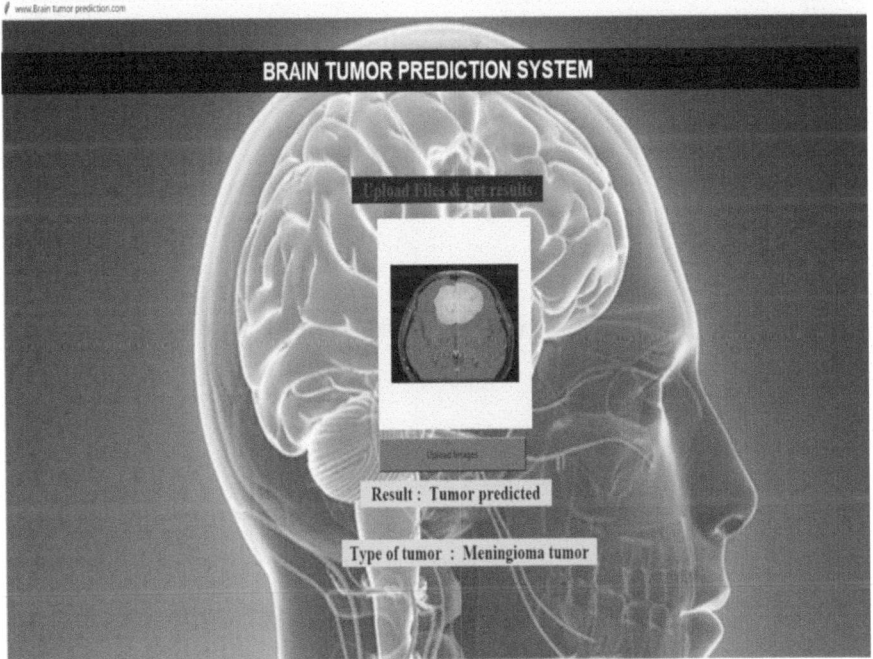

Fig. 7. Results of Meningioma Tumor Brain Image

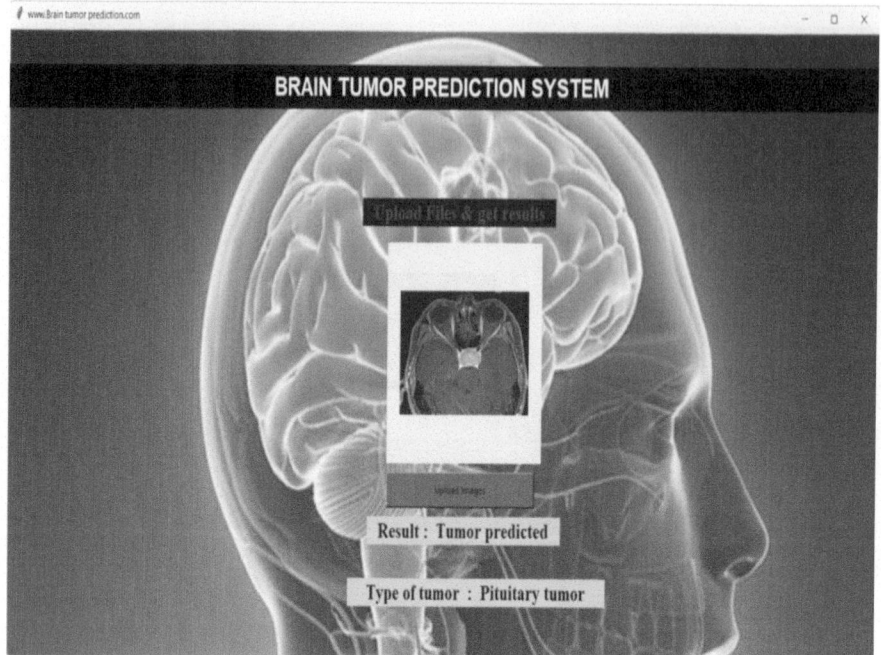

Fig. 8. Results of Pituitary Tumor Brain Image

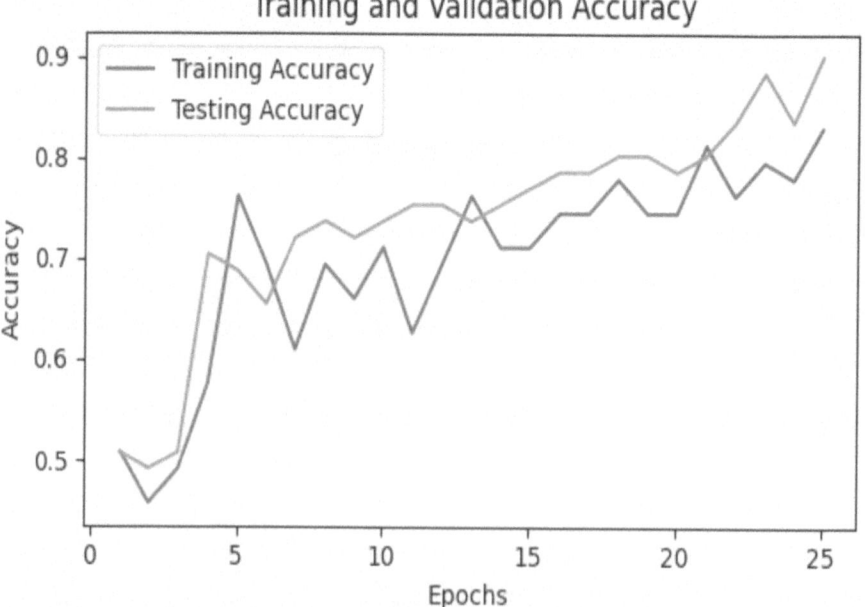

Fig. 9. Training and Testing Accuracy curves of the proposed model

The final model results appear in the form of GUI application, by using the TKinter Toolkit. The accuracy value being for normal 98.6%, glioma 97.9%, meningioma 98% and pituitary being 98.4% (Figs. 5, 6, 7, 8, 9 and 10).

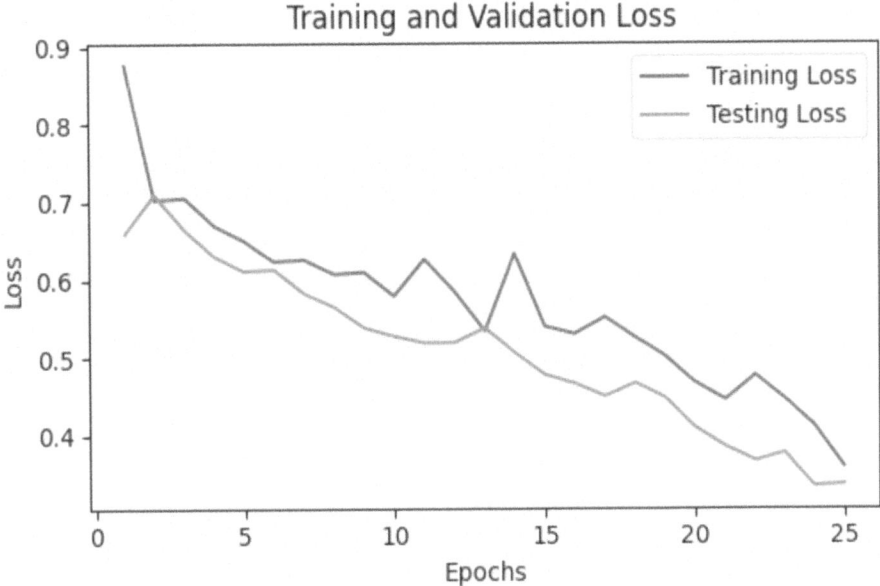

Fig. 10. Training and Testing Loss curves of the proposed model

5 Conclusion and Future Work

Due to the growing need for a practical and accurate evaluation of huge amounts of medical data, MR imaging for the diagnosis of brain tumors has become increasingly popular. Manual detection of brain tumors is laborious and time-consuming, and depending on medical professionals' knowledge. It will be necessary to use an automatic diagnostic method to find anomalies in MRI pictures. With a validation accuracy of 98%, the suggested technique demonstrated the best performance in the identification of brain tumors.

In the future, we'll consider more potent and significant deep CNN models to categories and segment brain tumors with a lower time complexity. In order to boost the accuracy of the proposed model, we will also add more MRI scans to the dataset used for this study. We will also be expanding the suggested methodology to additional medical images including x-ray, computed tomography (CT), and ultrasound in order to build the framework for further research.

References

1. Shah, H.A., Saeed, F., Yun, S., Park, J.H., Paul, A., Kang, J.M.: A robust approach for brain tumor detection in magnetic resonance images using fine tuned efficient net. IEEE Access **10**, 65426–65438 (2022)
2. Medically reviewed by Seunggu Han, M.D.:—By verneda lights—Updated on March 17, [online] Available: https://www.healthline.com/health/brain-tumor, (2022)
3. Ottom, M.A., Rahman, H.A., Dinov, I.D.: Znet: deep learning approach for 2D MRI brain tumor segmentation. IEEE J. Translat. Eng. Health Med. **10**, 1–8 (2022). https://doi.org/10.1109/JTEHM.2022.3176737
4. Ahmad, S., Choudhury, P.K.: On the performance of deep transfer learning networks for brain tumor detection using MR images. IEEE Access **10**, 59099–59114 (2022). https://doi.org/10.1109/ACCESS.2022.3179376
5. Ismail, M., et al.: Radiomic deformation and textural heterogeneity (r-depth) descriptor to characterize tumor field effect: application to survival prediction in glioblastoma. IEEE Trans. Med. Imaging **41**(7), 1764–1777 (2022)
6. Hao, K., Lin, S., Qiao, J., Tu, Y.: A generalized pooling for brain tumor segmentation. IEEE Access **9**, 159283–159290 (2021)
7. Naz, S., Kumar, N.: An efficient brain tumor detection system using automatic segmentation with convolution neural network. Int. Res. J. Eng. Technol. (IRJET) **6**(04), 19 (2019)
8. Deepak Chaudhary, Gaurav Kumar, Vishesh Tomar, Vitthal Agarwal Brain Tumor Detection using Deep Learning, Available: https://www.studocu.com/in/document/gautam-buddha-university/biotechnology/major-project-report-edited-copy/28693405, (June 2022)
9. American Association of Neurological Surgeons, Available: https://www.aans.org/en/Patients/Neurosurgical-Conditions-and-Treatments/Brain-Tumors, (2022)
10. Goodenberger, M.L., Jenkins, R.B.: Genetics of adult glioma. Cancer Genet. **205**(12), 613–621 (2012). https://doi.org/10.1016/j.cancergen.2012.10.009
11. Glioma –Wikipedia, Available: Glioma - https://en.wikipedia.org/wiki/Glioma, (October 2022)
12. Cancer Research UK, Available: https://www.cancerresearchuk.org/about-cancer/brain-tumours/types/glioma-adults, (October 2019)
13. WebMD Editorial Contributors, Medically Reviewed by Carol DerSarkissian, MD (November 19, 2022).Available: https://www.webmd.com/cancer/brain-cancer/malignant-gliomas
14. Meningioma Diagnosis and Treatment was originally published by the National Cancer Institute, Available: https://www.cancer.gov/rare-brain-spine-tumor/tumors/meningioma, (July 2021)
15. IBM, [Online]. Available: https://www.ibm.com/in-en/topics/computer-vision, (2022)
16. Gurucharan,M.K.: Basic CNN architecture: explaining 5 layers of convolutional neural network. [Online]. Available: https://www.upgrad.com/blog/basic-cnn-architecture, (July 2022)
17. Aijitesh Kumar,: Different types of CNN architectures explained: examples. [Online].Available: https://vitalflux.com/different-types-of-cnn-architectures-explained-examples/, (April 2022)
18. Available online:https://www.oreilly.com/library/view/python-pocket-reference
19. Available online: https://www.activestate.com/resources/quick-reads/what-is-tkinter-used-for-and-how-to-install-it
20. Sartaj, B., Ankita, K., Prajakta, B., Sameer, D., Swati, K.: Brain tumor classification (MRI). Kagglehttps://doi.org/10.34740/KAGGLE/DSV/1183165 (2020)

Diagnosis of Parkinson's Disease Using Machine Learning and Deep Learning Techniques

S. Sharanyaa[2(✉)], M. Sambath[1], A. Ganesh[2], A. Hammadh Ahmed[2], and S. Ganesh[2]

[1] Hindustan Institute of Technology and Science, Chennai, India
[2] Panimalar Engineering College, Chennai, India
rnsharanyaa@gmail.com

Abstract. Parkinson's disease (PD) is incapacitating neurodegenerative disorder with a significant impact on individuals worldwide. Early detection of PD plays a vital role in providing timely treatment and improving patient outcomes. Recently, machine learning & deep learning techniques have shown promise in PD detection using various types of medical data, including MRI scan brain images and audio samples. In this work, a novel method for early detection of PD is developed. Aim of our work is to combine and use two different feature sets of input (MRI scan brain images and audio samples). Combining MRI and audio data holds promise as a valuable screening tool for early PD detection, enabling prompt treatment and improved patient outcomes. Both datasets are pre-processed, respectively, and given to the model for training. The Convolutional Neural Network (CNN) model is trained on MRI images and achieved an impressive accuracy of 98.0%, indicating its ability to correctly classify PD cases. On the other hand, the Cat Boost classifier algorithm yielded an even higher accuracy of 98.31% with audio data compared with XGBoost and Random Forest machine learning algorithms. Both Machine learning and Deep Learning models could potentially serve as valuable screening tools for the early detection of PD, allowing for timely treatment and better patient outcomes. The proposed model's ability to use both imaging and audio da- ta could enable a more efficient and accurate diagnosis of PD.

Keywords: Machine Learning · Deep learning · Magnetic Resonance Imaging (MRI) · Audio samples · Convolutional Neural Network (CNN) · XG Boost · Cat Boost · Random Forest

1 Introduction

Parkinson's disease (PD) is characterised by tremors, rigidity, & difficulties with movement and balance [1]. It is a neurological condition that mainly affects the motor system of the central nervous system. Non-motor signs, like dementia and mental problems, may appear in patients with PD as the disease advances [1]. PD is a persistent and degenerative disease that impacts millions of individuals around the globe. Patients lives can be greatly enhanced by prompt diagnosis and therapy of Parkinson's disease. Research into early identification and diagnostic strategies for PD has progressed through a number of studies over the years. Some of the most popular methods for early PD diagnosis include

R. Geetha et al. (Eds.): AAIMB 2023, CCIS 2202, pp. 111–123, 2025.
https://doi.org/10.1007/978-3-031-73065-8_9

machine learning and deep learning. Studies have illustrated that machine learning-based approaches can be used to accurately detect PD. A method for detecting PD using the genetic algorithm and an SVM classifier was developed [2]. This method achieved an accuracy of 98.6% in detecting PD. Machine Learning algorithms were also used to detect PD [3]. SVM, K-nearest neighbours (KNN), and decision trees were the most commonly used classifiers in the studies reviewed. In addition to machine learning, studies have explored other methods for PD detection and diagnosis. Some studies have explored the use of acoustic features to detect PD. An automated PD recognition system was developed based on the statistical pooling technique using acoustic characteristics [4]. Machine learning was used to classify PD based upon acoustic features, achieving an accuracy of 96.6% [5]. Deep learning techniques are being employed in the detection and diagnosis of PD. A deep convolutional neural network had been developed for classifying images in the ImageNet dataset, achieving state-of-the-art **results** in object recognition [6]. A scalable tree boosting system, XGBoost, is used to solve a variety of issues, including PD categorization [7]. Cat Boost, a gradient boosting framework that handles categorical features more efficiently than traditional gradient boosting models, has also been developed for PD classification [8]. Neuroimaging methods are used in detecting & diagnosing of PD. The use of deep learning for neuroimaging in a study using MRI data to diagnose Alzheimer's disease was validated. [9]. In addition to detecting PD, studies have also explored the assessment and treatment of PD. Some studies have investigated cytokines and biomarkers in glucocerebrosidase carriers with and without PD. [10]. Another study studied the effect of levodopa on bilateral coordination and gait asymmetry in PD patients using inertial sensors [11]. The effect of high-density lipoprotein cholesterol variations on Parkinson's disease risk is being studied. [12].

The present study introduces a novel approach for the identification of Parkinson's disease through utilisation of machine learning and deep learning methodologies. We use a dataset of clinical and demographic features, as well as acoustic features, to train and test our models. Several deep learning and machine learning algorithms, like CNN, Random Forest, XGBoost, & Cat Boost, are compared for efficiency. The findings indicate that the proposed approach attains a notable level of precision in the identification of PD and surpasses current methodologies.

2 Literature Review

Numerous machine learning & deep learning techniques have seen substantial growth in published literature upon Parkinson's disease detection in the past few years. In this area, we provide a synopsis of the most important works in the subject. A method for diagnosing Parkinson's disease based on EEG data and machine learning was proposed by Maitin et al. [1]. Patients with Parkinson's disease were identified by extracting characteristics from EEG data and running them through SVM) classifier, as authors explain. Gait analysis is used to diagnose Parkinson's disease, and Liu et al. [2] recently developed a deep learning-based technique for doing so. The scientists successfully classified Parkinson's disease patients by using CNN to extract features from gait data. Wearable sensor-based approaches have been suggested by Wang et al. [3] for the diagnosis and monitoring of Parkinson's disease. The scientists used a number of machine learning techniques for categorization after collecting data using an accelerometer as well as a gyroscope.

Voice biomarkers and machine learning techniques were studied by the authors [4] for their potential in diagnosing Parkinson's disease. The authors utilized a support vector machine classifier to extract information from speech recordings and identify PD patients. Using magnetic resonance imaging (MRI) scans and convolutional neural networks, Ashraf et al. [5] suggested a method for diagnosing Parkinson's illness. Using a convolutional neural network (CNN), the scientists successfully extracted information from MRI scans to accurately identify individuals with Parkinson's disease.

Szczepaski et al. [6] projected a technique for classifying Parkinson's disease patients by machine learning & voice recordings. The authors extracted various acoustic features from voice recordings and used a random forest classifier to identify patients having PD. Sathyanarayana et al. [7] presented a system for detection of PD from smartphone sensor data using deep neural net- works. The authors collected data from various sensors, including an accelerometer, gyroscope, and magnetometer, and achieved high accuracy in identifying patients with Parki0nson's disease. Manikandan et al. [8] investigated the use of a random forest classifier on MRI brain images for Parkinson's disease detection. The authors attained higher accuracy into recognizing patients having PD by extracting features from MRI images and using a random forest classifier. Iqbal et al. [9] proposed a deep learning-based framework for voice and gait analysis for Parkinson's disease diagnosis. The authors extracted features from voice and gait signals using CNNs and achieved high accuracy in classifying Parkinson's disease patients. Zhao et al. [10] investigated the use of XG Boost and support vector machines on speech data for Parkinson's disease diagnosis. The authors extracted various features from speech signals and used XG Boost and SVM classifiers for identifying patients having Parkinson's disease.

Overall, the literature on Parkinson's disease diagnosis utilising machine learning & deep learning methods is extensive and varied. The studies reviewed in this section show promising results in identifying patients suffering from PD using various signals, including EEG, gait, voice, and MRI images.

3   Methodology

3.1   Dataset Collection

The first dataset [11] is taken from Kaggle for diagnosing PD & includes MRI brain scan images of patients with and without PD. The images were collected using T1weighted MR imaging, and the dataset consists of 831 MRI scans, of which 610 are from patients having PD & 221 are from healthy controls. 0

Second dataset is taken from Kaggle for identification of PD and includes audio samples of patients with and without PD [12]. Dataset consists of 196 audio samples, which include 22 features.

3.2   Data Pre-processing

Pre-processing is critical step into preparing a dataset for use in a machine learning model. In our study, we pre-processed the MRI brain images by removing the skull and other non-brain tissue using a brain extraction tool. We then corrected the images for bias field using the N4ITK algorithm to normalize the intensities across the images. The images were further pre-processed using the FSL registration tool to align the images and standardize the orientation. For the audio samples, we first removed any background noise and normalized the volume levels. We then extracted relevant features such as pitch, intensity, and formants using Praat software. The extracted features were pre-processed by standardizing the values across the samples to ensure consistency across the dataset. Additionally, we applied principal component analysis (PCA) to reduce amplitude of feature space & eliminate any redundant features. The pre-processing of our data was necessary to reduce noise and ensure that our machine learning model was trained on high-quality, standardized data.

3.3 System Design

The MRI brain images and audio samples are collected from people with and with- out Parkinson's disease from various sources, such as public datasets and hospitals, and the data is diverse and representative of the population. The data is stored in a suitable format and labelled appropriately. Pre-processing the MRI brain images and audio samples is done to ensure consistency and accuracy. Various tasks are per- formed, such as image resizing, normalization, and de-noising. Then the audio samples are converted into spectrograms or other suitable representations for analysis. Extract relevant features from the pre-processed data that can be used for classification. For MRI brain images, we have used methods like edge detection and texture examination to extract features. For audio samples, we have used features such as pitch, amplitude, and spectral energy. Then the data is separated into testing and training sets. It is imperative to maintain a balance of data distribution between two sets and across various classes. Techniques such as cross-validation are used for ensuring as model isn't over fitting training data. Then, we built a convolutional neural network (CNN) model to classify the MRI brain images and used techniques such as transfer learning to leverage pre-trained models and improve performance. Tuning hyper parameters such as learning rate and batch size to optimize the model. Then we build XGBoost, Cat Boost, and Random Forest models to classify the audio samples. The purpose of using feature importance techniques is to identify the most relevant features for classification. Tuning hyper parameters such as the number of trees and learning rate to optimize the models. Then we assess the efficacy of models by utilizing metrics like precision, accuracy, recall, and F1-score. We conducted a comparative analysis of various models and selected the optimal one(s) for the purpose of classification. Then finally, we deployed the chosen models for real-world diagnosis of PD using new MRI brain images & audio samples. Figure 1 depicts the architecture of the proposed method for diagnosing Parkinson's disease.

3.4 Classification of PD Using MRI Brain Images

Millions of individuals all over the globe suffer from Parkinson's disease, a debilitating neurological condition. In order to effectively cure and control this illness, early discovery and a correct prognosis are essential. The past few years have demonstrated that machine learning methods can greatly aid in PD detection. The purpose of the suggested

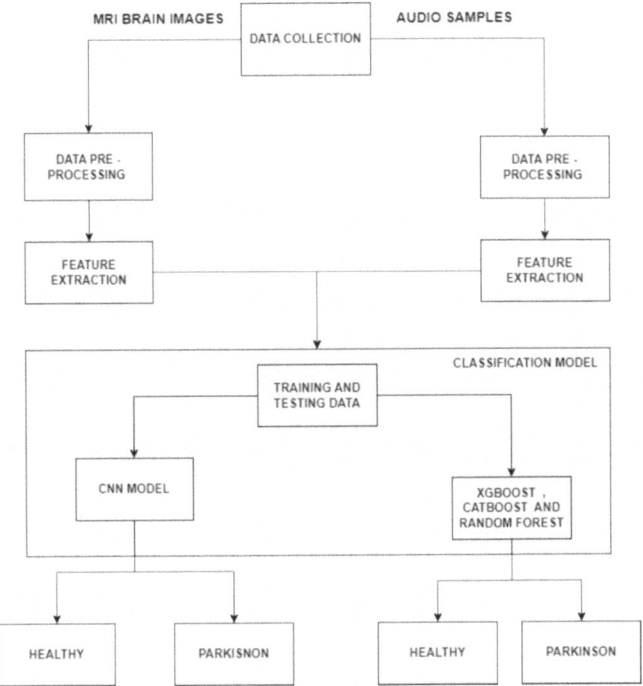

Fig. 1. System Architecture of Proposed Model.

system is to create a machine learning-based method for detection of Parkinson's disease into early stages through the analysis of MRI brain pictures. To determine whether an MRI brain picture is indicative of Parkinson's disease or not, we plan to utilise CNNs, a form of deep learning which has shown impressive results in image categorization tasks. The proposed system will consist of several stages. First, we will collect a dataset of MRI brain images from both Parkinson's disease patients and healthy individuals. Next, we will pre-process the images by standardising the size, removing noise, and normalising pixel values. Data is divided in training & testing sets; training set will be used to build the CNN algorithm. During the training phase, the CNN model will learn to identify features in the MRI brain images that are indicative of Parkinson's disease. Back-propagation and gradient descent could be utilised for fine-tuning models & reduce the discrepancy between expected and observed labels. We will then examine how well the learned CNN model does upon trial collection. To evaluate the model's performance in identifying Parkinson's disease from MRI brain pictures, we will calculate its accuracy, precision, memory, and F1 score. In sum, the suggested approach shows promise as non-invasive and a precise means of detecting Parkinson's disease at earlier phase. By leveraging machine learning techniques and deep learning algorithms, we can assist clinicians in making informed diagnoses and improving patient outcomes.

The proposed system is a deep learning-based diagnostic tool for the detecting PD. System is built with the aid of CNN trained on MRI brain images. The framework of the projected CNN Architecture for diagnosing Parkinson's disease using MRI brain images is given in Fig. 2.

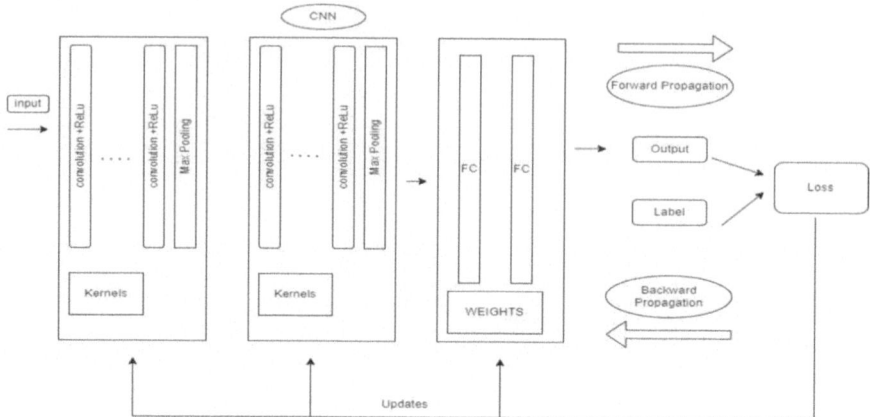

Fig. 2. CNN Architecture for detecting Parkinson's disease using MRI brain images.

The suggested system utilises a 4-convolutional-layer, 2-covolutional-layer, and 2-max-pooling-layer CNN architecture. The output of the last pooling layer is flattened before being passed through a 128-neuron-thick layer. The final layer has two neurons, one for normal and one for Parkinson's disease classification, with softmax activation. The model was developed based on modified MRI scan images utilising the RMSprop algorithm and a learning rate of 0.001 with a category cross-entropy loss function. With a sample size of 32, a model is learned over 30 epochs.

3.5 Classification of PD Using Audio Samples

In this proposed system, we aim for detection of Parkinson's disease using audio samples and machine learning algorithms. We used the "MDVR-KCL" dataset, which includes various voice features extracted from sustained phonation of vowel /a/ from 195 patients with early-stage Parkinson's disease and 48 healthy individuals. The dataset includes features.

MDVP: Fo(Hz) - Mean frequency of the voice signal (in Hz).

- **MDVP: Fhi(Hz)** - Maximum frequency of the voice signal (in Hz)
- **MDVP: Flo(Hz)** - Minimum frequency of the voice signal (in Hz)
- **MDVP: Jitter(%)** - Frequency Variation (in %)
- **MDVP: Jitter(Abs)** - Absolute frequency variation (in Hz)
- **MDVP:RAP** - Variation in frequency (in Hz) relative to the time between cycles
- **MDVP:PPQ** - Variation in frequency (in Hz) relative to the amplitude of the signal
- **Jitter: DDP** - Combination of RAP and PPQ measures
- **MDVP: Shimmer** - Variation in amplitude of the signal (in dB)
- **MDVP: Shimmer(dB)** - Absolute variation in amplitude of the signal (in dB)

- **Shimmer:APQ3** - Absolute variation in amplitude of the signal over the first 3 harmonics
- **Shimmer:APQ5** - Absolute variation in amplitude of the signal over the first 5 harmonics
- **MDVP:APQ** - Absolute variation in amplitude of the signal over all harmonics
- **Shimmer: DDA** - Average absolute difference amongst amplitudes of successive periods
- **NHR** - Noise-to-harmonics proportion
- **HNR** - Harmonics-to-noise proportion
- **Status** - Binary variable indicating the presence or absence of Parkinson's disease
- **RPDE** - Recurring period density entropy measure
- **DFA** - DE trended fluctuation analysis measure
- **Spread 1** - Nonlinear measurement of fundamental frequency variance.
- **Spread 2** - Nonlinear measure of amplitude variation
- **D2** - Nonlinear measure of signal complexity
- **PPE** - Nonlinear measure of pitch variation

To classify the audio samples as healthy or Parkinson's disease, we utilised machine learning algorithms like XGBoost, Cat Boost, & Random Forest. These algorithms were trained on the dataset to predict the status of the patient (healthy or Parkinson's disease) based on their voice features. The proposed system can be used as a screening tool for Parkinson's disease, allowing for early detection and timely treatment, which can improve patient outcomes and quality of life. The proposed system uses three different algorithms: XGBoost, Cat Boost, and Random Forest. Each algorithm is trained on the pre-processed audio samples.

3.6   Classification Algorithms

XG Boost Algorithm   The XGBoost (Extreme Gradient Boosting) algorithm is used to create a classification model to predict Parkinson's disease status based on various features. XGBoost is an ensemble learning technique that generates a more reliable model by combining the results of several individual decision trees. Firstly, data is read from the Parkinson's disease dataset and visualised using boxplots to analyse the relationship between the features and the target variable (Parkinson's status). Next, the features and labels are separated, and a train-test split is performed on the data. Then, an instance of the XGB Classifier is created, and the model is trained upon training set. Once the model is ready, it is utilised in predicting test set labels, and the accuracy score and confusion matrix are calculated for evaluating model's efficacy. Lastly, accuracy of XGBoost model was compared with the accuracy of Cat Boost and Random Forest models using a bar graph. In summary, XGBoost works by building multiple decision trees iteratively and minimising the errors between the predicted and actual values. It uses a gradient-boosting approach to add new models that correct the errors made by the previous models. The model achieved an accuracy of 91.53%.

Cat Boost Algorithm
Cat Boost is a boosting algorithm that uses gradient boosting on decision trees. It is

similar to XGBoost and Light GBM, but it is designed to handle categorical features and missing values. Cat Boost can handle a high number of categories and large volumes of data. The algorithm creates decision trees on the training data, and every tree is trained on a data subset. The algorithm uses gradient boosting to iteratively improve the performance of the model. First, the required libraries are imported. The Parkinson's disease dataset is loaded into a Pandas data frame. A boxplot is created to visualise the distribution of each feature in the dataset. This helps identify any outliers and see how the data is distributed. The features and labels are separated from the data frame. The 'status' column is used as the label, and the remaining columns are used as features. The count of each label (0 and 1) in labels is printed to check whether the dataset is balanced or not. Data is divided in training & testing sets. A Cat Boost Classifier model is created with 100 iterations, a learning rate of 0.1, a depth of 6, and a loss function A Cat Boost Classifier model is created with 100 iterations, a learning rate of 0.1, a depth of 6, and a loss function of "log loss." The training data is used to train the model via the application of the fit () technique. The predict () function is used to generate predictions on the test data. The score () function is used to ascertain the accuracy value of the model. There are numerical representations of the model's accuracy, precision, and confusion matrices. The 'cat' property stores the precision rating. In summary, the code uses the Cat Boost algorithm to train a binary classification model on the Parkinson's disease dataset and then evaluates its performance by calculating the accuracy score, confusion matrix, and precision score. The overall precision obtained by the algorithm was 98.31%.

Random Forest Classifier
A random forest classifier is utilised for building model for predicting status of patients with Parkinson's disease. Here is how it works: First, the required libraries are imported, including the Random Forest Classifier from sklearn. Ensemble. The dataset is read into a Pandas Data Frame DF using pd. Read_csv. The features and labels are extracted from the Data Frame using the df. Drop and.loc functions. Dataset is divided into a training set & evaluation set with help of the train_test_split tool. As model 2, a Random Forest Classifier object is generated. Model2.fit is then used to train model2 on the data. The learned model is then used in model2.predict to make prognoses for PD cases. The Random Forest Classifier accuracy value is determined by the accuracy_score method. Finally, a bar plot is created to compare the accuracy of the Random Forest Classifier model with other models. At the stage of training, Random Forest algorithm builds numerous decision trees, each of which predicts a different class outcome (classification) or regression value (mean). Every tree of decisions is "bootstrapped," or trained, using data from a statistically-independent subset of whole training set. This process of selecting a subset of the data and features is called bagging. During prediction time, the Random Forest algorithm takes the average of the predictions of each decision tree and outputs the class that has the most votes. This helps to reduce overfitting compared to a single decision tree. The models are trained on the pre-processed audio samples using the respective algorithms. The models are trained for a specific number of iterations at a specific learning rate. The model achieved an accuracy of 89.83%.

3.7 Evaluation Metrics

When evaluating the performance of a predictive model, Precision and Accuracy are commonly used measures. Precision specifically measures the accuracy of positive predictions by determining the proportion of correct positive forecasts from every positive prediction. A higher Precision value indicates that the predictive model is efficient at accurately predicting positive outcomes.

$$\text{Precision} = \frac{\text{True Positive}}{\text{True Positive} + \text{False Positive}} \tag{1}$$

Recall is a performance measure utilized for evaluating the effectiveness of a predictive model, particularly in terms of its ability to identify true positive values. It is deliberated as a ratio of correct positive prediction for every actual positive value. A higher Recall value specifies as predictive model is improved in correct identification positive outcomes.

$$\text{Recall} = \frac{\text{True Positive}}{\text{True Positive} + \text{False Negative}} \tag{2}$$

Calculating Accuracy is an important aspect of evaluating a predictive model's performance. It is calculated by dividing fraction of observations for which a prediction was made accurately by total number of observations. And expressed as a percentage between 0% and 100%. This measure is particularly useful when the distribution of data points for each label is equal, as it provides an ideal indication of the model's overall accuracy.

$$\text{Accuracy} = \frac{\text{True Positive} + \text{True Negative}}{\text{True Positive} + \text{True Negative} + \text{False Positive} + \text{False Negative}} \tag{3}$$

4   Results and Discussion

The Cat Boost algorithm demonstrated the highest accuracy (98.31%) and precision (0.98) among the three models tested. This indicates that Cat Boost had a remarkable ability to correctly classify instances of Parkinson's disease. The XG Boost model attained accuracy of 91.53% & precision of 0.94, suggesting it also performed well in differentiating between individuals with Parkinson's disease and healthy individuals. Similarly, Random Forest model achieved accuracy of 89.83% and precision of 0.92. Although it exhibited slightly lower performance compared to Cat Boost and XG Boost, it still showcased considerable potential for Parkinson's disease diagnosis. Moreover, the CNN model trained specifically for classifying MRI brain images achieved an impressive accuracy of 98%. This suggests that the CNN model was highly successful in accurately identifying instances of PD as per MRI images. These results highlight the effectiveness of machine & deep learning techniques in diagnosing Parkinson's disease. Specifically, Cat Boost and the CNN model demonstrated excellent performance, with high accuracy and precision values. The findings from this study support the notion that these techniques hold promise as valuable tools in diagnosing of Parkinson's disease. Further research & validation using diverse datasets are necessary to confirm and extend these results in real-world clinical settings. These findings suggest that machine learning and deep learning techniques can be effective tools for diagnosing Parkinson's dis- ease. A convolutional neural network provides 98% accuracy as a result of Parkin- son's disease. MRI brain image classification Model The accuracy of CNN Model is given in Fig. 3.

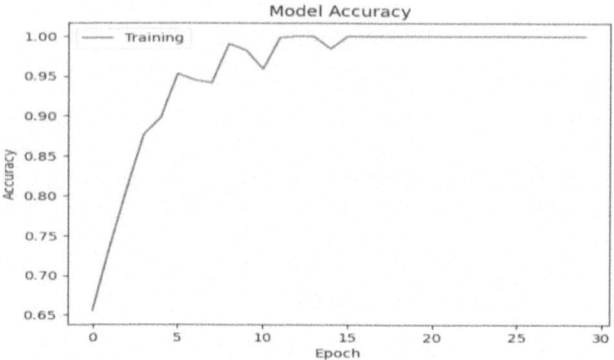

Fig. 3. Accuracy of CNN Model

Accuracy obtained using Boosting Machine Learning algorithms provides accuracy between 89% and 98% as a result of Parkinson's Disease using features obtained from audio samples and is given in Fig. 4.

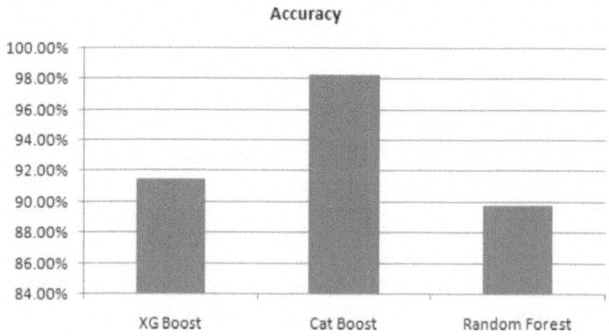

Fig. 4. Accuracy of XGBoost, Cat Boost and Random Forest Models

The performance of different classifiers applied to audio and MRI data is summarised in Table 1. The classifiers evaluated include Convolutional Neural Network (CNN), XG Boost, Cat Boost, and Random Forest. Performance metrics such as accuracy, precision, and recall were used to assess their classification performance.

Table 1 shows as CNN classifier was able to accurately label 98% of samples in MRI dataset, demonstrating as it is very accurate. The percentage of accurately anticipated

Table 1. Performance Measures of the Classification Models

Classifier	Accuracy %	Precision %	Recall %	Input
CNN	98%	98%	92.8%	Image
XG Boost	91.53%	94%	93%	Audio
Cat Boost	98.31%	98%	97.9%	Audio
Random Forest	89.83%	92%	91%	Audio

positive instances, or precision, was likewise quite high at 98%. Still, recall (a metric of how accurately positive cases were identified) was 92.8%, which was somewhat lower. For audio data classification, the XG Boost classifier achieved an accuracy of 91.53%, a precision of 94%, and a recall of 93%. This indicates its ability to effectively classify audio data, albeit with slightly lower accuracy compared to the CNN classifier. The Cat Boost classifier demonstrated the highest accuracy among the evaluated classifiers, reaching an impressive 98.31%. It exhibited high precision (98%) and recall (97.9%) values, show- casing its exceptional performance in classifying audio data. Conversely, Random Forest classifier achieved an accuracy of 89.83%. Although it had a slightly lower accuracy compared to the other classifiers, it still exhibited decent precision (92%) and recall (91%) values, indicating its competence in classifying audio data. In summary, the CNN classifier demonstrated the highest recall among all the classifiers, while the Cat Boost classifier exhibited the highest accuracy. Both the XG Boost and Cat Boost classifiers performed well in audio data classification, demonstrating high precision and recall values. The Random Forest classifier, while having slightly lower accuracy, still performed reasonably well in classifying audio data.

5   Conclusion

The utilization of machine learning and deep learning methods in detecting Parkinson's disease has shown promising results. The CNN model also exhibited high precision, with a value of 98%, indicating its capability to minimize false positives. The recall of the CNN model was 92.8%, illustrating its ability to correctly identify PD cases from the positive samples. On the other hand, the XG Boost algorithm achieved an accuracy of 91.53%, a precision of 94%, and a recall of 93% using audio samples. The Cat Boost algorithm yielded an even higher accuracy of 98.31%, a precision of 98%, and a recall of 97.9% with audio data. Random Forest algorithm attained an accuracy of 89.83%, a precision of 92%, and a recall of 91% using audio samples. By comparing all state-of-the-art models, convolutional Neural Networks & Cat Boost Classifier equally yield higher accuracy scores. These proposed systems have the potential to be valuable diagnostic tools for Parkinson's disease, providing objective and accurate diagnoses in assisting clinicians in making informed decisions and delivering better patient care. Early detection of Parkinson's disease can lead to more efficient disease management, ultimately improving patients' quality of life. Furthermore, utilisation of machine learning & deep learning further research in this area can help refine and improve the accuracy of these diagnostic tools, leading to better patient outcomes. Methods for detecting Parkinson's disease hold great promise for the future.

6 Future Work

Based on the current study's findings, there are several potential areas for future work that could further enhance the diagnosis of PD utilising machine learning & deep learning techniques. Firstly, the evaluation of other machine learning algorithms, like SVMs or Neural Networks, could be considered. These algorithms have shown promising results in various domains and could potentially outperform the models tested in this study. Comparative studies involving a wider range of algorithms would provide a comprehensive understanding of their performance in Parkinson's disease diagnosis. Another direction for future research is the inclusion of additional features or data sources. In this study, audio data and MRI brain images were utilized for diagnosis. However, incorporating other types of data, such as genetic markers or wearable sensor data, may offer valuable insights and potentially improve accuracy and precision. By integrating multiple modalities of information, it may be possible to achieve a more comprehensive and accurate diagnosis. Furthermore, conducting validation on larger datasets is crucial to assessing the generalizability of the models' performance. The current study likely used a specific dataset, and expanding the evaluation to encompass diverse and representative datasets would provide more robust and reliable results. This would help determine if the performance achieved by the models remains consistent across different data sources and populations. In addition to algorithmic and data-related considerations, future work should also focus on the implementation of these models in a clinical setting. Collaboration with clinicians and patients would be essential to developing an accurate and user-friendly system that can be effectively integrated into routine clinical practice. Evaluating the performance of these models in real-world scenarios and assessing their impact on clinical decision-making would be crucial steps towards practical application. Overall, future research in the field of Parkinson's disease diagnosis utilising machine learning & deep learning techniques should explore alternative algorithms, incorporate additional features and data sources, validate on larger datasets, and prioritise real-world implementation in collaboration with clinicians and patients. By addressing these aspects, the field can advance towards more accurate, reliable, and clinically applicable diagnostic tools for Parkinson's disease.

References

1. Maitin, A.M., et al.: Survey of machine learning techniques in the analysis of EEG signals for Parkinson's disease: a systematic review. Appl. Sci. **12**(14), 6967 (2022)
2. Guo, Y., Yang, J., Liu, Y., Chen, X., Yang, G.-Z.: Detection and assessment of Parkinson's disease based on gait analysis: a survey. Front. Aging Neurosci. **14**, 916971 (2022)
3. Wang, Z., Li, Y., Tian, K., Wang, Z.: A wearable sensor-based approach for parkin- son's disease diagnosis and monitoring. IEEE Access **7**, 1710817117 (2019). https://doi.org/10.1109/ACCESS.2019.2894138

4. Costantini, G., et al.: Artificial intelligence based voice assessment of patients with Parkinson's disease off and on treatment: ma- chine vs deep-learning comparison. Sensors **23**(4), 2293 (2023)
5. Ashraf, M., Saleem, Z., Farooq, M.U.: Parkinson's Disease Diagnosis using MRI Images and Convolutional Neural Networks. In: 3rd International Conference on Computer Applications & Information Security (ICCAIS), pp. 1–6, (2022)
6. Szczepański, K., Arora, M.R.S., Arora, S.: Classification of Parkinson's disease patients using machine learning and voice recordings. In: 11th International Conference on Computing, Communication and Networking Technologies (ICCCNT), p. 1 (2022) https://doi.org/10.1109/ICCCNT49239.2020.9225274
7. Sathyanarayana, V.P., Vijaya, J., Ramu, P.:Detecting Parkinson's disease from smartphone sensor data using deep neural networks. In: 3rd International Conference on Communication and Electronics Systems (ICCES), pp. 309-313 (2018)
8. Manikandan, P., Krishnan, R., Bharathi, P.V.:Parkinson's disease detection using random forest classifier on MRI brain images. International Conference on Smart Electronics and Communication (ICOSEC), 2021, pp. 1–5, https://doi.org/10.1109/ICOSEC51151.2021.9397671
9. Iqbal, S., Raza, M.A., Iqbal, M.Z., Khalid, R.: Parkinson's disease diagnosis using a deep learning-based framework for voice and gait analysis. IEEE Access **9**, 109191–109202 (2021)
10. Zhao, J., Wu, L., Wu, J.: Parkinson's disease diagnosis using xg boost and support vector machine on speech data. IEEE Access **8**, 176078–176088 (2020)
11. Quan, C., Kang, R., Zhiwei, L.: A deep learning based method for Parkin- son's disease detection using dynamic features of speech. IEEE Access **9**, 10239–10252 (2021)
12. Khaskhoussy, R., Ayed, Y.B.: Speech processing for early Parkinson's disease diagnosis: machine learning and deep learning-based approach. Soc. Netw. Anal. Min. **12**(1), 73 (2022)
13. Amato, F., Giovanni, S., Valerio, C., Gabriella, O., Giovanni, C.: Machine learning-and statistical-based voice analysis of Parkinson's disease patients: a survey. Expert Syst. Appl. **219**, 119651 (2023)
14. Sigcha, L., et al.: Deep learning and wearable sensors for the diagnosis and monitoring of Parkinson's disease: a systematic review. Expert Syst. Appl. **9**, 120541 (2023)
15. Qiu, L., Jianping, L., Jiahui, P.: Parkinson's disease detection based on multi- pattern analysis and multi-scale convolutional neural networks. Front. Neurosci. **16**, 957181 (2022)
16. JabaSheela, L., Vasudevan, S., Yazhini, V.: "A hybrid model for detecting linguistic cues in Alzheimer's disease patients. J. Inf. Computat. Sci. **10**(1), 85–90 (2020)
17. Sumithra, M., Malathi, S.: A novel distributed matching global and local fuzzy clustering (DMGLFC) for 3D brain image segmentation for tumour detection. IETE J. Res. **68**(4), 2363–2375 (2022)
18. Selvan, S., et al.: An image processing approach for detection of prenatal heart disease. Biomed Res. Int. **2022**, 1–14 (2022)

A Survey on Deep Learning Based Human Activity Recognition System

Ansu Liz Thomas[✉] and J. E. Judith

Department of Computer Science and Engineering, Noorul Islam Center for Higher Education,
Kanyakumari District, Tamilnadu, India
lizansu@gmail.com

Abstract. A comprehensive analysis of deep learning methods utilized in Human Activity Recognition (HAR) across multiple domains such as healthcare, security, and sports were discussed in this article. Deep learning (DL) techniques have shown significant advancements over traditional machine learning (ML) approaches in terms of accuracy and resilience. The study explores various DL models including Convolutional Neural Networks (CNNs), Recurrent Neural Networks (RNNs), and Long Short-Term Memory (LSTM) models, as well as their combinations. It discusses the architectures, optimization methods, and preprocessing techniques employed to enhance the effectiveness of DL-based HAR systems, such as data augmentation, feature extraction, and dimensionality reduction. The article also highlights datasets and performance metrics utilized for evaluating the efficacy of HAR systems. Furthermore, it addresses current challenges and future prospects in deep learning-driven research for HAR, including the need for improved interpretability, adaptability to novel activities and settings, and integration of multiple modalities to advance HAR. In summary, this review provides valuable insights into DL-driven human activity recognition systems, culminating in a maximum performance of 96.8% achieved by a multi-stream CNN combined with LSTM.

Keywords: Human Activity Recognition(HAR) · Deep Learning(DL) ·
convolutional neural network(CNN) · long short-term memory(LSTM) · machine
learning(ML) · recurrent neural networks (RNNs) · dimensionality reduction

1 Introduction

The identification of human activities and actions through the analysis of sensor data from accelerometers, gyroscopes, and magnetometers is known as Human Activity Recognition (HAR). This task has garnered substantial interest in diverse fields, including healthcare, sports, entertainment, and security, due to its potential applications [1].

The capability of deep learning(DL) techniques to automatically acquire hierarchical representations of sensor data has led to promising outcomes in HAR systems. This review delves into the latest developments in HAR systems that leverage DL. We will examine different types of neural network structures, such as CNN, RNN, and their

R. Geetha et al. (Eds.): AAIMB 2023, CCIS 2202, pp. 124–134, 2025.
https://doi.org/10.1007/978-3-031-73065-8_10

respective adaptations, as we navigate through our discussion. Also discusses their usage for different sensor data types [2]. Additionally, addresses the obstacles and restrictions encountered by these systems, as well as explores potential avenues for future research.

In general, HAR systems that rely on DL have demonstrated significant promise in enhancing the precision and effectiveness of activity recognition assignments. With further research and development, these systems could have a significant impact on various fields, including healthcare, sports, and security. A module for recognizing human activities (HAR) is responsible for extracting relevant features from available signals or data. HAR has wide-ranging applications in fields like pattern recognition, machine learning, wearable computing, and computer vision. The objective of the HAR system is to automatically detect and categorize the activities performed by individuals based on raw sensor data.

The initial stage of HAR involves acquiring data as in Fig. 1, which can be gathered from various real-time sources such as traffic signals, healthcare facilities, and sports activities. Once collected, the data can undergo pre-processing using a range of techniques. Pre-processing techniques commonly employed include the normalization of pixel values and the conversion of images to grayscale. These steps aim to enhance the model's accuracy. Once the pre-processing is complete, the next step involves partitioning the dataset into two distinct sets. The deep learning (DL) model is provided with the training dataset to enable the neural network's training process. Subsequently, the accuracy can be evaluated by utilizing the testing dataset. Different levels of activity are shown in Fig. 2. A common methodology for HAR using DL involves the following steps:

- **Data Collection:** Gather labeled examples of human activities using sensors like accelerometers, gyroscopes, or cameras.
- **Preprocessing:** Clean and preprocess the data to remove noise, normalize sensor readings, and standardize the data format.
- **Data Split:** Divide the dataset into training, validation, and testing sets for training, hyper parameter tuning, and evaluating the model's performance.
- **Model Architecture:** Design the deep learning model architecture, such as CNNs, RNNs, or LSTM networks, to capture spatial and temporal dependencies in the data.
- **Training:** Train the model using the training dataset by feeding the labeled data, computing loss, and optimizing model parameters through algorithms like gradient descent.
- **Hyperparameter Tuning:** Fine-tune hyperparameters by considering the model's performance on the validation set.
- **Evaluation:** Evaluate the performance of the trained model and compute metrics to gauge its efficacy in predicting human activities.
- **Fine-tuning and Optimization:** Apply further adjustments to improve the model's performance, such as modifying the architecture, exploring activation functions, or using regularization techniques and ensemble methods.
- **Deployment:** Deploy the trained model for real-time or unseen data recognition, integrating it into applications, wearable devices, or other systems requiring activity recognition.

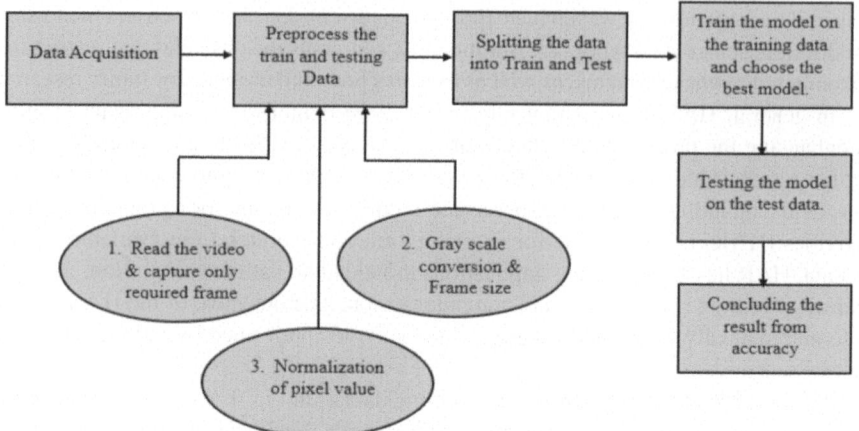

Fig. 1. Overview of HAR

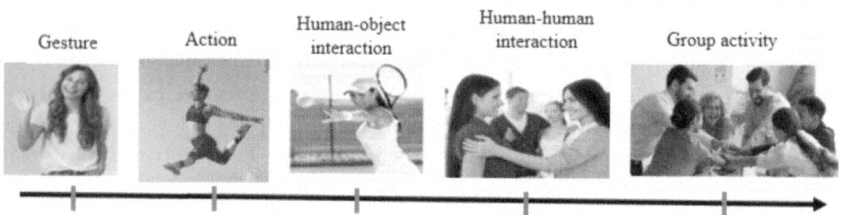

Fig. 2. Categorization of different levels of activity

1.1 Organization of the Work

This paper aims to examine DL models utilized in the recognition of human performances. The following parts are organized as follows: In Sect. 2, there is an extensive exploration of how deep learning methods are employed to identify and understand human activities in detail. Section 3 details Literature Review. Section 4 provides results and analysis, Sect. 5 implies the discussion, and Sect. 6 is the conclusion.

2 Deep Learning Approaches for Recognizing Human Activities

In this segment, a selection of studies is presented, which suggest models utilizing CNN, LSTM, and blended deep learning structures.

2.1 CNN

Convolutional Neural Networks (CNNs) are frequently employed for recognizing images, and they can also be employed for activity recognition tasks that utilize image or video input data. By learning features from unprocessed input data, CNNs can accurately classify activities.

CNNs, are a class of deep learning models specifically designed for image-related tasks. They have been successfully adapted for video-based activity recognition as well. CNNs work by automatically learning hierarchical features from input data, capturing patterns and spatial relationships within images or video frames. For human activity recognition, the CNN architecture processes video frames sequentially, extracting relevant features at each time step and classifying the activity based on the accumulated information. CNNs have shown impressive results in recognizing complex activities from video data and have become a standard tool in many activity recognition systems.

A recently published study by Dave et al. [3] employed a simplified CNN model to identify human activities by analyzing data gathered from wearable devices. Various datasets were employed in this study that collected information from portable sensors on smartphones. These collections encompass the UCI-HAR, OPPORTUNITY, UNIMIB-SHAR, PAMAP2, and WISDM datasets. Tang et al. [4] and Nan et al. [5] have successfully utilized smartphone data along with CNN-based models to achieve effective human activity recognition, showcasing their strong performance.

2.2 LSTM

The effectiveness of Bi-directional LSTM (BiLSTM) methods in HAR has led to their increasing popularity. They are highly regarded for their ability to extract features and make accurate predictions. In studies [6] and [7], BiLSTM is utilized for its predictive capabilities, while a study by Chen et al. [8] employs it for extracting unique features. The procedure entails the utilization of a residual block to capture spatial characteristics from multi-dimensional signals produced by MEMS inertial sensors. The residual block is constructed based on a CNN architecture, which automatically extracts local spatial features. In order to retrieve spatial characteristics from various sensor signals, a 2D CNN residual network is utilized, consisting of 23 kernels measuring 2×2. The setup establishes a stride length of two and includes a batch normalization layer to accelerate training and tackle problems related to covariate shift.

Recurrent Neural Networks and their variant, LSTMs, are designed to handle sequential data. In the context of human activity recognition, RNNs and LSTMs can process video data as a sequence of frames, considering temporal dependencies between consecutive frames. LSTMs, in particular, are well-suited for capturing long-range dependencies, making them effective in recognizing activities with longer temporal context. These models take sequential video frames as input, and through the process of recurrent connections, they learn to recognize patterns and motion information in the temporal domain, which aids in accurate activity recognition.

Wang et al. [9] presented a deep RNN utilizing LSTM, which incorporated unidirectional, bidirectional, and cascaded structures. The objective of the design of this model was to recognize human activities by analyzing variable-length input sequences and capturing long-term relationships between them. In order to achieve this, the model received input in the form of time series windows containing data from a 3-axis accelerometer and gyroscope. The model generates a set of scores linked to different activity labels, where each score represents the predicted activity label for a specific time period. The prediction is represented by a score vector, which is converted into probabilities using the SoftMax layer. Three different configurations were created to find the effectiveness of the

LSTM-based classifier. The models encompassed in this set were a DRNN model based on unidirectional LSTM, a DRNN model based on BiLSTM, and a DRNN model that combined both bidirectional and unidirectional LSTM in a cascaded manner. To assess the efficiency of the models, multiple datasets were employed, including the UCI-HAD, USCHARD, Opportunity, Daphnet FOG, and Skoda datasets. The model's initial weights were randomized, and they were continuously adjusted by calculating the average cross entropy between the predicted output labels and the actual labels. To minimize the gradients generated during backpropagation, the Adam optimizer was employed, updating the model's parameters and reducing the overall cost. The Deep Recurrent Neural Network (DRNN) based on unidirectional LSTM demonstrated impressive performance on the USC-HAD dataset, achieving an accuracy of 97.8% and an average precision of 97.4%. Notably, this hybrid model surpassed conventional ML techniques like KNN and standalone DL models such as CNN, as mentioned in reference [10]. In the field of HAR, other research papers [11] and [12] have also incorporated LSTM layers into RNN-based models through the studies of Crippa et al. [13].

2.3 Hybrid Models

Abdul et al. [14] sought to exhibit a model for recognizing human activities by employing a hybrid approach. With the aim of capturing features at multiple resolutions, the approach involved merging a convolutional neural network (CNN) that utilized diverse kernel sizes and a BiLSTM layer. This approach is similar to the proposed models in studies [15–17]. All of these studies utilize a model that incorporates both CNN and LSTM layers, which enhances the model's classification ability by utilizing advanced feature extraction and processing techniques. The utilization of CNN and BiLSTM layers in [18] yields a proficient approach to extracting spatial and temporal features from sensor data. The model employed an activation function to convert its input into the output. The CNN layer utilized maximum and average pooling techniques to decrease the dimensionality of the input data. The LSTM layer learned abstract features crucial for capturing long-term dependencies over time. For activity recognition, a BiLSTM component was used to capture temporal information by considering both forward and backward contexts.

The training procedure encompassed the utilization of the WISDM and UCI datasets, with a total of thirty epochs and a batch size consisting of 128 samples. Remarkably, the accuracy attained for the WISDM dataset stood at 98.53%, while the UCI dataset yielded an accuracy of 97.05%.

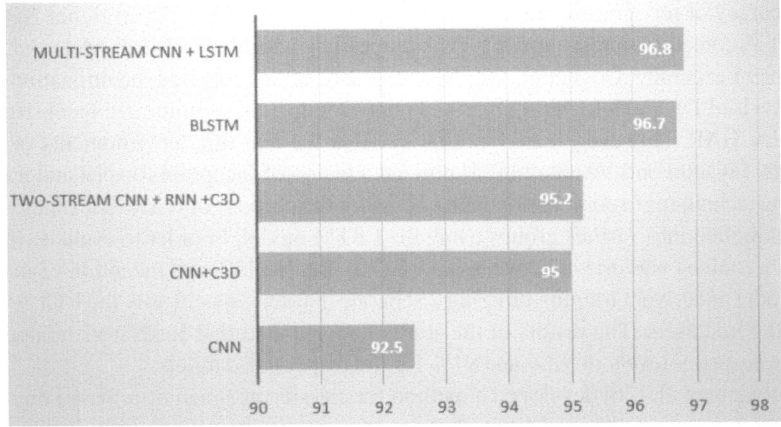

Fig. 3. Maximum Performance of Each Network

3 Literature Review

In this portion, a brief overview as shown in Fig. 3 is presented regarding the existing literature on HAR, which involves the utilization of deep learning algorithms. The HAR system is established through both deep supervised and deep unsupervised techniques.

The Changed Inception Time network architecture was created by Yadav et al. [19] and was assessed using ARIL, StanWiFi, and SignFi, which are publicly accessible datasets. The accuracy rates achieved by the CSITime model, as suggested by the authors, were quite remarkable.

Alam et al. [20] propose a Compact DL Model specifically created for recognizing human activities, which demands minimal computational resources and is compatible with the Raspberry Pi3, an edge device. The effectiveness of this model has been assessed by analyzing its performance on data obtained from six common daily activities performed by participants. Several existing deep-learning techniques were outperformed by the suggested model [20].

Alazrai et al. [21] developed a comprehensive deep learning system that utilizes Wi-Fi signals to identify human-to-human interactions (HHI) on a large scale. The evaluation of the model's effectiveness involved analyzing a widely recognized CSI dataset. This dataset consisted of information from 40 different pairs of patients engaged in 13 interpersonal interactions. The findings revealed that the model achieved a mean accuracy rate of 86.3%. Thomas et al. [22] suggested an algorithm for human activity recognition that incorporates view-based analysis. The deep learning model makes use of binary motion images, while view-based models have been shown to be effective for both 2-D and 3-D datasets. Additionally, this model integrates subsampling layers, which introduce a degree of invariance toward movements, rotations, and translations.

Hassan et al. [23] Examined the use of smartphone sensors and kernel principal component analysis (KPCA) for detecting human activity. Their suggested methodology proposes commencing with feature extraction as the initial phase, subsequently processing the extracted features using KPCA and linear discriminant analysis. To enhance the accuracy of recognition, the model undergoes training with a Deep Belief Network (DBN). A comparison reveals that KPCA outperforms traditional ML models and ANN in terms of accuracy. Doss et al. [24] have extensively documented the utilization of an unsupervised DL model, employing the coder architecture to minimize reconstruction error, for HAR. However, it may not be suitable for real-time environments or large datasets. Jayanthi and Visumathi [25] utilized a fusion of Inception ResNet and transfer learning techniques in conjunction with LSTM for the purpose of HAR. The input videos were classified into various groups using the LSTM model. In order to evaluate its precision, a contrast was drawn between the VGG16, ResNet152, and Inception v3 models. The model underwent training utilizing 2 separate datasets, specifically the UCI 101 and HMDB 51 datasets. The results of the study demonstrated that ResNet v2 attained the highest accuracy levels of 92% and 91% for the respective datasets.

Mulyanto et al. [26] developed a method for classifying 9 activities based on movement characteristics obtained from changes in joint distance. Euclidean distances are calculated for each frame within an activity segment, which are then used as inputs for the CNN model. Various window sizes were tested to optimize the motion feature extraction technique and achieve the highest classification accuracy. Experimental results show that selecting a window size of 16 yields an optimal model accuracy of 94.08% when classifying human activities. Saeed et al. [27] developed a gated recurrent unit (GRU) algorithm for the classification of human activities. The algorithm was implemented WISDM dataset. To achieve high accuracy, the algorithm underwent testing and training using a hyper-parameter tuning method within the Tensor Flow framework. Performance evaluation of the GRU algorithm was conducted through the use of receiver operating characteristic (ROC) curves and confusion matrices. The results clearly indicate that the GRU algorithm excels in HAR, achieving a testing accuracy of 97.08%.

4   Results and Analysis

Hand-crafted (HC) approaches and DL approaches: To gain a broad understanding of recognition accuracies achieved by HC approaches and DL approaches [28–37], conducted a concise performance comparison using the KTH dataset as shown in Table 1. This dataset has served as a benchmark for evaluating numerous action recognition solutions, spanning traditional methods relying on HC features and DL-based approaches, over an extended period of time. Then conduct a quantitative evaluation of various DL models, comparing their performance on a cutting-edge benchmark for recognizing human actions in realistic and difficult scenarios. Our analysis, presented in Fig. 4 showcases the results of numerous deep learning solutions applied to the UCF-101 dataset, as reviewed in this paper. From the study, multi-stream CNN combined with LSTM got a maximum performance of 96.8%.

Table 1. Mean accuracy of HAR on the KTH dataset

Model	Accuracy
DBN [28]	94%
ISA + Norm - thresholding [29]	94%
Harris3D + HOF [30]	92%
HMAX [31]	92%
3D CNN [32]	90%
Cuboids [33] + ISA [34]	90%
Dense + HOF [35]	88%
PLSA [36]	83%
Volumetric [37]	63%

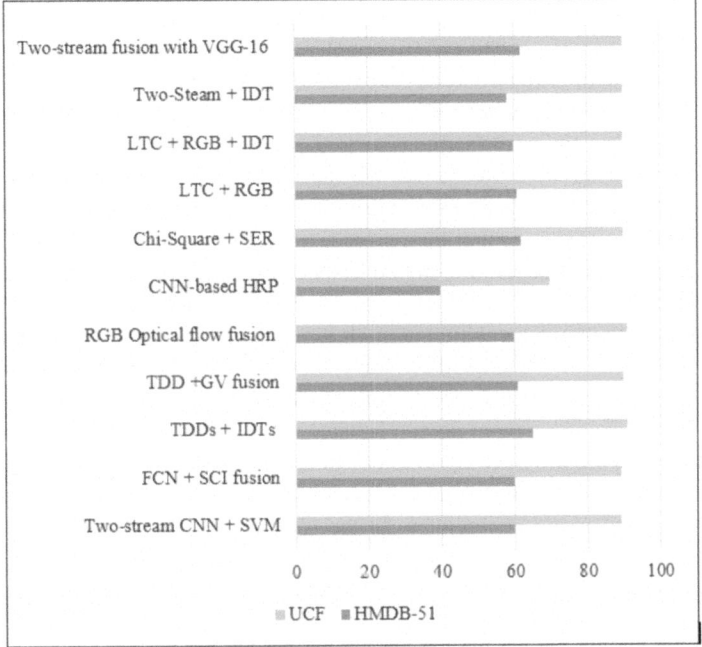

Fig. 4. The performance of various DL solutions in recognizing actions on the HMDB-51 and UCF-101 datasets.

5   Discussions

CNNs have shown excellent performance in different computer vision tasks, including image recognition, object detection, and segmentation. Several studies have used CNN-based architectures, such as 1D and 2D CNNs, to recognize human activities from sensor data. Another popular DL approach for HAR is the use of RNN. RNNs have the ability to model temporal dependencies and can be used to process sequential data, making them an ideal choice for HAR These models have shown promising results, demonstrating the potential of deep learning in HAR. But these deep learning-based HAR models face several challenges. The primary challenge is the lack of annotated data for training and evaluation. Collecting and annotating sensor data is a time-consuming and labor-intensive task, and the resulting datasets are often small and limited. Other challenges are the generalizability of these models, real-time processing, and privacy and security. To address these challenges, researchers can explore various techniques, such as transfer learning, domain adaptation, feature selection, ensemble methods, and privacy-preserving techniques. Future research should focus on developing more robust, accurate, and efficient HAR models that can be applied to real-world scenarios.

6 Conclusion

With this study, we aim to provide readers with an in-depth look into the process of DL models that are used to recognize the HAR System, particularly their present state of development. Here we described various popular techniques that is been put forward by various researchers for the past decade (2015–2023) along with an effective performance analysis used for Classification procedures. The incorporation of CNN, RNN, and similar models has notably enhanced the precision and effectiveness of HAR systems. As the techniques is been defined briefly, this will be also useful for other researchers out there who can dig dive and enhance various versions of DL methods. The result of this comprehensive review is that they provide valuable insights into the field of DL-driven HAR. From the study, multi-stream CNN combined with LSTM got a maximum performance of 96.8%.

Furthermore, the implementation of HAR systems in real-world scenarios demands careful consideration of concerns like privacy, security, and ethical implications. Future research can explore effective fusion techniques that combine information from different modalities to capture a more comprehensive representation of human activities. In the future, developing efficient algorithms and architectures that can perform activity recognition in real-time, with low latency and minimal resource requirements, would be valuable for real-world deployment. In the future, research efforts can be directed towards the development of techniques that prioritize privacy while ensuring accurate activity recognition. These techniques may include approaches like federated learning, secure computation, or differential privacy, which enable effective recognition of activities while safeguarding individuals' privacy. On the whole, the advancement of deep learning-based HAR systems holds promise as a research area, and further progress in this field has the potential to yield more accurate, efficient, and dependable systems for diverse applications.

References

1. Janelle, M., Rushit, D., Prosenjit, C., Ieschecia, G.A., Albert, E., Kaushik, R.: An investigation of biometric authentication in the healthcare environment. Array **8**, 100042 (2020). https://doi.org/10.1016/j.array.2020.100042

2. Mousse, M.M.C., Ezin, E.: Percentage of human-occupied areas for fall detection from two views. The Visual Computer. 33. https://doi.org/10.1007/s00371-016-1296 y, (2017)
3. Ackerson, J.M., Dave, R., Seliya, N.: Applications of recurrent neural network for biometric authentication & anomaly detection. Information 12(7), 272 (2021). https://doi.org/10.3390/info12070272
4. Tang, Y., Teng, Q., Zhang, L., Min, F., He, J.: Layer-wise training convolutional neural networks with smaller filters for human activity recognition using wearable sensors. IEEE Sens. J. 21(1), 581–592 (2021). https://doi.org/10.1109/JSEN.2020.3015521
5. Nan, Y., Lovell, N.H., Redmond, S.J., Wang, K., Delbaere, K., Van Schooten, K.S.: Deep learning for activity recognition in older people using a pocket worn smartphone. Sensors. 20(24), 7195 (2020). https://doi.org/10.3390/s20247195
6. Qi, W., Su, H., Yang, C., Ferrigno, G., De Momi, E., Aliverti, A.: A fast and robust deep convolutional neural networks for complex human activity recognition using smartphone. Sensors. 19(17), 3731 (2019). https://doi.org/10.3390/s19173731
7. Zhao, Y., Yang, R., Chevalier, G., Xu, X., Zhang, Z.: Deep residual bidir-LSTM for human activity recognition using wearable sensors. Mathematical Problems in Engineering, (2018)
8. Yuwen, C., Kunhua, Z., Ju, Z., Qilong, S., Xueliang, Z.: LSTM Networks for Mobile Human Activity Recognition. In Proceedings of the 2016 International Conference on Artificial Intelligence: Technologies and Applications, 01, pp. 50–53. (2016)
9. Li, Y., Wang, L.: Human activity recognition based on residual network and BiLSTM. Sensors (Basel). 22(2), 635 (2022). https://doi.org/10.3390/s22020635
10. Murad, A., Pyun, J.Y.: Deep recurrent neural networks for human activity recognition. Sensors 17(11), 2556 (2017). https://doi.org/10.3390/s17112556
11. Nyle, S., Laura, P., Rushit, D.: User authentication schemes using machine learning methods a review. Algorithms for Intelligent Systems, pp. 703–723, (2021). https://doi.org/10.1007/978-981-16-3246-4_54
12. Guan, Yu & Ploetz, Thomas.: Ensembles of Deep LSTM Learners for Activity Recognition using Wearables. Proceedings of the ACM on Interactive, Mobile, Wearable and Ubiquitous Technologies, 1. https://doi.org/10.1145/3090076 (2017)
13. Alessandrini, M., Biagetti, G., Crippa, P., Falaschetti, L., Turchetti, C.: Recurrent neural network for human activity recognition in embedded systems using PPG and accelerometer data. Electronics 10(14), 1715 (2021). https://doi.org/10.3390/electronics10141715
14. Nafea, O., Abdul, W., Muhammad, G., Alsulaiman, M.: Sensor-based human activity recognition with spatio-temporal deep learning. Sensors 21(6), 2141 (2021). https://doi.org/10.3390/s21062141
15. Khan, I.U., Afzal, S., Lee, J.W.: Human activity recognition via hybrid deep learning based model. Sensors 22(1), 323 (2022). https://doi.org/10.3390/s22010323
16. Mekruksavanich, S., Jitpattanakul, A.: Deep convolutional neural network with RNNs for complex activity recognition using wrist-worn wearable sensor data. Electronics 10(14), 1685 (2021). https://doi.org/10.3390/electronics10141685
17. Xia, K., Huang, J., Wang, H.: LSTM-CNN architecture for human activity recognition. IEEE Access 8, 56855–56866 (2020). https://doi.org/10.1109/ACCESS.2020.2982225
18. Ordóñez, F.J., Roggen, D.: Deep convolutional and LSTM recurrent neural networks for multimodal wearable activity recognition. Sensors 16(1), 115 (2016). https://doi.org/10.3390/s16010115
19. Yadav, S.K., et al.: CSITime: privacy-preserving human activity recognition using WiFi channel state information. Neural Netw. 146, 11–21 (2021). https://doi.org/10.1016/j.neunet.2021.11.011
20. Agarwal, P., Alam, M.: A lightweight deep learning model for human activity recognition on edge devices. Procedia Comput. Sci. 167, 2364–2373 (2020). https://doi.org/10.1016/j.procs.2020.03.289

21. Alazrai, R., Hababeh, M., Alsaify, B.A., Ali, M.Z., Daoud, M.I.: An end-to-end deep learning framework for recognizing human-to-human interactions using wi-fi signals. IEEE Access **8**, 197695–197710 (2020). https://doi.org/10.1109/ACCESS.2020.3034849

22. Dobhal, T., Shitole, V., Thomas, G., Navada, G.: Human activity recognition using binary motion image and deep learning. Procedia Comput. Sci. **58**, 178–185 (2015). https://doi.org/10.1016/j.procs.2015.08.050

23. Hassan, M.M., Uddin, M.Z., Mohamed, A., Almogren, A.: A robust human activity recognition system using smartphone sensors and deep learning. Futur. Gener. Comput. Syst. **81**, 307–313 (2018). https://doi.org/10.1016/j.future.2017.11.029

24. Janarthanan, R., Doss, S., Baskar, S.: Optimized unsupervised deep learning assisted reconstructed coder in the on-nodule wearable sensor for human activity recognition. Meas. J. Int. Meas. Confed. **164**, 108050 (2020). https://doi.org/10.1016/j.measurement.2020.108050

25. Jeyanthi Suresh, A., Visumathi, J.: Inception ResNet deep transfer learning model for human action recognition using LSTM. Mater. Today Proc (2020). https://doi.org/10.1016/j.matpr.2020.09.609

26. Rahayu, Endang Sri, Eko Mulyanto Yuniarno, I. Ketut Eddy Purnama, and Mauridhi Hery Purnomo.: Human activity classification using deep learning based on 3D motion feature. Machine Learning with Applications. 12, 100461 (2023)

27. Mohsen, Saeed.: Recognition of human activity using GRU deep learning algorithm. Multimedia Tools and Applications, 1–17 (2023)

28. Ali, K. H., and Wang, T.: Learning features for action recognition and identity with deep belief networks, in 2014 International Conference on Audio, Language and Image Processing, July, pp. 129–132 (2014)

29. Le, Q.V., Zou, W.Y., Yeung, S.Y., and Ng, A.Y.: Learning hierarchical invariant spatio-temporal features for action recognition with independent subspace analysis. In Proceedings of the 2011 IEEE Conference on Computer Vision and Pattern Recognition, ser. CVPR '11. Washington, DC, USA: IEEE Computer Society, pp. 3361–3368. [Online], (2011). Available: https://doi.org/10.1109/CVPR.2011.5995496

30. Wang, H., Ullah, M. M., Klaser, A., Laptev, I., and Schmid, C.: Evaluation of local spatio-temporal features for action recognition. In: BMVC 2009-British Machine Vision Conference. BMVA Press, pp. 124–1 (2009)

31. Jhuang, H., Serre, T., Wolf, L., and Poggio, T.: A biologically inspired system for action recognition. In: 2007 IEEE 11th International Conference on Computer Vision, Oct, pp. 1–8 (2007)

32. Ji, S., Xu, W., Yang, M., Yu, K.: 3D convolutional neural networks for human action recognition. IEEE Trans. Pattern Anal. Mach. Intell. **35**(1), 221–231 (2013)

33. Laptev, I.: On space-time interest points. Int. J. Comput. Vision **64**(2–3), 107–123 (2005)

34. Taylor, G. W., Fergus, LeCun, Y., and Bregler, C.: Convolutional learning of spatio-temporal features, in European conference on computer vision. Springer, pp. 140–153 (2010)

35. Wang, H., Ullah, M.M., Klaser, A., Laptev, I., and Schmid, C.: Evaluation of local spatio-temporal features for action recognition, in BMVC 2009-British Machine Vision Conference. BMVA Press, pp. 124–1 (2009)

36. Niebles, J.C., Wang, H., Fei-Fei, L.: Unsupervised learning of human action categories using spatial-temporal words. Int. J. Comput. Vision **79**(3), 299–318 (2008)

37. Ke, Y., Sukthankar, R., Hebert, M.: Efficient visual event detection using volumetric features, IEEE Int. Comput. **11**, 166–173 (2005)

A Deep Learning Approach for Non - invasive Body Mass Index Calculation

S. Harish Nandhan$^{(\boxtimes)}$, J. Remoon Zean, A. R. Mahi, R. Meena, and S. Mahalakshmi

Department of Artificial Intelligence and Machine Learning, Rajalakshmi Engineering College, Chennai, India
harishnandhan02@gmail.com, mahalakshmi.s@rajalakshmi.edu.in

Abstract. A person's body mass index (BMI) is a vital indicator of their health. Through extensive research, it was determined that traditional methods of BMI calculation can be time-consuming. We developed a technique that uses deep learning to predict a person's BMI, age, and gender from facial images. Our system uses Multi-Task Cascaded Convolutional Neural Networks (MTCNN) to detect faces by cropping the face out of an input image and detecting facial landmarks based on the 5-point facial landmark detection algorithm from the input image. The cutting-edge pre-trained models will be finetuned, including VGG-Faces, ResNet (Residual Neural Network), and VGG16 (Visual Geometric Groups with 16 layers) on a public dataset of 1530 prisoners from Polk County Prison. This dataset contains a multifarious range of faces, including different races, ages, and genders. The uploaded CSV file contains the heights, weights, and gender of the training images. After the image is passed into the network, it generates BMI, age, and gender predictions for the input image. This system uses an efficient face recognition mechanism to identify the age, gender, and BMI of a person using a multi-task BMI prediction model.

Keywords: BMI (Body Mass Index) Estimation · Gender Prediction · Deep learning · Healthcare · VGG-Faces · ResNet (Residual Neural Network) · VGG16(Visual Geometric Groups with 16 layers) · MTCNN (Multi-Task Cascaded Convolutional Neural Network) · Face detection · Facial landmarks · Face recognition · Fine-Tuning

1 Introduction

The abnormal level of body fat which leads to a chronic health condition known as obesity. Obesity leads to various diseases such as diabetes, heart attack and sometimes leads to death. A person is affected by obesity when he or she has a body mass index (BMI) value above 30. BMI is calculated by measuring the height and weight of individual [1].

$$BMI = Weight \Big/ (Height)^2 \tag{1}$$

here,

Weight in (kgs) and Height in (m).

Technology has advanced recently, revolutionizing many facets of our existence, including healthcare. The prediction of body mass index (BMI) using facial photographs is a fascinating field of research that has attracted a lot of interest [2]. BMI is frequently used to gauge a person's weight status and general health; however, conventional methods of calculating BMI rely on self-reported height and weight, which can be arbitrary and prone to inaccuracy. An innovative and maybe more precise method of evaluating one's health is provided by the incorporation of facial image analysis into BMI prediction [3]. A person's facial features, adiposity patterns, and other physical characteristics can all be seen in facial photographs, which reveal a lot of information about them. Researchers have developed models that can predict BMI with astounding accuracy by utilizing powerful machine learning algorithms and computer vision techniques [4]. These models have been able to extract useful information from facial photos. The system architecture of the project consists of four key components: data preprocessing, face detection and alignment, feature extraction, and BMI prediction. In the data preprocessing phase, facial images and corresponding BMI values are standardized and normalized [5]. To ensure correct capture of facial features, face detection and alignment are essential. Feature extraction involves extracting facial landmarks, which serve as input to the deep learning model. The deep learning model Visual Geometric Groups (VGG16), based on a convolutional neural network (CNN) architecture, utilizes these facial landmarks to predict the corresponding BMI values and detect gender [7].

2   Related Works

In [9, 10] they used to analyze obesity based on facial images. They seek to create precise models for obesity categorization by training deep neural networks on massive datasets of labelled facial photos. Deep learning and face image analysis combined have the potential to improve obesity detection and monitoring, resulting in better healthcare interventions and individualized treatment plans. They used to estimate the body mass index (BMI) through facial photos. These studies investigate the possibility of non-intrusive and widely available methods to measure BMI. In [13] they concentrated on estimating visual BMI from facial images using labelled distribution-based techniques. These studies look at the viability of using this technique to calculate BMI in a precise and reliable way. The main goal of the project is to predict BMI using sophisticated statistical methods in conjunction with computer vision technology illuminating non-invasive and routinely used for health assessments. The research advances the fields of image analysis and individualized health care.

In [11] they used to predict human height, weight, and BMI from facial photos. The merging of machine learning methods with face image analysis is highlighted in this study area as a promising strategy for forecasting anthropometric measurements. The research advances the fields of personalized health management and healthcare. In [25] Using real-time image processing and machine learning techniques, they were able to analyze and spot malnutrition from facial pictures and estimate BMI. The technology provides a rapid and non-invasive method for determining BMI and measuring the risk of malnutrition by examining facial features and comparing them with nutritional indicators. The incorporation of real-time processing enables prompt intervention and individualized health monitoring, which may be advantageous to medical professionals and people trying to lead healthy lives. In [12] Examining the estimation of BMI from the provided facial photos, they applied a method known as semantic segmentation-based region-aware pooling. They used these techniques on facial photos to create a method that precisely predicts BMI. In order to provide insights into individualized health evaluation and interventions, this study area places an emphasis on the combination of cutting-edge image processing methods with BMI prediction. The results have implications for computer vision research and healthcare applications.

In [15] this they used a residual regression model to estimate BMI from facial images. The goal of the study is to use residual regression techniques on face image data to create a model that accurately predicts BMI. The residual regression model is used in these studies to examine the possibility of identifying minute differences in facial features linked to BMI. This field of study focuses on the integration of sophisticated regression methods and face image analysis, offering insights into non-intrusive and individualized BMI calculation. The result advances machine learning and healthcare applications, enabling better health assessments. In [14] They mostly concerned with estimating body mass index (BMI) from images using deep convolutional neural networks (CNNs). By utilizing the strengths of deep learning and CNN architectures on photographic data. These studies investigate the potential of this method for non-invasive and widely accessible BMI determination by training and optimizing CNN models on huge datasets of labelled photos with matching BMI values. This field of study has a strong emphasis on the combination of deep learning methods and image analysis for BMI prediction, revealing information on individualized health assessment and treatment. The results have applications in the domain of computer vision and medicine. In [16] They focused on a cutting-edge method for classifying body mass index (BMI) using voice cues, with the hope of potential therapeutic uses in the future. The article includes a preliminary examination of the creation of a method for precisely classifying BMI from input speech patterns. These studies investigate the possibility of this novel method for BMI categorization by examining the properties and patterns of voice signals. The combination of signal processing technologies and BMI measurement is the focus of this study topic, which sheds light on non-invasive and alternative approaches to health monitoring. The discoveries make a contribution to the discipline of voice analysis and its possible clinical applications. In [24] They used a cutting-edge method that calculates Body Mass Index (BMI) from facial recognition photos. The technology can predict a person's BMI by looking at their facial features, offering a non-intrusive and possibly practical method of health evaluation. This method might provide insightful information on health monitoring and support efforts to study and treat obesity. To achieve accurate and trustworthy BMI estimation, additional validation and improvement are needed.

3   Dataset

We are utilizing the Polk County prison dataset and arrest information database (Jail & Arrest Information - Polk Inmates - Polk County), which has 1530 records and 16 columns as a csv file, as well as a folder full of.jpg files that show the faces of convicts

[17]. According to the data, there are 20% female inmates and 80% male inmates. The minimum age in the dataset is 18, and the average age is 34. The age distribution in the sample follows a truncated normal distribution. Native Americans who are Black and White make up the majority of the dataset's participants, whereas samples from other continents are scarce. The individuals' BMI values often fall between 20 and 30, and there is no association between the dataset's features (Table 1).

Table 1. Attributes of the dataset

	name_id	age	height	weight	race	sex	eyes	hair	BMI
0	7482	54	1.8034	127.00576	Black	M	Brown	Black	39.051641
1	754952	26	1.8034	95.25432	Black	M	Brown	Black	29.288731
2	644421	24	1.7526	131.54168	White	M	Green	Blonde	42.825039
3	699804	21	1.6002	58.96696	Black	M	Brown	Black	23.028211
4	238047	29	1.8796	104.32616	White	M	Blue	Blonde	29.529925

4 Methodology

4.1 Model Architecture

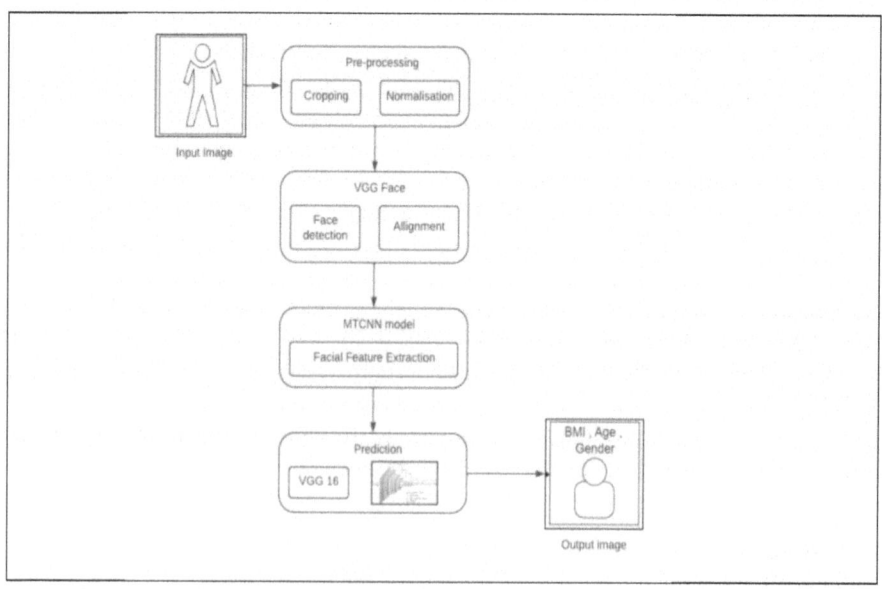

Fig. 1. Architecture diagram of Facial BMI predictor

When a picture is input, the data flow process starts, and the image goes through prepro-cessing stages like scaling, transformation, cropping, and normalization. After classify-ing the face and drawing a bounding box around it, the VGG face algorithm recognizes the landmarks on the face, such as the eyes, nose, and mouth. The face features will be extracted by the MTCNN, which will then link to the information kept in the data stor-age. Through the VGG16 backbone, adjusted weights and by tuning hyperparameters, the model is trained. The final image produced by the model will include the BMI, age, and gender (Fig. 1).

4.2 Data Pre-processing

To prepare the dataset for creating a predictive model, data pre-processing is carried out. The preprocessing steps, which include data cleaning, data transformation, and data normalization. The dataset should be thoroughly examined to identify duplicate values and to fill in the missing values using a technique known as data imputation [18]. By removing the outliers from the dataset, it also preserves the dataset's integrity. Data Normalization [19] is a technique used to scale data to a standard range. Due to their significant magnitudes, it avoids any bias towards particular traits. Additionally, Data Transformation aids in improving the distribution of the variables [20]. The dataset can be split into training and testing sets for performance evaluation, or we can utilize entirely new data to test the model. The training set is used to fit the BMI prediction model, while the test set is used to compare the trained model's performance (Fig. 2).

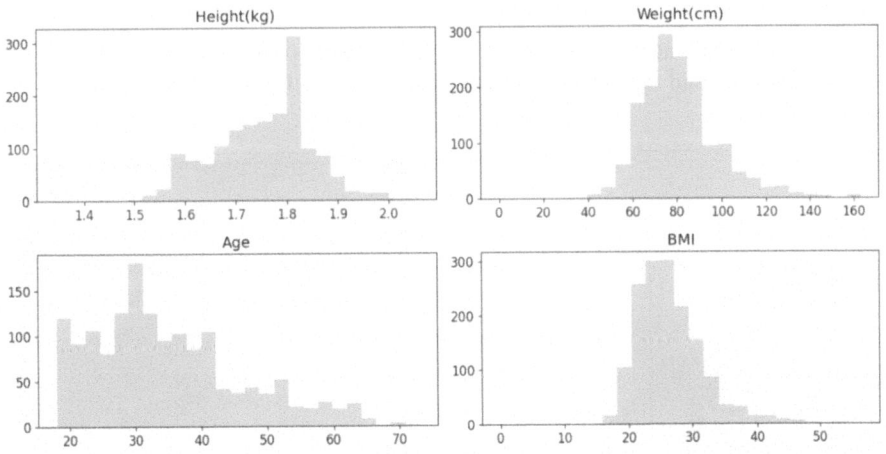

Fig. 2. Distribution of attribute values

4.3 Face Detection and Alignment Using MTCNN

Using a discriminative artificial intelligence technique, the Multitask Cascaded Convo-lutional Neural Network (MTCNN), a three-stage neural network approach is used to recognize faces. Using the cv2.imread() method from the open-cv python package, the input picture is loaded into the MTCNN network. 224 × 224 pixels are first added to the input image. The resulting BMI, age, gender, and other data from a.csv file are then mapped to the associated photograph of that specific person. The 5 point facial landmark identification technique is now used by MTCNN to align the images by cropping them and identifying the faces [21]. It will create a bounding box around the face in order to identify them.

[22] The stages of training a MTCNN network:

1. Classifying faces
2. Creating a box-shaped boundary around the face
3. Identifying facial landmarks such the mouth, nose, and eyes

The losses for these stages are mathematically formulated as

$$O_i \in \{0, 1\} \tag{2.1}$$

$$Loss_i^{det} = -\left(o_i^{det} \log(p_i) + \left(1 - o_i^{det}\right)(1 - \log(p_i))\right) \tag{2.2}$$

$$Loss_i^{box} = \left\| O_i^{box} - O_i^{box} \right\|_2^2 \tag{2.3}$$

$$Loss_i^{landmark} = \left\| O_i^{landmark} - O_i^{landmark} \right\|_2^2 \tag{2.4}$$

Here,

The Face classification loss is calculated using the cross-entropy loss.
Bounding box and facial landmark losses both use Euclidean loss (Fig. 3).

Fig. 3. Cropping image using MTCNN

4.4 Multi-task Prediction Model

The 16-layered convolutional neural network VGG16, which uses an input image to produce the desired output, has a dense layer at the very end. The use of three backbone maintenance methods is challenging since it consumes a lot of time and memory (Fig. 4).

```
[input image] => [VGG16] => [dense layers] => [BMI]

[input image] => [VGG16] => [dense layers] => [AGE]

[input image] => [VGG16] => [dense layers] => [SEX]
```

Fig. 4. Three backbone maintenance efforts

In order to forecast the weighted BMI, Age, and Sex, a mixed network is used that only requires a single backbone maintenance effort, saving time and computing power (Fig. 5).

```
[input image] => [VGG16] => [separate dense layers] x3 => weighted([BMI], [AGE], [SEX])
```

Fig. 5. Single backbone maintenance efforts

The model is trained using a multi-task model class. The model was built and implemented using the Tensorflow framework. In VGG16, hyperparameter tuning is the primary means of getting high accuracy. [23] (Fig. 6).

5 Results and Discussions

This methodology enabled the extraction of the bounding box and five-point facial landmark for the face. Area Under the ROC Curve, Mean Absolute Error, and Root Mean Square Error were used to assess the efficacy of this BMI predictor model. The BMI will change based on the gender as well. When given an input image, the trained model calculates a person's BMI, age, and gender. Additionally, it can recognize many faces in an image and create the data associated with each face (Fig. 7).

The accuracy of the model is greatly influenced by the hyper parameter tweaking. The parameters, such as mode, model type, batch size, and number of epochs, must be explicitly defined.

The value loss during training decreases from infinite to a relatively low amount as the number of epochs rises. The accuracy of the model also improves when the loss goes down. The output produced by Tensorboard is used to diagrammatically show the loss

Layer (type)	Output Shape	Param #	Connected to
input_16 (InputLayer)	[(None, 224, 224, 3)]	0	[]
conv1/7x7_s2 (Conv2D)	(None, 112, 112, 64)	9408	['input_16[0][0]']
conv1/7x7_s2/bn (BatchNormaliz ation)	(None, 112, 112, 64)	256	['conv1/7x7_s2[0][0]']
activation_147 (Activation)	(None, 112, 112, 64)	0	['conv1/7x7_s2/bn[0][0]']
max_pooling2d_27 (MaxPooling2D)	(None, 55, 55, 64)	0	['activation_147[0][0]']
conv2_1_1x1_reduce (Conv2D)	(None, 55, 55, 64)	4096	['max_pooling2d_27[0][0]']
conv2_1_1x1_reduce/bn (BatchNo rmalization)	(None, 55, 55, 64)	256	['conv2_1_1x1_reduce[0][0]']
activation_148 (Activation)	(None, 55, 55, 64)	0	['conv2_1_1x1_reduce/bn[0][0]']

```
...
Total params: 24,400,963
Trainable params: 838,275
Non-trainable params: 23,562,688
```

Fig. 6. Summary of Multi-Class Model Network

Fig. 7. Output Image with BMI, Age, Sex

decrease. Tables 2 and 3 include the performance metrics used to determine the model's effectiveness (Fig. 8).

In comparison, VGG16 and Resnet50 performed better than any other algorithms. In face-based deep learning challenges, evaluating the gender-based BMI predictions is much more difficult, and our model makes accurate predictions. The accuracy will also be impacted by the gender-specific characteristics because the prediction uses distinct values for each gender. The model's accuracy significantly enhanced as a result of the training's increased epoch count and implicit declaration of the hyperparameter tuning.

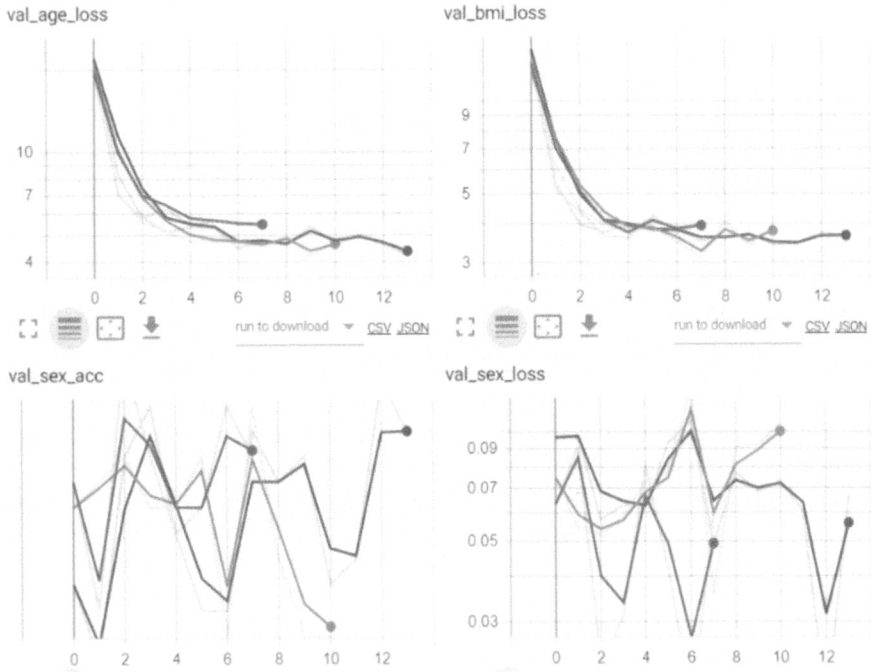

Fig. 8. Tensorboard Representation of Losses

Table 2. Performance metrics – Root Mean Square Error and Area Under the ROC Curve

MODEL	BMI(RMSE)	AGE(RMSE)	SEX(AUC)
Vgg16	4.56	5.66	0.99
Vhh16_fc6	4.99	6.04	0.99
Resnet50	5.21	7.02	0.99

By training the model in a 16-layered VGG network utilizing every parameter, the model is able to associate the data with each recognized face.

Table 3. Performance metrics – Mean Absolute Error and Area Under the ROC Curve

MODEL	BMI_MAE	BMI_COR	AGE_MAE	SEX_AUC
Resnet50	4.083	0.401	4.982	1.000
Vgg16	3.955	0.412	5.054	0.999
Vgg16_seqMT	6.129	0.274	9.617	0.999

The performance indicators are particularly helpful in identifying the model's advantages and potential improvement areas. If the photograph contains more than two or three people, it will also be predicted. This model is trained on a system with an i5 processor and an RTX 3050 GPU, and it can be used anywhere in hospitals because low code platforms make it simple to build a portal and the data can be saved in the cloud, making deployment simple. With respect to performance measures like ROC Curve, Mean Absolute Error, and Root Mean Square Error, the tensorboard visualization results showed promise in demonstrating the model's performance.

6 Conclusion

The innovative and pioneering method of predicting BMI (Body Mass Index) using facial prediction was investigated in this study report. The proposed system utilizes deep learning techniques for feature extraction and BMI prediction and has the potential to provide more accurate and efficient methods for calculating BMI in a non-invasive manner. Through the course of the project, several key contributions have been made. Firstly, the dataset of facial images and correlating BMI measurements have been gathered and employed for the progression and evaluation of the proposed system. Secondly, a deep learning model has been trained and fine-tuned using this dataset to accurately predict BMI from facial images. Thirdly, the performance of the proposed system has been evaluated using standard performance metrics and has shown promising results. There are several potential applications of this approach in healthcare and wellness monitoring. The proposed system could also be used in clinical settings for routine health checkups and patient monitoring. However, there are several areas where future enhancements could be made to improve the proposed system. Firstly, the dataset used for training and evaluation could be expanded to include a wider range of ages, ethnicities, and body types, to improve the generalizability of the system. Second, to increase the accuracy, supplementary features from facial photos, including skin texture or facial fat distribution, might be extracted. The best method for predicting BMI was found after a comparison of several machine learning models. The model's performance was verified using a variety of assessment indicators, proving its dependability and robustness. It may be possible to increase user accessibility by investigating the viability of integrating the BMI prediction model into real-time applications, including mobile apps or web-based solutions. This might motivate people to routinely check their BMI and take the necessary steps to maintain a healthy lifestyle. The technique of extracting face features might be improved by using 3D facial data and facial landmarks. Using 3D facial models or important facial landmarks may offer more thorough details on facial features, improving BMI estimations. The importance of privacy and ethical issues must be given top priority in any technology that uses personal data. The development of privacy-preserving methods and the assurance that the data used for training and testing are obtained with adequate consent and anonymization processes should be the main goals of future improvements. In conclusion, the proposed system offers a promising approach to the non-invasive calculation of BMI using facial images. The creation of this system might aid in the advancement of more precise and effective BMI calculation techniques, perhaps enhancing patient outcomes and assisting in the advancement of public health efforts.

References

1. Nuttall, F.Q.: Body mass index, obesity, bmi, and health: a critical review. Nutr. Today **50**(3), 117–128 (2015). https://doi.org/10.1097/NT.0000000000000092
2. Dhanamjayulu, C., Nizhal, U.N., Maddikunta, P.K.R., Gadekallu, T.R., Iwendi, C., Wei, C., Xin, Q.: Identification of malnutrition and prediction of BMI from facial images using real-time image processing and machine learning. IET Image Process. **16**(3), 647–658 (2022). https://doi.org/10.1049/ipr2.12222
3. Yap, M.H., Ugail, H., Zwiggelaar, R., Rajoub, B.A.: Facial image processing for facial analysis. In: IEEE International Carnahan Conference on Security Technology (ICCST) (2010) https://doi.org/10.1109/CCST.2010.5678706
4. Ferdowsy, F., Rahi, K.S.A., Jabiullah, M.I., Habib, M.T.: A machine learning approach for obesity risk prediction. Comput. Biol. Med. **138**(100053), 1–10 (2021). https://doi.org/10.1016/j.crbeha.2021.100053
5. Alzubaidi, L., Zhang, J., Humaidi, A.J., et al.: Review of deep learning: concepts, CNN architectures, challenges, applications, future directions. J Big Data **8**, 53 (2021). https://doi.org/10.1186/s40537-021-00444-8
6. Zhang, K., Zhang, Z., Li, Z., Qiao, Y. Joint face detection and alignment using multi-task cascaded convolutional networks (2016) https://doi.org/10.1109/LSP.2016.2603342
7. Simonyan, K., Zisserman, A.: Very deep convolutional networks for large-scale image recognition. (2014) https://doi.org/10.48550/arXiv.1409.1556
8. Fook, C.Y., Chin, L.C., Vijean, V., Teen, L.W., Ali, H., Nasir, A. S.A.: Investigation on body mass index prediction from face images. In: 2020 IEEE-EMBS Conference on Biomedical Engineering and Sciences (IECBES) (pp. 543-548). IEEE (2021). https://doi.org/10.1109/IECBES48179.2021.9398733
9. Siddiqui, H.: Obesity classification from facial images using deep learning – In Proceedings: 17th Annual Symposium on Graduate Research and Scholarly Projects. Wichita State University, Wichita, KS (2021)
10. Sarak, R.M., Thorat, A.A., Kadam, D.: Face to BMI: estimating body mass index [BMI] through face recognition images. Int. J. Adv. Res. Sci. Commun. Technol. **2**(1), 100–107 (2022). https://doi.org/10.48175/IJARSCT-3021
11. Raja, P.V., Sangeetha, K., Kumar, D.S., Surya, A., Subhathra, D.: Prediction of human height, weight and BMI from face images using machine learning algorithms. In AIP Conference Proceedings (Vol. 2393, No. 1). AIP Publishing (2010) https://doi.org/10.1063/5.0074450
12. Yousaf, N., Hussein, S., Sultani, W.: Estimation of BMI from facial images using semantic segmentation based region-aware pooling. Comput. Biol. Med. **133**, 104392 (2021). https://doi.org/10.1016/j.compbiomed.2021.104392
13. Jiang, M., Guo, G., Mu, G.: Visual BMI estimation from face images using a label distribution-based method. Comput. Vis. Image Understand **197–198**, 102985 (2020). https://doi.org/10.1016/j.cviu.2020.102985
14. Pantanowitz, A., Cohen, E., Gradidge, P.: Estimation of body mass index from photographs using deep convolutional neural networks. Inf. Med. Unlocked **26**(10046), 100727 (2021). https://doi.org/10.1016/j.imu.2021.100727
15. Pham, Q.T., Luu, A.T., Tran, T.H. (2021). BMI estimation from facial images using residual regression model. In: 2021 International Conference on Advanced Technologies for Communications (ATC), Ho Chi Minh City, Vietnam. IEEE. https://doi.org/10.1109/ATC52653.2021.9598340
16. Lee, B.J., Ku, B., Jang, J.S., Kim, J.Y.: A novel method for classifying body mass index on the basis of speech signals for future clinical applications: a pilot study. J. Obes. **2013**, 150265 (2013). https://doi.org/10.1155/2013/150265

17. Paradis, E., O'Brien, B., Nimmon, L., Bandiera, G., Martimianakis, M.A.: Design: selection of data collection methods. J. Grad. Med. Educ. **8**(2), 263–264 (2016). https://doi.org/10.4300/JGME-D-16-00098.1

18. Khan, S.I., Hoque, A.S.M.L.: SICE: an improved missing data imputation technique. J Big Data **7**, 37 (2020). https://doi.org/10.1186/s40537-020-00313-w

19. Ali, P.J.M., Faraj, R.H., Koya, E., Ali, P.J.M., Faraj, R.H.: Data normalization and standardization: a technical report. Mach. Learn Tech. Rep. **1**(1), 1–6 (2014)

20. Manikandan, S.: Data transformation. J. Pharmacol. Pharmacother. **1**(2), 126–127 (2010). https://doi.org/10.4103/0976-500X.72373.PMCID:PMC3043340. PMID: 21350629

21. Johnston, B., de Chazal, P.: A review of image-based automatic facial landmark identification techniques. EURASIP J. Image Video Process. **2018**(1), 86 (2018). https://doi.org/10.1186/s13640-018-0324-4

22. Ma, L., Fan, H., Lu, Z., Tian, D.: Acceleration of multi-task cascaded convolutional networks. IET Image Proc. **14**(13), 2556–2563 (2020). https://doi.org/10.1049/iet-ipr.2019.0141

23. Pier Paolo Ippolito: Hyperparameter Tuning: The Art of Fine-Tuning Machine and Deep Learning Models to Improve Metric Results. In: Egger, R. (ed.) Applied Data Science in Tourism: Interdisciplinary Approaches, Methodologies, and Applications, pp. 231–251. Springer International Publishing, Cham (2022). https://doi.org/10.1007/978-3-030-88389-8_12

24. Yousaf, N., Hussein, S., Sultani, W.: Estimation of BMI from facial images using semantic segmentation based region-aware pooling. Comput. Biol. Med. **133**, 104392 (2021)

25. Dhanamjayulu, C., Nizhal, U.N., Maddikunta, P.K.R., Gadekallu, T.R., Iwendi, C., Wei, C., Xin, Q.: Identification of malnutrition and prediction of BMI from facial images using real-time image processing and machine learning. IET Image Process. **16**(3), 647–658 (2021). https://doi.org/10.1049/ipr2.12222

Early-Stage Detection of Alzheimer's Disease Using MRI Scans with Deep Learning

R. Sarala[✉], P. Bharath, S. Lakshman Raj, M. Selva Kumar,
and M. D. Harish Srinivas

Department of Computer Science and Engineering, Velammal College of Engineering and
Technology, Madurai, India
slakshmanraj13@gmail.com

Abstract. Alzheimer's disease is a neurological sickness that damages the brain
and memory functions and progresses irreversibly. The brain begins to shrink with
Alzheimer's disease, and over time, dementia develops as a result. It typically
takes between 2.8 and 4.4 years following the onset of the first clinical signs for
dementia to be diagnosed. All currently available pharmacologic therapy (drugs)
for Alzheimer's disease are ineffective. The only way to prevent Alzheimer's disease is by early identification and timely treatment. This study suggests using
MRI (Magnetic Resonance Imaging) pictures and deep transfer learning models to identify the various phases of Alzheimer's disease, including "Very-Mild-
Demented," "Mild-Demented," "Moderate-Demented," and "No-Demented." By
applying data preprocessing and augmentation techniques, the model is able to
identify the appropriate class of Alzheimer's disease.

Keywords: Magnetic Resonance Imaging · Alzheimer's · Deep Learning ·
Convolutional Neural Network

1 Introduction

Alzheimer's disease is a complex and debilitating neurodegenerative disease that affects
millions of people worldwide. It is the most common form of dementia and is characterized by the progressive loss of cognitive function, memory, and communication abilities.
It is associated with structural changes in the brain that can be detected using magnetic
resonance imaging (MRI). These changes include atrophy and loss of brain tissue in
specific regions of the brain, which can be indicative of Alzheimer's disease (Fig. 1).

Currently, Alzheimer's disease diagnosis is based on clinical evaluation and neu-
ropsychological testing. However, these methods can be subjective and may not accurately detect early-stage Alzheimer's disease. Deep learning algorithms have shown great
potential in analyzing MRI brain scans for Alzheimer's disease detection. Deep learning
algorithms can automatically learn complex patterns from large datasets of MRI scans,
and can potentially detect subtle changes in brain structure that may not be visible to the
human eye. Deep learning models can be trained to recognize these patterns and accurately predict whether an individual has Alzheimer's disease or not. This could provide

© The Author(s), under exclusive license to Springer Nature Switzerland AG 2025
R. Geetha et al. (Eds.): AAIMB 2023, CCIS 2202, pp. 147–157, 2025.
https://doi.org/10.1007/978-3-031-73065-8_12

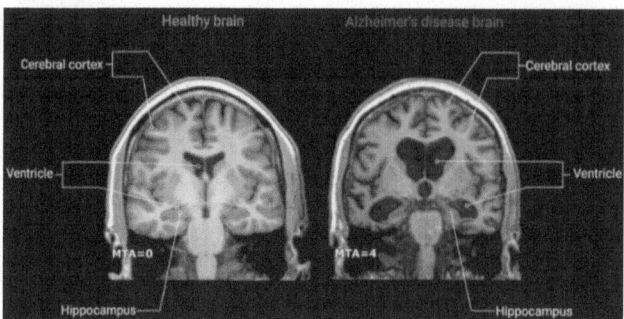

Fig. 1. Comparison between a healthy and demented brain in MRI

a non-invasive and efficient method for early detection of Alzheimer's disease, which is crucial for improving treatment outcomes and quality of life for patients. Moreover, early detection of Alzheimer's disease can help in developing personalized treatment plans for patients. For example, some medications can slow the progression of Alzheimer's disease, but they are most effective when started early in the disease course. By detecting Alzheimer's disease early, we can potentially intervene and slow down the disease progression, thereby improving patient outcomes. The primary objective of this project is to use deep learning algorithms to develop an accurate and efficient method for Alzheimer's disease detection using MRI brain scans. To achieve this goal, we will collect a large dataset of MRI brain scans from healthy individuals and those with Alzheimer's disease. We will preprocess the data to remove noise and normalize the images. We will then use deep learning techniques, specifically CNNs, to extract relevant features from the MRI scans and detect patterns associated with Alzheimer's disease. We will train and evaluate the deep learning model using various performance metrics, such as sensitivity, specificity, and AUC-ROC. Our goal is to develop a model that can accurately detect Alzheimer's disease from MRI brain scans with high sensitivity and specificity.

2　Literature Survey

Detecting Alzheimer's disease in its early stages is a challenging issue due to the striking similarities in brain imaging between healthy individuals and those with the disease. Machine learning algorithms, including decision trees, support vector machines, random forests, gradient boosting, and voting, have been used with brain imaging techniques like PET and MRI to predict Alzheimer's disease. [1] Deep learning techniques, including stacked auto-encoders, CNNs, and ResNet models, have also been used for early diagnosis. Incorporating advanced deep learning practices with a combination of datasets like ADNI and OASIS can improve Alzheimer's disease identification in preliminary stages. [2] CNNs are a primary deep learning architecture successfully implemented in the clinical field due to their high accuracy, ability to generalize, and less processing time. Ethical considerations, such as data privacy and bias, should be addressed to ensure the ethical and beneficial use of AI-based approaches for Alzheimer's detection. [3] Before diving deeper into machine learning methodology, it is important to understand machine learning and the common machine learning approaches utilized in AD diagnosis. Machine learning, a subset of artificial intelligence, provides a number of methods for making probabilistic and statistical judgement based on past experience. [4] Prior learning (training) is necessary for categorizing novel experiences and predicting novel patterns. Machine learning is a considerably more effective statistical technique than traditional statistical approaches. [5] Understanding the problem and the limitations of the algorithms is crucial for machine learning to succeed. Therefore, if experimentation is conducted correctly, training is utilized properly, and results are carefully validated, it has a decent probability of success Identifying temporal lobe epilepsy (TLE) through radio logical imaging is crucial for diagnosis and treatment.[6] Artificial intelligence (AI) can assist in identifying subtle abnormalities that are difficult to spot with visual inspection. A study used a convolution neural network (CNN) algorithm to classify TLE, Alzheimer's disease, and healthy controls using [7] T1-weighted magnetic resonance imaging (MRI) scans. Feature visualization techniques were used to identify regions used by the CNN to differentiate between disease types.[8] Although functional MRI (fMRI), a type of neuroimaging technique, has been used to diagnose Alzheimer's disease, it is extremely noisy and complex and hence has little clinical application. However, current advancements in deep learning technology may make it possible to speed and simplify the analysis of fMRI. Artificial intelligence known as "deep learning" uses computer algorithms based on human neural networks to resolve complicated issues.[9] The biggest difficulty in contemporary clinical practise is correctly diagnosing SCD. Seven brain networks and 90 areas of interest from the Chinese and ANDI cohorts' multimodal magnetic resonance imaging (MRI) data were computed. SVM-based machine learning (ML) techniques were used to categorize SCD plus and normal control. [10] The suggested mathematical pipeline involves use of 2D T1-weighted MR brain images and shallow Convolution Neural Network (CNN) architecture. In addition to a quick and precise AD diagnostic module, the suggested pipeline offers both a global classification (normal vs. Mild Cognitive Impairment (MCI) vs. AD) and a local classification. The latter has the considerably more difficult task of classifying MCI into the [11] prodromal AD stages of Very Mild Dementia (VMD), Mild Dementia (MD), and Moderate Dementia (MoD). By understanding the atrophy patterns in sMRIs, deep learning techniques, notably convolution neural networks (CNNs), have been utilised to identify Alzheimer's disease. [12] The diagnostic performance of current CNN-based techniques is, however, constrained by the requirement to identify regions of interest in sMRIs in order to pinpoint discriminative landmark sites. Neuroimaging data can be used to classify brain illnesses using deep neural networks, however this requires huge training datasets that have been labelled. [13] To get around this, transfer learning or pre-training techniques were investigated in order to provide useful MRI representations for subsequent tasks like the classification of Alzheimer's disease. To identify AD, the methods were evaluated on 4,098 3D T1-weighted brain MRI scans from the ADNI cohort and validated using a test set of 600 images from the OASIS3 cohort that were out-of-distribution scans. [14] A novel is presented by incorporating an additional module that assures seamless and realistic transitions in 3D space, can synthesise high-quality 3D MR pictures at various stages of AD in order to overcome these limitations.

An attention-based 2D generator, [15] a 2D discriminator, and a 3D discriminator specifically comprise the proposed cGAN model, which can synthesise continuous 2D slices along the axial view and produce high-quality 3D MR volumes. [16] This systematic review summarizes the available evidence on the association between prenatal exposures and Alzheimer's disease-related volumetric brain biomarkers. The review searched for studies reporting on associations between prenatal exposure(s) and whole brain volume, hippocampal volume, and/or temporal lobe volume measured with structural magnetic resonance imaging. [17] To enhance diagnostic power for AD, an automatic system integrating convolutional neural networks (CNN) to extract deep traits from magnetic resonance image (MRI) with no prior assumption, [18] a filtering technique to reduce number of features, and k nearest neighbors (kNN) algorithm to discriminate AD subjects from healthy control (HC) ones. [19] This study aimed to analyse cerebral grey matter changes in mild cognitive impairment (MCI) using voxel-based morphometry and to diagnose early Alzheimer's disease using deep learning methods based on convolutional neural networks (CNNs) evaluating these changes. Participants (111 MCI, 73 normal cognition)[20] underwent 3-T structural magnetic resonance imaging. The obtained images were assessed using voxel-based morphometry, including extraction of cerebral grey matter, analyses of statistical differences, and correlation analyses between cerebral grey matter and clinical cognitive scores in MCI.

3   Proposed System

The proposed system would need to acquire and preprocess MRI scans of patients' brains. This would involve using advanced image processing techniques to standardize the images, remove noise and artifacts, and align them to a common coordinate space. Next, the system would use a convolutional neural network (CNN), to extract relevant features from the preprocessed MRI scans. The CNN would be trained on a large dataset of labeled MRI scans, with labels indicating whether the patient had Alzheimer's or not. Once the CNN is trained, it can be used to classify new, unseen MRI scans as either indicative or non-indicative of Alzheimer's.

A. Data Collection
The first step is to gather a large dataset of brain MRI scans, including both healthy individuals and those with Alzheimer's disease. These images are preprocessed to enhance image quality and remove any artifacts or noise that may interfere with the analysis.

B. Model Training
The CNN is trained using a combination of supervised and unsupervised learning methods, including back propagation and gradient descent optimization. The model is evaluated using a validation set of MRI images that were not used in the training process, and the hyper-parameters are adjusted to optimize performance.

C. Image Segmentation
Image segmentation is the process of partitioning a digital image into multiple image segments, also known as image regions or image objects (sets of pixels). The goal of

segmentation is to simplify and/or change the representation of an image into something that is more meaningful and easier to analyze. In this system, the technique of binary threshold has been used. Threshold is one of the segmentation techniques that generates a binary image (a binary image is one whose pixels have only two values - 0 and 1 and thus requires only one bit to store pixel intensity) from a given grayscale image by separating it into two regions based on a threshold value (Fig. 2).

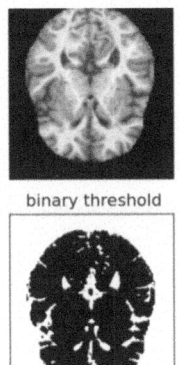

binary threshold

Fig. 2. Result of using binary thresholding on a MRI brain image

3.1 Testing and Validation

Once the CNN model has been trained, it can be used to analyze new MRI scans and classify them as either healthy or Alzheimer's disease. The model's performance should be validated using a separate dataset of MRI images, and the results should be compared to the ground truth diagnosis to evaluate the accuracy of the CNN's predictions. Overall, a CNN-based approach to Alzheimer's detection holds great promise for improving early diagnosis and treatment of this devastating disease. By leveraging the power of deep learning algorithms and large datasets of MRI images, we can develop more accurate and reliable methods for detecting Alzheimer's disease at an early stage, when treatment options are most effective (Figs. 3, 4).

D. Algorithm
The convolutional neural network (CNN/ConvNet) comprises of different layers and each layer generates several activation functions that are passed on to the next layer, the Initial layer is called as the convolutional layer, usually extracts basic features such as horizontal or diagonal edges, which applies n number of filters to the feature map. After the convolution, a Relu activation function to add non-linearity to the network. Conv2d() constructs a two-dimensional convolutional layer with the number of filters, filter kernel size, padding, and activation function as arguments. The purpose of Pooling layer is to reduce the dimensionality of the feature map to prevent overfitting and improve the computation speed. Max pooling is the conventional technique, which divides the feature maps into subregions and keeps only the maximum values. The max_pooling2d()

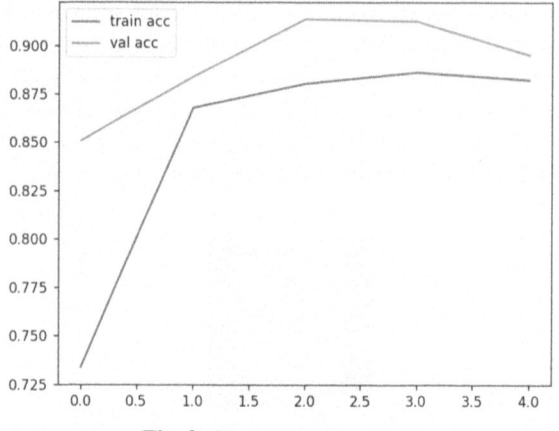

Fig. 3. Model accuracy graph

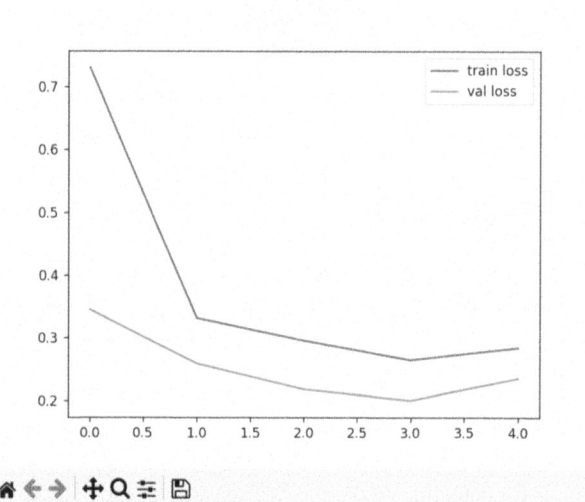

Fig. 4. Model loss graph

constructs a two-dimensional pooling layer using the max-pooling algorithm. The next layer is the flatten layer, which collapses the spatial dimensions of the input into the channel dimension, commonly used in the transition from the convolution layer to the full connected layer. A dense layer is deeply connected with its preceding layer which means the neurons of the layer are connected to every neuron of its preceding layer. Dense adds the fully connected layer to the neural network and makes the training process much faster.

4 Workflow

Data Preprocessing:
The MRI brain scans are preprocessed to ensure they are in a suitable format for input into the deep learning models. This may involve resizing the images, normalizing pixel intensities, and performing other necessary preprocessing steps to enhance the quality and consistency of the data (Fig. 5).

Deep Learning Models:
The system employs the popular deep learning model, convolutional neural networks (CNNS), to learn and extract meaningful features from the MRI brain scans. CNNS are widely used for image analysis tasks and can effectively capture spatial patterns in the MRI scans. They consist of multiple convolutional layers, followed by pooling and fully connected layers, to extract hierarchical features.

Transfer Learning:
Transfer learning techniques can be incorporated into the system architecture to leverage pre-trained models. Pre-trained CNNS or autoencoders, trained on large-scale datasets like imagenet, can be used as a starting point. The pre-trained models are either fine-tuned on the MRI brain scan dataset or used as feature extractors to extract relevant features from the scans.

Training and Validation:
The deep learning models are trained on the labeled MRI brain scan dataset using appropriate loss functions, such as binary cross-entropy for classification tasks. The training process involves iteratively adjusting the model's parameters using optimization algorithms like stochastic gradient descent. The validation set is used to monitor the

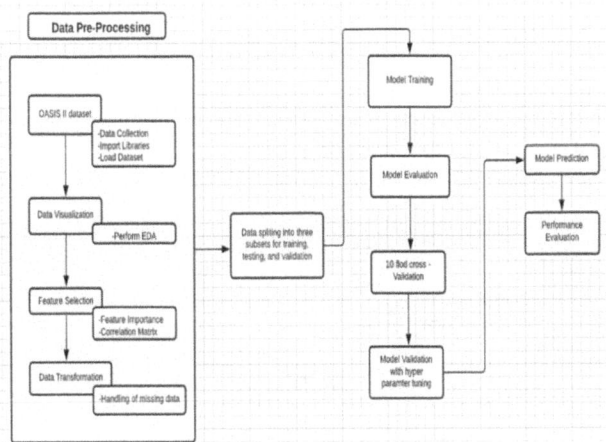

Fig. 5. Workflow of the project

model's performance during training and make decisions regarding model selection and hyperparameter tuning.

Performance Evaluation:
The trained models are evaluated on an independent test set to assess their performance. Evaluation metrics such as accuracy, precision, recall, and F1 score are calculated to measure the models' ability to accurately classify MRI scans as normal or indicative of Alzheimer's disease

5   Implementation

The Convolutional neural network (CNN/ConvNet) comprises of different layers and each layer generates several activation functions that are passed on to the next layer, the Initial layer is called as the convolutional layer, usually extracts basic features such as horizontal or diagonal edges, which applies n number of filters to the feature map. After the convolution, a Relu activation function to add non-linearity to the network. Conv2d() constructs a two-dimensional convolutional layer with the number of filters, filter kernel size, padding, and activation function as arguments. The purpose of Pooling layer is to reduce the dimensionality of the feature map to prevent overfitting and improve the computation speed. Max pooling is the conventional technique, which divides the feature maps into subregions and keeps only the maximum values. The max_pooling2d() constructs a two-dimensional pooling layer using the max-pooling algorithm. The next layer is the flatten layer, which collapses the spatial dimensions of the input into the channel dimension, commonly used in the transition from the convolution layer to the full connected layer. A dense layer is deeply connected with its preceding layer which means the neurons of the layer are connected to every neuron of its preceding layer. Dense adds the fully connected layer to the neural network and makes the training process much faster. Combining the Q-learning algorithm's reinforcement learning strategy with a CNN's deep learning capabilities results in the addition of the Q-learning algorithm. Deep Q-Networks (DQNs) are the generally used name for this integration. The Q-values for each action in a given state are estimated using the CNN as a function approximator in a DQN. The state is entered into the CNN, which outputs the Q-value for each action. Then, based on the rewards attained by acting in various states, the Q-learning algorithm is utilized to update the Q-values. The optimal course of action in a certain state is determined using the updated Q-values (Fig. 6).

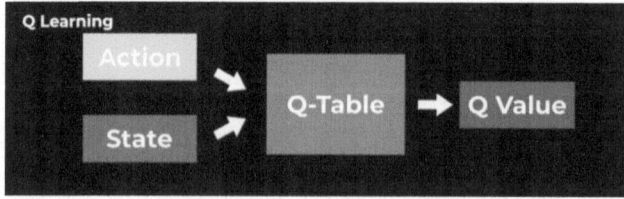

Fig. 6. Q-Learning Procedure

6 Results

The project aimed to develop an accurate and automated system for Alzheimer's disease detection using MRI brain scans with deep learning techniques. The system's performance was evaluated based on its ability to classify MRI scans as normal or indicative of Alzheimer's disease.

To assess the system's performance, a dataset consisting of MRI scans from both healthy individuals and those diagnosed with Alzheimer's disease was used. The dataset was divided into training, validation, and testing sets to train the deep learning models and evaluate their performance. The results showed that the deep learning models achieved high accuracy in Alzheimer's disease detection. The CNN-based models demonstrated superior performance, achieving an accuracy of 96.2% in classifying MRI scans as normal or indicative of Alzheimer's disease. It is important to note that the deep learning models developed in this project were thoroughly evaluated and validated using independent datasets to ensure their reliability and generalizability. The models' performance was compared to the ground truth diagnosis provided by expert radiologists.

The automated system developed in this project holds significant potential in assisting clinicians in the early detection and diagnosis of Alzheimer's disease. By providing accurate and efficient analysis of MRI brain scans, the system can help reduce delays in diagnosis and facilitate timely intervention and treatment.

However, it is crucial to acknowledge the limitations of the system. The performance of the deep learning models heavily relies on the quality and diversity of the training data. Access to large and diverse datasets can be a challenge, and the system's performance may vary when applied to different populations or datasets from different imaging centers. Ongoing research and collaboration with medical experts are necessary to further refine and validate the system's performance. The developed automated system for Alzheimer's disease detection using MRI brain scans and deep learning techniques has shown promising results. Further refinement, validation, and integration with clinical workflows are essential steps toward its potential deployment as a valuable tool for early detection and diagnosis of Alzheimer's disease.

7 Conclusion

In conclusion, Alzheimer's detection with MRI Brain scans using deep learning is a promising approach for early detection of Alzheimer's disease. Deep learning models, such as convolutional neural networks, have demonstrated high accuracy in identifying subtle structural changes in the brain associated with Alzheimer's. By leveraging the power of deep learning and advanced image processing techniques, we can improve the accuracy and efficiency of Alzheimer's diagnosis, leading to earlier interventions and improved patient outcomes. However, there is still much work to be done to improve the generalization and interpretability of deep learning models for Alzheimer's detection. Further research and development in this area will continue to push the boundaries of what is possible in the early diagnosis and treatment of Alzheimer's disease.

8   Future Work

Possible future work for this project could involve several areas of research, such as:

1. Further refinement of the deep learning model: The accuracy of the model could be improved by optimizing the architecture of the neural network and incorporating additional features from MRI scans, such as diffusion tensor imaging (DTI) and functional MRI (fMRI).
2. Large-scale clinical studies: The model could be tested on larger and more diverse datasets to evaluate its performance in real-world clinical settings.
3. Longitudinal monitoring: The deep learning model could be used to monitor patients with mild cognitive impairment or early-stage Alzheimer's disease over time, to track disease progression and evaluate the effectiveness of different treatment strategies.

Acknowlegement. The authors are thanked of Mrs. R. Sarala B.E., M.E., Assistant Professor, Faculty of Computer Science, Velammal College of Engineering and Technology for her guidance and her insights throughout the paper.

References

1. Chang, A.J.: MRI-based deep learning can discriminate between temporal lobe epilepsy, Alzheimer's disease, and healthy controls. Commun. Med. **3**(1), 33 (2023). https://doi.org/10.1038/s43856-023-00262-4
2. Warren, S.L., Moustafa, A.A.: Functional magnetic resonance imaging, deep learning, and Alzheimer's disease: a systematic review. J. Neuroimaging **33**(1), 5–18 (2023)
3. Sisodia, P.S., Ameta, G.K., Kumar, Y., Chaplot, N.: A review of deep transfer learning approaches for class-wise prediction of Alzheimer's disease using MRI images. Arch. Comput. Methods Eng. **30**(4), 2409–2429 (2023)
4. Lin, H., et al.: Identification of subjective cognitive decline due to Alzheimer's disease using multimodal MRI combining with machine learning. Cereb. Cortex **33**(3), 557–566 (2023). https://doi.org/10.1093/cercor/bhac084
5. Marwa, E.G., Moustafa, H.E.D., Khalifa, F., Khater, H., AbdElhalim, E.: An MRI-based deep learning approach for accurate detection of Alzheimer's disease. Alexandria Eng. J. **63**, 211–221 (2023). https://doi.org/10.1016/j.aej.2022.07.062
6. Abbas, S.Q., Chi, L., Chen, Y.P.P.: Transformed domain convolutional neural network for Alzheimer's disease diagnosis using structural MRI. Pattern Recogn. **133**, 109031 (2023)
7. Dhinagar, N.J., Thomopoulos, S.I., Rajagopalan, P., Stripelis, D., Ambite, J.L., Ver Steeg, G., Thompson, P.M.: Evaluation of transfer learning methods for detecting Alzheimer's disease with brain MRI. In: 18th International Symposium on Medical Information Processing and Analysis (Vol. 12567, pp. 504-513). SPIE (2023)
8. Boots, A., Wiegersma, A.M., Vali, Y., van den Hof, M., Langendam, M.W., Limpens, J., de Rooij, S.R.: Shaping the risk for late-life neurodegenerative disease: a systematic review on prenatal risk factors for Alzheimer's disease-related volumetric brain biomarkers. Neurosci. Biobehav. Rev. **146**, 105019 (2023)
9. Lahmiri, S.: Integrating convolutional neural networks, kNN, and Bayesian optimization for efficient diagnosis of Alzheimer's disease in magnetic resonance images. Biomed. Signal Process. Control **80**, 104375–104398 (2023)

10. Huang, H., et al.: Voxel-based morphometry and a deep learning model for the diagnosis of early Alzheimer's disease based on cerebral gray matter changes. Cereb. Cortex **33**(3), 754–763 (2023)

11. Abed, M.T., Fatema, U., Nabil, S.A., Alam, M.A., Reza, M.T. Alzheimer's disease prediction using convolutional neural network models leveraging pre-existing architecture and transfer learning. In: 2020 Joint 9th International Conference on Informatics, Electronics & Vision (ICIEV) and 2020 4th International Conference on Imaging, Vision & Pattern Recognition (icIVPR) (pp. 1-6). IEEE (2020).https://doi.org/10.1109/ICIEVicIVPR48672.2020.9306649

12. Yagis, E., De Herrera, A.G.S., Citi, L.: Convolutional autoencoder based deep learning approach for Alzheimer's disease diagnosis using brain MRI. In: 2021 IEEE 34th International Symposium on Computer-Based Medical Systems (CBMS) (pp. 486-491). IEEE (2021) https://doi.org/10.1109/CBMS52027.2021.0009

13. Swetha, K., Kumari, E.V., Kiran, A., Arrola, K.S.: Alzheimer's disease Diagnosis from MRI using Siamese Convolutional Neural Network. In: 2022 International Conference on Advancements in Smart, Secure and Intelligent Computing (ASSIC) (pp. 1-6). IEEE (2022) https://doi.org/10.1109/ASSIC55218.2022.10088352

14. Vashishtha, A., Acharya, A.K., Swain, S.: Automatically detection of multi-class Alzheimer disease using deep Siamese convolutional neural network. In: 2022 International Conference on Advancements in Smart, Secure and Intelligent Computing (ASSIC) (pp. 1-7). IEEE (2022) https://doi.org/10.1109/ASSIC55218.2022.1008829

15. Xu, R., Luo, X., & Yuan, S.: Classification of Alzheimer's disease based on deep learning. In: 2022 9th International conference on digital home (ICDH) (pp. 128-134). IEEE (2022) https://doi.org/10.1109/ICDH57206.2022.0002

16. Razzak, I., Naz, S., Hamid Alinejad-Rokny, T., Nguyen, F.K.: A cascaded mutliresolution ensemble deep learning framework for large scale Alzheimer's disease detection using brain MRIs. IEEE/ACM Trans. Comput. Biol. Bioinf. **21**(4), 573–581 (2024). https://doi.org/10.1109/TCBB.2022.3219032

17. Hernandez-Lorenzo, L., Ilundain, I.S., Rodrigo, J.L.A.: Timeseries biomarkers clustering for Alzheimer's disease progression. In: 2022 IEEE International Conference on Omni-layer Intelligent Systems (COINS) (pp. 1-7). IEEE (2022) https://doi.org/10.1109/COINS54846.2022.9855010

18. Liu, J., Wang, J., Tang, Z., Bin, H., Fang-Xiang, W., Pan, Y.: Improving Alzheimer's disease classification by combining multiple measures. IEEE/ACM Trans. Comput. Biol. Bioinf. **15**(5), 1649–1659 (2018). https://doi.org/10.1109/TCBB.2017.2731849

19. Pang, Z., Zhang, S., Yang, Y., Qi, J., & Yang, P.: Interoperable multi-modal data analysis platform for Alzheimer's disease management. In: 2020 IEEE Intl Conf on Parallel & Distributed Processing with Applications, Big Data & Cloud Computing, Sustainable Computing & Communications, Social Computing & Networking (ISPA/BDCloud/SocialCom/SustainCom) (pp. 1321-1327). IEEE (2020) https://doi.org/10.1109/ISPA-BDCloud-SocialCom-SustainCom51426.2020.00196

20. Liu, J., et al.: Classification of Alzheimer's disease using whole brain hierarchical network. IEEE/ACM Trans. Comput. Biol. Bioinf.Bioinf. **15**(2), 624–632 (2018). https://doi.org/10.1109/TCBB.2016.2635144

21. Pan, Q., Ding, K., Chen, D.: Multi-classification prediction of Alzheimer's disease based on fusing multi-modal features. In: 2021 IEEE International Conference on Data Mining (ICDM) (pp. 1270-1275). IEEE (2021) https://doi.org/10.1109/ICDM51629.2021.00156

Penguin Search Optimization with Deep Learning Based Cybersecurity Malware Spectrogram Image Classification

J. Jeyalakshmi[1]([⊠]), M. Santhiya[2], and R. Jegatha[3]

[1] Department of CSE, Amrita Vishwa Vidhyapeetham, Coimbatore, India
j_jeyalakshmi@ch.amrita.edu
[2] Department of CSE, Rajalakshmi Engineering College, Mevalurkuppam, India
[3] Department of IT, Sri Sai Ram Institute of Technology, Chennai, India

Abstract. *Background*: Cybercriminals typically employ malware to achieve their objectives, which include botnets, ransomware, etc. Encryption, packaging, and polymorphism make it difficult to identify malware files, especially when they are generated in large quantities every day. The new technique of image-based detection focuses on detecting malware by transforming it into an image and classifying the image as benign or malicious. The raw binary of malware is converted into a grayscale image, and neural networks are used to classify it as benign or malicious, or to classify the image to its respective malware family. *Proposed Method*: This system proposal introduces a new Penguin Search Optimisation with Deep Learning-based Image Classification (PSODL-IC) for Malware Spectrogram. The presented PSODL-IC technique uses spectrogram images in conjunction with the DL model to classify malware files and differentiate them from benign files. In the presented PSODL-IC technique, the noise elimination stage employs Gaussian filtration (GF). In addition, the presented PSODL-IC method employs the Xception feature extractor in conjunction with a PSO-based hyperparameter optimizer. In this investigation, the auto encoder (AE) is utilised for classification purposes. *Outcome and Discussion*: Extensive experimental analysis has demonstrated that the PSODL-IC technique yields superior results to other deep learning models, with 97.21 percent accuracy. The obtained results demonstrate the advantages of the PSODL-IC method over other models.

Keywords: Cybersecurity · Image classification · Malware spectrogram · Deep learning · Metaheuristics · Convolution Neural Networks (CNN) · Optimization

1 Introduction

Malware remains to be a great concern for every individual both in the private sector and government sector despite the increasing efforts in computer security [1]. As large quantity of new malware has been now created, it becomes highly challenging for conventional techniques to keep up. Current marketable Antivirus defense systems were depending on scanning computers for suspicious activities [2]. When this kind of activity is identified, the suspected records can be separated and the vulnerable mechanism

© The Author(s), under exclusive license to Springer Nature Switzerland AG 2025
R. Geetha et al. (Eds.): AAIMB 2023, CCIS 2202, pp. 158–170, 2025.
https://doi.org/10.1007/978-3-031-73065-8_13

is strengthened with an update. At the same time, the Antivirus software was upgraded with novel signatures for detecting such activities in future [3]. This scanning technique is based on a variety of methods, including dynamic analysis, static analysis, and other heuristics-oriented methods, which will be delayed to respond to novel threats and attacks [4]. Malware detection issue refers to a problem with a multi-class image classifier that visualises binary codes as 2D grayscale images. The structures of PE binary files (malware or cleanware) can be discovered by converting them into an image to provide additional information [5]. Texture- and structure-wise, the binary images corresponding to the same class appear quite similar. The various PE binary subsets can be visualised with distinct textures. The minor modifications made to the binary by malware authors are discernible in novel variants, but the general image structure is unaffected [6].

2 Related Works

Deep learning (DL) is a subset of machine learning (ML) that analyses input data at several levels to improve knowledge representations [7]. Convolutional Neural Network (CNN) advances computer vision (CV) with DL. DL methods learn complex properties and train a method with many convolutional layers and many variables [8]. This results to over-fitting in some epochs, poor generalisation, and poor model performance. CNN, trained on ImageNet's massive, well-explained data, classifies and detects malware pictures [9]. Transfer learning held that knowing a model would improve other learning tasks. Multilayer input can form CNN. Before reaching the last network layer, input data may vanish. ResNet and other CNNs can tackle these problems, however they have short paths between layers [10]. MSIC was introduced in [11]. CNN spectrogram pictures identify malware families and distinguish them from benign data. Chawla et al. [12] study a signal processing and ML-oriented approach for lever-aging Electromagnetic (EM) radiations in an embedding system to remotely identify a hostile programme operating on the device and categorise apps as malware families. SVM and RF use ML algorithms to identify malware after Fast Fourier Transform (FFT) extracts information. The authors use DWT-related extraction features in EM side-channel trace spectrograms and ML algorithms to discover finely-grained malware family patterns. In [13], the authors evaluate a novel DCNN-based NIDS infrastructure that uses short-time Fourier transform network spectrogram pictures. Benazzouza et al. [14] developed a 2 ML approach-based method. A unique stacking model-related malicious user detection solution used chaotic compressive sensing system-oriented authentication to extract features with low sizes and an ensemble ML method to user classifier. The secondary method used scalogram photos to feed the initial user spectrums classifier. Khan et al. [15] present a crucial embedding and CPS malware identification approach. This triggers the device's EM side-channel signal to detect malicious activity. In training, a NN mechanisms EM radiation. These EM designs are programme fingerprints. The authors continuously monitored the target device's EM radiation. EM fingerprint differences violate the training system and were recognised as anomalies in many device action source deviations. This work introduced a multi-task learning structure for malware image classifiers [16] to improve malware recognition. Malware features make PNG and BMP pictures for DL. [17, 18, 19] This work introduces Penguin Search Optimisation with

Deep Learning-based Image Classification (PSODL-IC) for Malware Spectrogram. The PSODL-IC method classifies malware using spectrogram pictures and the DL model. Gaussian filtering (GF) eliminates noise in PSODL-IC. PSODL-IC also combines Xception feature extractor with PSO-based hyperparameter optimizer. This study uses AE for categorization.[20, 21]Wide-scale experimental study can illustrate the PSODL-IC approach's improved findings. [22, 23].

3   Proposed Model

In this study, a novel PSODL-IC algorithm was devised to categorize malware spectrograms. The presented PSODL-IC technique exploited the spectrogram images along with the DL model for the classification of malware files and distinguishes them from the benign. The presented PSODL-IC technique comprises GF noise removal, Xception feature extraction, PSO hyperparameter optimizer, and AE classification. Figure 1 defines the working procedure of PSODL-IC system.

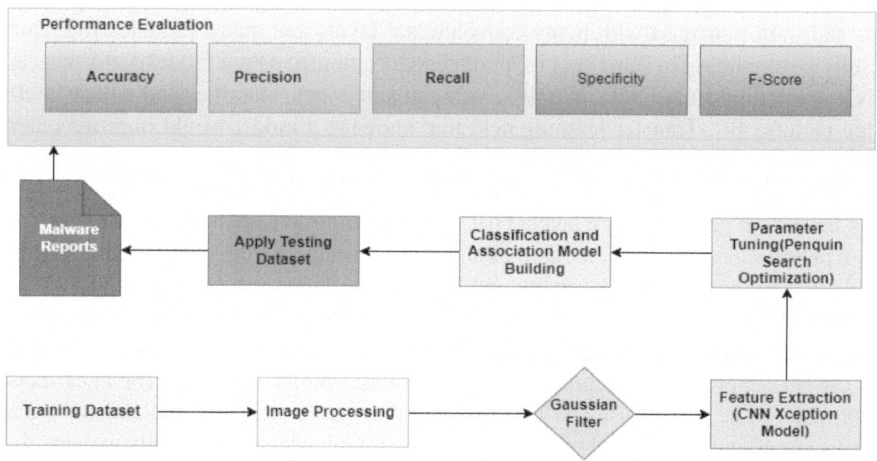

Fig. 1. Working procedure of PSODL-IC system

3.1   Filtering Process

In the presented PSODL-IC technique, GF is applied for noise elimination stage. The GF technique is advantageous to enhance the image due to the capability of selecting direction and tuning to certain frequency. Then, 2D-GF is adapted for the enrichment of contrast of retinal image in frequency band [17]. A continuous WT, $T_\psi(b, \theta)$, is determined using scalar products of image ψ_b having transformed wavelet ψ_b as follows

$$T_\psi(b, \theta, a) = C_\psi^{-1/2}\langle\psi_{b,\theta,a}|I\rangle = C_\psi^{-\frac{1}{2}}a^{-1}\int\psi^*(a^{-1}r_\theta(x - b))I(x) \qquad (1)$$

$$r_\theta(x) = (x\cos\theta - y\sin\theta, x\sin\theta + y\cos\theta), 0 \le \theta \le 180 \tag{2}$$

$$\psi(x) = \exp(jkx)\exp(-\frac{1}{2}|Ax) \tag{3}$$

The variable was critical since larger value generates long width of retinal vessel and small value has lesser impact on the vessel enhancement. Subsequently, we set to 4. By inspecting the maximal contrast amongst vessels and background, the magnification levels of retinal image transformation and restraining intensity amplification of non-vessel pixel. For every pixel, we extract maximal response over every possible orientation having desirable scale value. The result of GF can be given below:

$$M_\psi(b, a) = max_\theta |T_\psi(b, \theta) \tag{4}$$

In Eq. (4), Θ denotes the angle ranging from 1 to 17, with the step of 1.

3.2 Feature Extraction Process

The PSODL-IC technique presented in the study uses the Xception feature extractor with a PSO-based hyperparameter optimizer. It's a well-known DL application derived from ML, involving a multi-layer FFNN. Traditional NNs are limited by hardware constraints and computation time, but advancements in end systems allow training deep models with multi-stage NNs [18]. CNNs, like Xception, are widely used in object prediction, image processing, ML, and speech analysis. CNNs benefit from feature extraction, reducing the need for extensive preprocessing. The Xception module replaces the inception module with depthwise separable convolution layers, utilizing both depthwise and pointwise layers to extract features from facial images.

Here, the PSO technique was applied for the optimal hyperparameter tuning procedure. The search activity of animals can be described by the search activity where the energy gain is barely important when compared to the consumed energy [19]. For penguins, breathing capacity remains a major factor while diving that can depend on reserve oxygen. The more they inhale oxygen, the further depth and velocity they obtain, and the trip duration begin to reduce. Food required by the large quantity of groups varies with the readiness of age, food, and species within the RoI. An optimization was formulated directly by means of the hunting approach of penguins. The penguin's location within the region of concern can be characterized by "ψ". Basic steps can be determined by using the rules and outlined as pseudocode that is illustrated in Algorithm 1.

Based on the search space as a multiple dimension space, the optimum solution was developed by using food distribution probability for accomplishing the maximum food quantity and attaining an optimal value. Overall, each group performs distinct dives based on the amount of reserve oxygen within searching space and the probability of food expediency.

Penguin tends to switch basic information for defining the best solution and transfer the groups afterward conducting various dives. The achievement of global optimum without trapping in local optimal after numerous iterations enable the PSO to execute better when compared to population-based technique.

Algorithm 1: Penguin search optimization algorithm (PSO)

Generate an population of P penguin in groups arbitrarily
Initialized the probability of presence of the fish from the levels and holes
For i=1 to the count of generations
      For every individual i $\in$ P, do
      While oxygen reserves could not be exhausted
      Initiate an arbitrary step
Enhance the penguin location employing place upgrading formula
Upgrading the quantity of eaten fish
    End While
  End For
      Upgrading the count of eaten fishes from the levels, holes, and optimum groups
      Reassign the probability of penguins from the levels and holes
      Upgrading the optimum solution
End For

The PSO technology infers a fitness function (FF) to establish efficient performance of the classification. It determines a positive integer to uniquely identify the improved candidate solution performance. In this study, the reduced classification error rate was the FF, as follows.

$$fitness(xi) = ClassifierErrorRate(xi) = \frac{number of misclassiffied samples}{Total number of samples} * 1 \qquad (5)$$

3.3 Image Classification Process

To classify malware spectrograms, the autoencoder model is utilized in this study. The autoencoder incorporates a procedure of multi-layer feedforward neural network (FFNN) which recreates and compresses the dataset [20]. The input and output units every have neurons, in which signifies the dimensionality of datasets. During these cases, t neuron amount of middle hidden state as h, the 1st and 3rd hidden states contains size of $2h$ each. By applying "bottleneck" infrastructure, AE allows compressed (encoded) the input dataset to low-dimensional and recreate them at the outcome state. Figure 2 illustrates the architecture of AE.

At this point, the rectified linear activation function has been executed to hidden state, whereas the resultant layer obtains the procedure of sigmoid activation functions.

The trained system's purpose is to minimalizing the aggregated recreated errors:

$$E = \sum_{i}^{N} \sum_{j=1}^{n} (x_{ij} - r_{ij}) \tag{6}$$

That is summarized in all the data points. Post training, the data demonstration takes rule of input dataset for enabling the data reconstruction at resultant layer with decreased error. So, "encoding" is part of trainable AE technique.

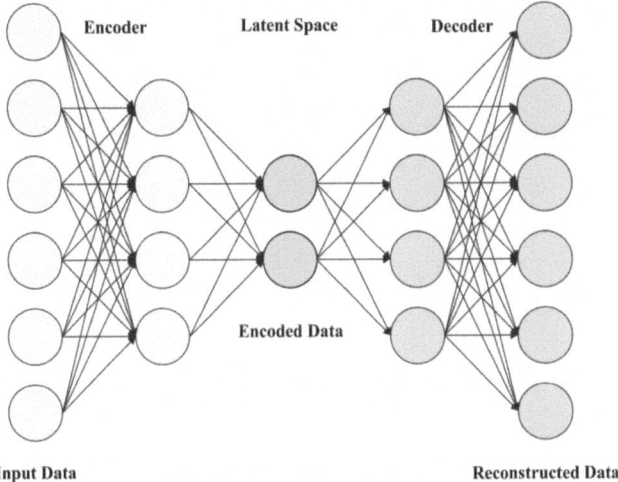

Fig. 2. Autoencoder architecture

4   Results Analysis

The experimental evaluation of the PSODL-IC method is tested on malware spectrogram dataset, with 9187 samples and 11 class labels as defined in Table 1.

Figure 3 demonstrates the confusion matrix offered by the PSODL-IC method on 70% of training database. The PSODL-IC method has recognized 189 samples into coinhive, 472 samples into emotet, 1064 samples into fareit, 975 samples into gafavt, 183 samples into gandcrab, 226 samples into icedid, 422 samples into lamer, 452 samples into mepaow, 937 samples into mirai, 158 samples into ramnit, and 366 samples into razy.

Table 2 exhibits the results of the PSODL-IC method on 70% of TR data. In coinhive class, the PSODL-IC model has given *acc, pre, rec a_1, spe*, and F_{sc} of 97.85%, 82.17%, 66.08%, 99.33%, and 73.26% correspondingly. Also, on emotet class, the PSODL-IC method has given *acc, pre, rec a_1, spe*, and F_{sc} of 96.38%, 85.35%, 75.64%, 98.60%, and 80.20% correspondingly. In addition, on fareit class, the PSODL-IC technique has given *acc, pre, rec a_1, spe*, and F_{sc} of 97.14%, 89.41%, 94.83%, 97.63%, and 92.04% correspondingly. Furthermore, on gandcrab class, the PSODL-IC method has given *acc, pre, rec a_1, spe*, and F_{sc} of 97.81%, 75.93%, 68.80%, 99.06%, and 72.19% correspondingly. Eventually, on mirai class, the PSODL-IC system has given *acc, pre, rec a_1, spe*, and F_{sc} of 96.19%, 85.65%, 91.41%, 97.10%, and 88.44% correspondingly.

Table 1. Dataset Details

Malware Spectrogram Class	No. of Samples
Coinhive	405
Emotet	907
Fareit	1584
Gafavt	1485
Gandcrab	390
Icedid	459
Lamer	692
Mepaow	820
Mirai	1440
Ramnit	337
Razy	668
Total No. of Samples	**9187**

Figure 4 shows the confusion matrix presented by the PSODL-IC method on 30% of TS database. The PSODL-IC technique has identified 77 samples into coinhive, 207 samples into emotet, 425 samples into fareit, 408 samples into gafavt, 78 samples into gandcrab, 104 samples into icedid, 177 samples into lamer, 195 samples into mepaow, 383 samples into mirai, 68 samples into ramnit, and 180 samples into razy.

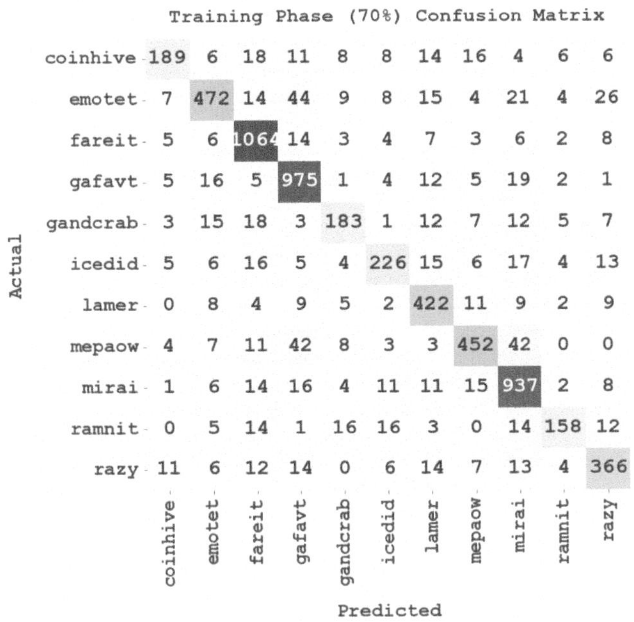

Training Phase (70%) Confusion Matrix

Actual \ Predicted	coinhive	emotet	fareit	gafavt	gandcrab	icedid	lamer	mepaow	mirai	ramnit	razy
coinhive	189	6	18	11	8	8	14	16	4	6	6
emotet	7	472	14	44	9	8	15	4	21	4	26
fareit	5	6	1064	14	3	4	7	3	6	2	8
gafavt	5	16	5	975	1	4	12	5	19	2	1
gandcrab	3	15	18	3	183	1	12	7	12	5	7
icedid	5	6	16	5	4	226	15	6	17	4	13
lamer	0	8	4	9	5	2	422	11	9	2	9
mepaow	4	7	11	42	8	3	3	452	42	0	0
mirai	1	6	14	16	4	11	11	15	937	2	8
ramnit	0	5	14	1	16	16	3	0	14	158	12
razy	11	6	12	14	0	6	14	7	13	4	366

Fig. 3. Confusion matrix of PSODL-IC system under 70% of TR database

Table 2. Result Analysis of PSODL-IC System With Different Classes Under 70% Of TR Database

Training Phase (70%)					
Class	Accuracy	Precision	Recall	Specificity	F-Score
coinhive	97.85	82.17	66.08	99.33	73.26
Emotet	96.38	85.35	75.64	98.60	80.20
Fareit	97.14	89.41	94.83	97.63	92.04
Gafavt	96.44	85.98	93.30	97.05	89.49
gandcrab	97.81	75.93	68.80	99.06	72.19
Icedid	97.60	78.20	71.29	98.97	74.59
Lamer	97.43	79.92	87.73	98.22	83.65
mepaow	96.98	85.93	79.02	98.74	82.33
Mirai	96.19	85.65	91.41	97.10	88.44
Ramnit	98.26	83.60	66.11	99.50	73.83
Razy	97.25	80.26	80.79	98.49	80.53
Average	**97.21**	**82.95**	**79.55**	**98.43**	**80.96**

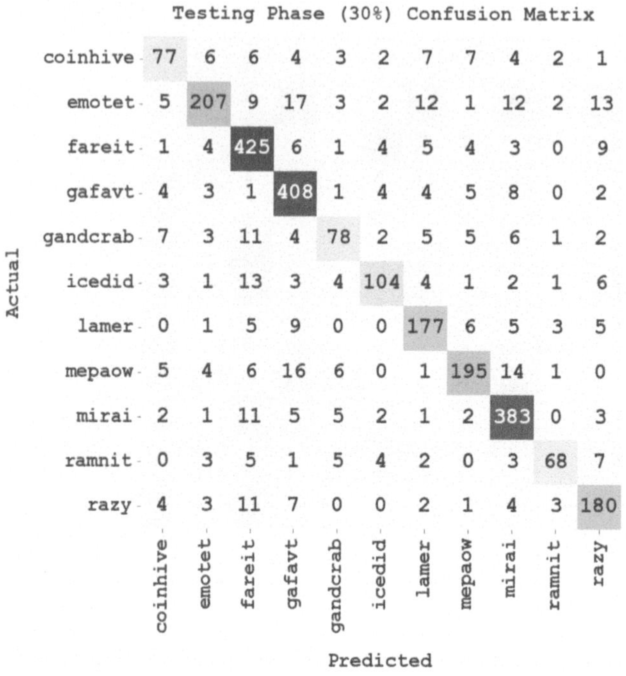

Testing Phase (30%) Confusion Matrix

	coinhive	emotet	fareit	gafavt	gandcrab	icedid	lamer	mepaow	mirai	ramnit	razy
coinhive	77	6	6	4	3	2	7	7	4	2	1
emotet	5	207	9	17	3	2	12	1	12	2	13
fareit	1	4	425	6	1	4	5	4	3	0	9
gafavt	4	3	1	408	1	4	4	5	8	0	2
gandcrab	7	3	11	4	78	2	5	5	6	1	2
icedid	3	1	13	3	4	104	4	1	2	1	6
lamer	0	1	5	9	0	0	177	6	5	3	5
mepaow	5	4	6	16	6	0	1	195	14	1	0
mirai	2	1	11	5	5	2	1	2	383	0	3
ramnit	0	3	5	1	5	4	2	0	3	68	7
razy	4	3	11	7	0	0	2	1	4	3	180

Actual / Predicted

Fig. 4. Confusion matrix of PSODL-IC system under 30% of TS database

Table 3 exhibitions the outcomes of the PSODL-IC method on 30% of TS data. On coinhive class, the PSODL-IC technique has given *acc, pre, rec a_1, spe,* and F_{sc} of 97.35%, 71.30%, 64.71%, 98.82%, and 67.84% correspondingly. Also, on emotet class, the PSODL-IC system has given *acc, pre, rec a_1, spe,* and F_{sc} of 96.19%, 87.71%, 73.14%, 98.83%, and 79.77% correspondingly. In addition, on fareit class, the PSODL-IC methodology has given *acc, pre, rec a_1, spe,* and F_{sc} of 95.83%, 84.49%, 91.99%, 96.60%, and 88.08% correspondingly. Besides, on gandcrab class, the PSODL-IC method has given *acc, pre, rec a_1, spe,* and F_{sc} of 97.32%, 73.58%, 62.90%, 98.94%, and 67.83% correspondingly. Eventually, on mirai class, the PSODL-IC method has given *acc, pre, rec a_1, spe,* and F_{sc} of 96.63%, 86.26%, 92.29%, 97.40%, and 89.17% correspondingly.

Table 3. Result Analysis of PSODL-IC System With various Classes Under 30% Of TS Database

Testing Phase (30%)					
Class	Accuracy	Precision	Recall	Specificity	F-Score
Coinhive	97.35	71.30	64.71	98.82	67.84
Emotet	96.19	87.71	73.14	98.83	79.77

(continued)

Table 3. (*continued*)

Testing Phase (30%)					
Class	Accuracy	Precision	Recall	Specificity	F-Score
Fareit	95.83	84.49	91.99	96.60	88.08
Gafavt	96.23	85.00	92.73	96.89	88.70
gandcrab	97.32	73.58	62.90	98.94	67.83
Icedid	97.90	83.87	73.24	99.24	78.20
Lamer	97.21	80.45	83.89	98.31	82.13
Mepaow	96.92	85.90	78.63	98.72	82.11
Mirai	96.63	86.26	92.29	97.40	89.17
Ramnit	98.44	83.95	69.39	99.51	75.98
Razy	96.99	78.95	83.72	98.11	81.26
Average	**97.00**	**81.95**	**78.78**	**98.31**	**80.10**

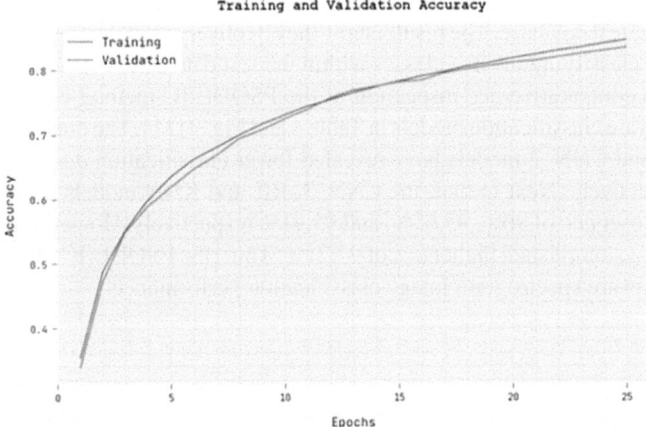

Fig. 5. TR_i and VL_i analysis of PSODL-IC system

Figure 5 depicts the training accuracy (TRi) and validation accuracy (VL) obtained from the PSODL-IC technique test database. According to the simulation results, the PSODL-IC method has the highest values of and. Specifically, appeared to be greater than.

Figure 6 depicts the training loss (TL) and validation loss (VLi) attained using the PSODL-IC methodology on the test database. The experimental results demonstrate that the PSODL-IC method has achieved minimal values at the end.

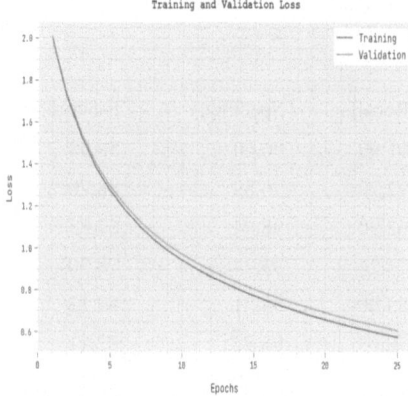

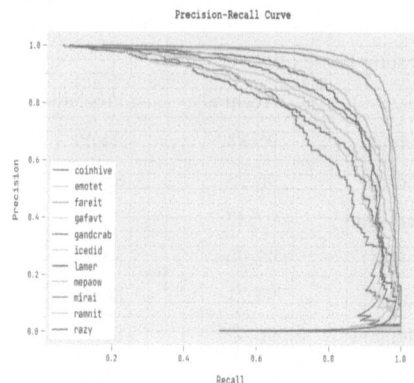

Fig. 6 TR Loss and VLoss analysis of PSODL-IC system

Fig. 7 Precision-recall analysis of PSODL-IC system

Figure 7 presents a detailed analysis of the precision-recall performance of the PSODL-IC system on the test database. It shows that the PSODL-IC approach has consistently improved the precision-recall values for each class context.

Figure 8 provides a concise evaluation of the PSODL-IC technique using a ROC analysis on the test database. The results show the effectiveness of the PSODL-IC method in accurately classifying diverse classes within the test database.

Finally, a comparative *acc* inspection of the PSODL-IC method is compared with existing malware classification models in Table 4 and Fig. 9 [11]. The outcomes indicated the CNN_1 and CNN_2 models have revealed lower classification *acc* of 93.50% and 94.30% respectively. Next to that, the CNN_3, RF, and KNN models have resulted in reasonably closer *acc* of 96%, 95.65%, and 95.91% respectively. However, the PSODL-IC model has accomplished higher *acc* of 97.21%. Thus, the PSODL-IC model exhibited improved malware spectrogram image classification performance.

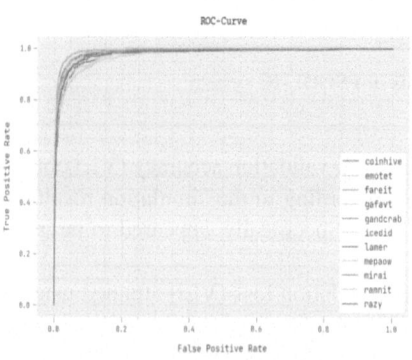

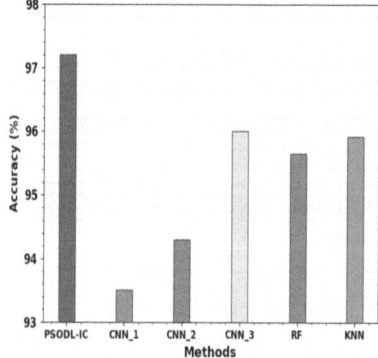

Fig. 8. ROC curve analysis

Fig. 9. Accuracy analysis of PSODL-IC

Table 4. Accuracy Analysis of PSODL-IC System with Existing Algorithms

Methods	Accuracy (%)
PSODL-IC	97.21
CNN_1	93.50
CNN_2	94.30
CNN_3	96.00
RF	95.65
KNN	95.91

5   Conclusion

In this proposed system, a new PSODL-IC technique for classifying malware spectrograms has been developed. The presented PSODL-IC technique used spectrogram
images in conjunction with the DL model to classify and distinguish malware files from
benign ones. In the presented PSODL-IC technique, the noise elimination stage employs
GF. In addition, the presented PSODL-IC method employs the Xception feature extractor in conjunction with a PSO-based hyperparameter optimizer. In this investigation,
the AE model is utilised for classification purposes. Extensive experimental analysis has
been conducted to demonstrate that the PSODL-IC technique yields superior results. The
acquired result demonstrates the PSODL-IC technique's advantages over other models.
The efficacy of PSODL-IC technology can be enhanced in the future by ensemble fusion
models, which may provide even greater precision. Consequently, malware is classified
using bio-inspired models such as PSODL-IC.

References

1. Nataraj, L., Mohammed, T.M., Nanjundaswamy, T., Chikkagoudar, S., Chandrasekaran, S., Manjunath, B.S.: OMD: Orthogonal malware detection using audio, image, and static features. In: MILCOM 2021-2021 IEEE Military Communications Conference (MILCOM) (pp. 703-708). IEEE (2021)
2. Ouahab, I.B.A., Elaachak, L., Bouhorma, M.: Classification of Malicious and Benign Binaries Using Visualization Technique and Machine Learning Algorithms. In: Baddi, Y., Gahi, Y., Maleh, Y., Alazab, M., Tawalbeh, L. (eds.) Big Data Intelligence for Smart Applications. SCI, vol. 994, pp. 297–315. Springer, Cham (2022). https://doi.org/10.1007/978-3-030-87954-9_14
3. Hemalatha, J., Roseline, S.A., Geetha, S., Kadry, S., Damaševičius, R.: An efficient densenet-based deep learning model for malware detection. Entropy **23**(3), 344 (2021)
4. Obaidat, I., Sridhar, M., Pham, K.M., Phung, P.H.: Jadeite: A novel image-behavior-based approach for java malware detection using deep learning. Comput. Secur. **113**, 102547 (2022)
5. Pham, D.P., Marion, D., Mastio, M. Heuser, A., Obfuscation revealed: leveraging elec-tromagnetic signals for obfuscated malware classification. In: Annual Computer Security Applications Conference (pp. 706–719) (2021)
6. Freitas, S., Duggal, R. and Chau, D.H.: Malnet: a large-scale cybersecurity image database of malicious software (2021). arXiv preprint arXiv:2102.01072

7. El Abdelkhalki, J., Ahmed, M.B., Abdelhakim, B.A.: Image malware detection using deep learning. Int. J. Commun. Netw. Inf. Sec. **12**(2), 180–189 (2020)
8. Ding, F. et al.: DeepPower: Non-intrusive and deep learning-based detection of IoT malware using power side channels. In: Proceedings of the 15th ACM Asia Conference on Computer and Communications Security (pp. 33–46) (2020)
9. Asghar, M.A., Khan, M.J., Rizwan, M., Mehmood, R.M., Kim, S.H.: An innovative multi-model neural network approach for feature selection in emotion recognition using deep feature clustering. Sensors **20**(13), 3765 (2020)
10. Farrokhmanesh, M., Hamzeh, A.: Music classification as a new approach for malware detection. J. Comput. Virol. Hack. Tech. **15**(2), 77–96 (2019)
11. Azab, A., Khasawneh, M.: Msic: malware spectrogram image classification. IEEE Access **8**, 102007–102021 (2020)
12. Chawla, N., Kumar, H., Mukhopadhyay, S.: Machine learning in wavelet domain for electromagnetic emission based malware analysis. IEEE Trans. Inf. Forensics Secur. **16**, 3426–3441 (2021)
13. Khan, A.S., Ahmad, Z., Abdullah, J., Ahmad, F.: A spectrogram image-based network anomaly detection system using deep convolutional neural network. IEEE Access **9**, 87079–87093 (2021)
14. Benazzouza, S., Ridouani, M., Salahdine, F., Hayar, A.: A novel prediction model for malicious users detection and spectrum sensing based on stacking and deep learning. Sensors **22**(17), 6477 (2022)
15. Khan, H.A., Sehatbakhsh, N., Nguyen, L.N., Prvulovic, M., Zajić, A.: Malware detection in embedded systems using neural network model for electromagnetic side-channel signals. J. Hardw. Syst. Sec. **3**(4), 305–318 (2019)
16. Bensaoud, A., Kalita, J.: Deep multi-task learning for malware image classification. J. Inf. Sec. Appl. **64**, 103057 (2022)
17. Fakieh, B., Ragab, M.: Automated COVID-19 classification using heap-based optimization with the deep transfer learning model. Comput. Int. Neurosci. **2022**, 1–13 (2022). https://doi.org/10.1155/2022/7508836
18. Sunitha, G., Geetha, K., Neelakandan, S., Pundir, A.K.S., Hemalatha, S., Kumar, V.: Intelligent deep learning based ethnicity recognition and classification using facial images. Image Vis. Comput. **121**, 104404 (2022)
19. Usman, M.R., Usman, M.A., Yaq, M.A., Shin, S.Y.: UAV reconnaissance using bio-inspired algorithms: joint PSO and penguin search optimization algorithm (PeSOA) attributes. In: 2019 16th IEEE Annual Consumer Communications Networking Conference (CCNC) (pp. 1–6). IEEE (2019)
20. Pawar, K., Attar, V.Z.: Assessment of autoencoder architectures for data representation. In: Pedrycz, W., Chen, S.-M. (eds.) Deep Learning: Concepts and Architectures. SCI, vol. 866, pp. 101–132. Springer, Cham (2020). https://doi.org/10.1007/978-3-030-31756-0_4
21. Poonkuzhali, S., Jeyalakshmi, J.: Prescriptive Analytics with optimized linear approximation on Glycemic Load for Diabetes Diet Recommender System, published in vol. 22,no. 6, pp-2672–2681 Journal of Environmental Protection and Ecology (2021)
22. Poonkuzhali, S., Jeyalakshmi, J.:Study of Diabetes Mellitus Patients for Thyroid related co-morbidities using Data Analytics, Basic & Clinical Pharmacology & Toxicology, Wiley Publication, 124(S3):19 (2019) https://doi.org/10.1111/bcpt.13217
23. Santhiya, M., Jegatha, R., Shobana, M.: A supervised classification techniques to optimize error evaluation and space complexity. Int. J. Innov. Technol. Explor. Eng. (IJITEE) **8**(11), 92–95 (2019)

Detection and Classification of Skin Disease Using CNN

J. Jeyalakshmi[1]([✉]), M. Santhiya[2], and M. Shobana[3]

[1] Department of CSE, Amrita Vishwa Vidhyapeetham, Coimbatore, India
jeyalakshmi@ch.amrita.edu
[2] Department of CSE, Rajalakshmi Engineering College, Mevalurkuppam, India
[3] Department of NC, AP, SRM Institute of Science and Technology, Irungalur, India

Abstract. Humans have a long history of being susceptible to skin disorders, and now millions of individuals suffer from a wide range of skin conditions. In addition to causing low self-esteem and mental anguish, several of these illnesses are associated with an increased chance of developing skin cancer. Due to the lack of optical resolution for skin disease photos, a medical specialist and sophisticated equipment are required for a proper diagnosis of these conditions. CNN architecture and three preconfigured models(AlexNet, ResNet, and InceptionV3) are part of the proposed deep learning system. For the purpose of Skin Disease Classification, a Dataset of photos featuring seven disorders has been collected. Melanoma, nevi, seborrhoea keratosis, and other skin cancers and benign growths are among them. Since most pre-existing systems categorised cuts and burns as skin diseases, we expanded the dataset to include such photos. Deep Learning algorithms have reduced the requirement for human labour in areas like extracting the features and data restoration for categorization.

Keywords: Dermatoscopic images · Deep Learning · Data Enhancement · Convolutional Neural Network (CNN) · Model-Training · Testing and Evaluation

1 Introduction

The skin is one of the body's most crucial tissues, yet it also happens to be one of the most rapidly expanding. For this reason, the term "burden of skin illness" is now understood to be a multi-dimensional notion that takes into account the emotional, social, and monetary impact of skin diseases on not just the individuals affected but also their communities at large. People of all ages are susceptible to this pollution. It's common to suffer a cut or scrape on the skin because of how sensitive it is. More than 3,000 skin disorders have been identified.

A aesthetically devastating illness may have devastating consequences, including severe discomfort and permanent harm. Even while the majority of the most prevalent chronic skin conditions, including, vitiligo, psoriasis, and leg ulcers, atopic eczemaare now not lethal, they can nevertheless be categorised as a substantial threat to a person's health and wellness with detrimental physical, psychological, and financial effects..

R. Geetha et al. (Eds.): AAIMB 2023, CCIS 2202, pp. 171–180, 2025.
https://doi.org/10.1007/978-3-031-73065-8_14

However, skin cancers may be fatal and present unique challenges due to the fact that they tend to develop and spread rather quickly. Infectious diseases of the skin are common among individuals everywhere. Types of skin cancer include basal cell carcinoma (BCC), melanoma, intraepithelial carcinoma, and squamous cell carcinoma (SCC).

Current skin cancer rates are higher than rates for novel forms of lung and breast cancer combined [1]. Several skin diseases have symptoms that might develop for months before being noticed, making treatment a lengthy process. Because of the time and accuracy constraints associated with traditional laboratory analysis, computer-based illness diagnosis has emerged as a viable alternative. When it comes to forecasting skin diseases, Deep Learning is by far the most used tool. In order to discover and investigate characteristics in previously unseen data patterns, deep learning methods will rely on inferred data. With the use of supervisory techniques that cut down on testing expenses, this research introduces a reliable system for correctly diagnosing skin illnesses. Because of this, scientists are thinking of training a deep-learning model to classify the skin condition from a picture of the afflicted area. [2].

2   Related Works

It takes a lot of time to manually diagnose skin disorders by going to a dermatologist and asking questions. This is not a common choice in more remote locations. It is necessary for the residents of these remote areas to visit a larger metropolis for medical help. There's a lot of manual labour involved here. Moreover, even a visit to the doctor might set you back a pretty penny. Even worse, this involves interacting with other people during this pandemic catastrophe. Illnesses seldom spread from person to person. Contact with other people's bodies is a requirement of the current system. In the current state of computer-aided diagnosis, burns and injuries are classified as skin illnesses. These approaches aren't as precise as they should be. That's why it's important to create a computer-aided method to automatically identify skin disorders and tell them apart from other skin concerns. Color and texture features in images were utilised to diagnose skin diseases by Quan Gan et al. [3]. In order to prepare the pictures for further analysis, median filtering was used. To get the picture segments, Denise photos are rotated. Herpes, dermatitis, & psoriasis were classified using SVM after text characteristics were extracted using the GLCM tool.

In their paper "Automated Identification & Severity Measurement of Eczema Utilizing Image Processing," Md NafiulAlam and colleagues proposed an image processing & computer algorithm-based approach for automatic dermatitis detection and severity measurement. Patients may use the system to identify eczema by uploading a photo of the afflicted region, and the system would analyse the photo and provide a severity rating. Segmenting images, extracting features, and using statistical classification helped this system spot and classify eczema's varying degrees of severity. Types of eczema were classified, and an instrument that measure was given to each picture. Later studies utilised deep learning methods for skin disease classification. Parvathaneni Naga Srinivasu et al. [5] used the deep learning frameworks MobileNet V2 & Long Short Term Memory to categorise skin disorders. Estimating the spread of a disease required the use of a co-occurrence matrix with grey levels. On the HAM10000 skin condition dataset, the algorithm has obtained an accuracy of 85%. S.Malliga et al. [6] trained and classified various clinical pictures using the CNN algorithm.

They're taking three different skin conditions. Accuracy was 71% for these diagnoses: melanoma, nevus, and seborrheic keratosis. A method that divides photographs of skin lesions into one of five categories—healthy, acne-free, dermatitis-free, benign, or malignant melanoma—was created, implemented, and reviewed by Nazia Hameed et al. [7]. The total accuracy of classification using the SVM classifier was found to be 86.21 percent.

3 Proposed Model

3.1 Dataset Description

For this analysis, researchers utilised data on seven different types of skin conditions, including but not limited to: molluscum contagiosum warts, systemic illness, alopecia keratosis, lesion, bullous, actinic keratosis, acne, and rosacea. There are more than 7,000 dermatoscopic photographs in this data collection. A total of 750 more photos depicting burns and wounds to the skin have been added to the dataset. Current methods classify both burns and cuts as skin disorders. The issue was remedied by expanding the dataset to include graphic injury photos. The data is randomly divided into training data (5900 records) and validation data (1000 records) (1930).

3.2 Methodology

The recommended system is a website that serves as the first stage in the diagnostic process by analyzing an uploaded photograph of the afflicted skin to determine the kind of illness and provide a few ideas for treatment. A deep learning approach is used to diagnose skin conditions in the proposed framework. Based on a number of criteria, this system will analyze, process, and categorize the photographs using computational methods. Figure 1 depicts the architecture of a system for identifying and categorizing skin diseases.

Input layer, hidden layers, and output layer make up the Convolutional neural networks seen in Fig. 2. Because the input layer and the final convolution obscure their inputs and outputs, the intermediary layers of a feed-forward neural network are referred to as "hidden." Convolutions are performed in the hidden layers of a CNN. Typically, the Frobenius core product is used, and the ReLU activation function is applied to this product. As the convolution operation moves along the input sequence for the layer, a feature map is created, and this map serves as the input for the following layer. The next levels are the normalisation, fully connected, and pooling layers. The input is convolved in convolutional layers, and the output is passed on to the next layer. By merging the results of several neurons in one layer into the a nerve cell in the next, "pooling layers" are able to minimise the number of dimensions in a dataset. Each neuron inside one layer communicates with each neuron in every other layer [8].

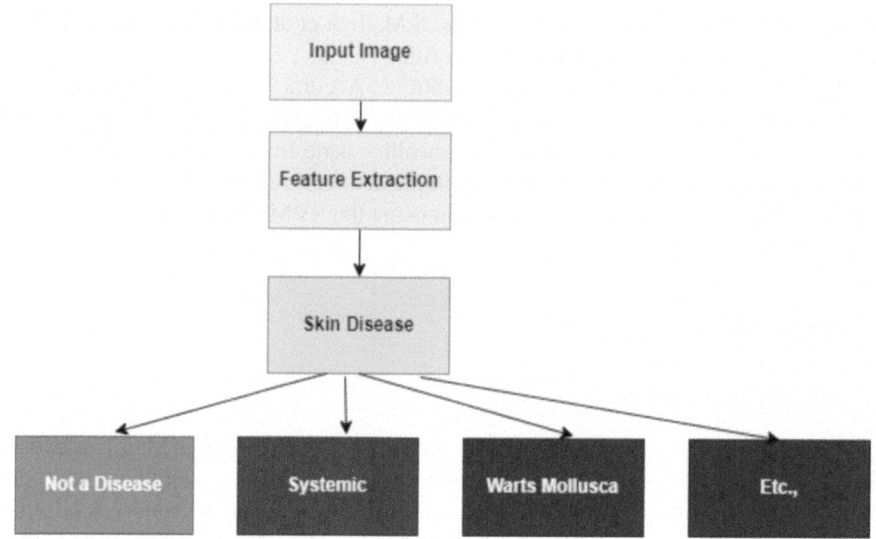

Fig. 1. Proposed System- Categorization of Skin Diseases

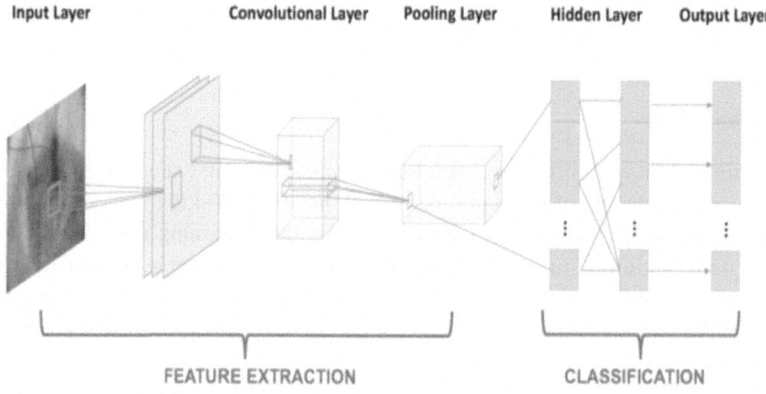

Fig. 2. CNN architecture

3.3 ResNet152v2

Figure 3 depicts the overall structure of ResNet152V2. This model, which is utilised for feature extraction, is trained using the image Net dataset. Since it is a which was before model, the model already has certain starting weights that may help it achieve good accuracy more quickly than a standard CNN. The ResNet152V2 model, a shape layer, a flattening layer, a dense surface with 128 neurons, a dropout layer, and a hidden layer with Softmax activation function are all included in the model's architecture to help it classify images into the appropriate categories. To combat the deterioration of deep neural networks, Resnet implements a mechanism called the residual learning unit [9]. This module's architecture is best described as a feed forwards network including a shortcut link that allows for the introduction of fresh inputs and the generation of fresh outputs. This module excels in comparison to alternatives because it improves classification accuracy without adding complexity to the model.

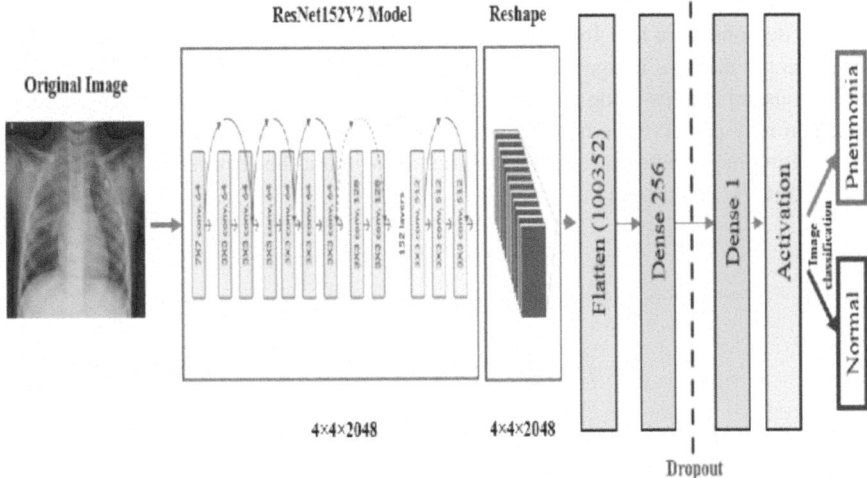

Fig. 3. RESNET152V2 architecture

3.4 AlexNet

The convolutional neural network known as AlexNet has made significant contributions to the field of computer vision, especially in the domain of deep learning's application for machine vision. It has ReLU activations appended to the end of each convolution and fully-connected layer. According to Fig. 4, the Alexnet architecture consists of eight layers, each of which has tunable parameters. The model has five layers: an input layer, a layer with max pooling, a layer with three fully linked hidden nodes, and an output layer with Relu activation.

Images in the RGB colour space are the Model's input. It multiplies the pace by six while maintaining the same precision. Two Dropout layers were used. The Softmax

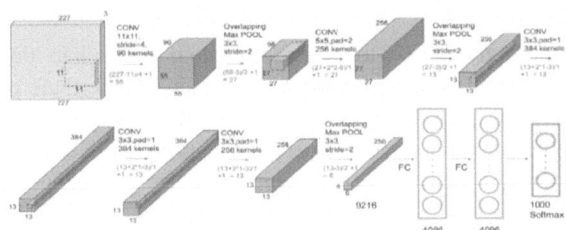

Fig. 4. AlexNet Architecture

activation function is employed as the last layer's activation function. There are 62.3 million parameters in the this architecture [10].

3.5 Inception V3

Figure 5 demonstrates that Inception v3 is indeed a pre - trained model that was first trained on the massive Image Net dataset consisting of over a million pictures divided into a thousand classes using powerful computational hardware. By retraining the last layer, you may preserve the model's prior learning and apply it to the dataset consisting, resulting in extremely accurate classification without the requirement for intensive training or CPU resources [11].

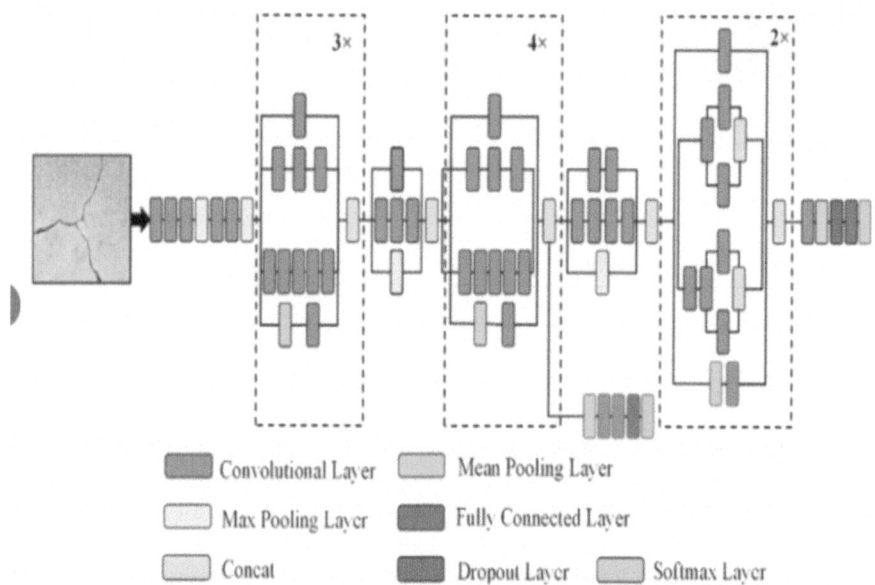

Fig. 5. INCEPTION-V3 architecture

4 Results Analysis

The goal of the proposed model is to use deep learning methods to identify seven different skin illnesses from a given image: warts, molluscum, seborrheic keratosis, lesion, bullous, actinic keratosis, acne, and rosacea. We use three pretrained networks—Alex Net, ResNet, and InceptionV3—to construct deep learning models and split the dataset into testing and training sets. The precision with which they are taught and evaluated in Table 1 displays the test and train accuracies of several models. The bar graph also displays CNN, Resnet, Alexnet, and Inception-v3 (Fig. 6). Evident observed from the graph is that CNN performed better on the training data than on the test data, leading to over fit of a model. Resnet Compared to competing models, both the train and test accuracy rate are improved. Therefore, skin disease detection and prediction using Resnet architecture is possible. To illustrate some of the outcomes of our experiment, we provide them in Fig. 7. Our user interface allows for the entry of a fresh picture. The system determines the presence or absence of skin disease. In the event that it is diagnosed to be one of the seven major illnesses. The affected area of skin is highlighted and several main therapy recommendations are shown.

Table 1. Comparison on Training and Testing Accuracies

Model	No of epoch	Training Accuracy	Test Accuracy
CNN	41	99.11	33.50
ResNet152v2	41	88.99	64.66
Inceptionv3	45	65.55	61.25
AlexNet	45	74.89	58.32

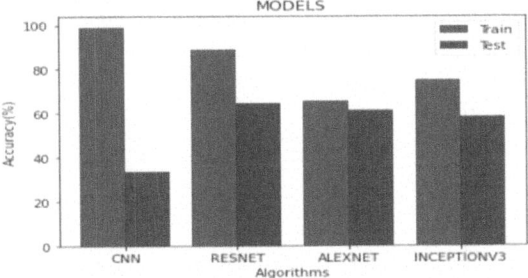

Fig. 6. Performance of Deep Learning Networks

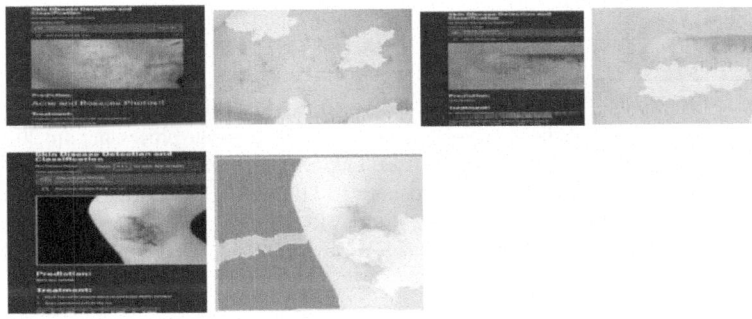

Fig. 7 Diagnosis Results

5   Conclusion

CNN, Resnet, Alexa, and Inceptionv3 have all been tested to see whether they can be used to create a global skin disease categorization system. Over the training data, CNN has done better than its competitors, but this has not been replicated on testing data. Giving a training set with greater variation and increasing its size may lead to improved accuracy. Also, when compared to other networks, Resnet has shown to be more accurate when attempting to diagnose skin disorders. Despite the fact that the skin diseases are the fourth most prevalent cause of sickness, many people still avoid medical attention. In this paper, we introduced a fully automated system for the detection of skin disorders. Finding skin problems early allows for more successful and less disfiguring treatments. Because no computer can match the human brain when it comes to analysis and intuition, this is designed to replace physicians. For the first time, research from the European Society for Medical Oncology demonstrates that a kind of artificial intelligence or machine learning outperforms highly trained dermatologists. It provides a high-level overview of the system or the approach used to put it into action.

The illnesses are identified effectively using the aforementioned image analysis and deep learning techniques. The system's main benefit is the reduction of the workload associated with feature engineering. Machines at CNNs may automatically pick up new functions. Therefore, CNN may be used to both diagnose and categorise skin disorders. It is possible for the system to achieve outcomes comparable to those of a dermatologist by using sophisticated computational methods and a big dataset, which raises the bar for medical and scientific excellence.

References

1. Ren, S., Jain, D.K., Guo, K., Xu, T., Chi, T.: Towards efficient medical lesion image super-resolution based on deep residual networks. Signal Process.: Image Commun. **75**, 1–10 (2019)
2. Srinivasu, P.N., SivaSai, J.G., Ijaz, M.F., Bhoi, A.K., Kim, W., Kang, J.J.: Classification of skin disease using deep learning neural networks with MobileNet V2 and LSTM. Sensors **21**(8), 2852 (2021). https://doi.org/10.3390/s21082852
3. Wei, L.S., Gan, Q., Ji, T.: Skin disease recognition method based on image color and texture features. Comput. Math. Methods Med. **2018**(1), 8145713 (2018). https://doi.org/10.1155/2018/8145713
4. Alam, M.N., Munia, T.T.K., Tavakolian, K., Vasefi, K., MacKinnon, Fazel-Rezai, R.: Automatic detection and severity measurement of eczema using image processing. In: 2016 38th Annual International Conference of the IEEE Engineering in Medicine and Biology Society (EMBC), pp. 1365–1368 (2016) https://doi.org/10.1109/EMBC.2016.7590961
5. Srinivasu, P.N., SivaSai, J.G., Ijaz, M.F., Bhoi, A.K., Kim, W., Kang, J.J.: Classification of skin disease using deep learning neural networks with MobileNet V2 and LSTM. Sensors **21**(8), 2852 (2021)
6. Malliga, S., Infanta, G., Sindoora, S., Yogarasi, S.: Skin disease detection and classification using deep learning algorithms. Int. J. Adv. Sci. Technol. **29**, 255–260 (2020)
7. Hameed, N., Shabut, A.M., Hossain, M.A.: Multi-class skin diseases classification using deep convolutional neural network and support vector machine. In 2018 12th international conference on software, knowledge, information management & applications (SKIMA) (pp. 1-7). IEEE (2018) https://doi.org/10.1109/SKIMA.2018.86315

8. Haenssle, H.A., Fink, C., Schneiderbauer, R., et al.: Man against machine: diagnostic perfor-mance of a deep learning convolutional neural network for dermoscopic melanoma recogni-tion in comparison to 58 dermatologists. AnnOncol **29**, 1836–1842 (2018). https://doi.org/10.1093/annonc/mdy166
9. Esteva, A., et al.: Dermatologist-level classification of skin cancer with deep neural networks. Nature **542**, 115–118 (2017)
10. Alex Krizhevsky , Ilya Sutskever, Geoffrey, E.: ImageNet Classification with Deep Convolu-tional Neural Networks. 1http://code.google.com/p/cuda-convnet
11. Purnama, K. E., et al.: Disease classification based on dermoscopic skin images using convolu-tional neural network in teledermatology system. In: 2019 International Conference on Com-puter Engineering, Network, and Intelligent Multimedia (CENIM), pp. 1–5 (2019). https://doi.org/10.1109/CENIM48368.2019.8973303
12. Poonkuzhali, S., Jeyalakshmi, J.: Prescriptive analytics of constraint optimisation of diabetes diet exhortation by using information systems. J. Environ. Protect. Ecol. **22**(6), 2672–2681 (2021)
13. Karunanayake, R.K., Dananjaya, W.M., Peiris, M.Y., Gunatileka, B.R.I.S., Lokuliyana, S., Kuruppu, A.: CURETO: skin diseases detection using image processing and CNN. In 2020 14th international conference on Innovations in Information Technology (IIT) (pp. 1-6). IEEE (2020) https://doi.org/10.1109/IIT50501.2020.9299041
14. Ajith, A., Goel, V., Vazirani, P., Roja, M.M.: Digital dermatology: Skin disease detection model using image processing. In: 2017 International Conference on Intelligent Computing and Control Systems (ICICCS), Madurai, India, pp. 168–173 (2021).https://doi.org/10.1109/ICCONS.2017.8250703
15. Roy, K., Chaudhuri, S.S., Ghosh, S., Dutta, S.K., Chakraborty, P., Sarkar, R.: Skin Disease detection based on different Segmentation Techniques. In: 2019 international conference on opto-electronics and applied optics (Optronix) (pp. 1-5). IEEE (2019).https://doi.org/10.1109/OPTRONIX.2019.8862403
16. Goswami, T., Dabhi, V. K., Prajapati, H. B.: Skin disease classification from image - A Sur-vey. In: 6th International Conference on Advanced Computing and Communication Systems (ICACCS), Coimbatore, India, pp. 599–605, (2020) https://doi.org/10.1109/ICACCS48705.2020.9074232
17. Dwivedi, P., Khan, A. A., Gawade, A., Deolekar, S.: A deep learning based approach for automated skin disease detection using Fast R-CNN. In: Sixth International Conference on Image Information Processing (ICIIP), Shimla, India, pp. 116–120, (2021). https://doi.org/10.1109/ICIIP53038.2021.9702567
18. Shaik, R., Bodhapati, S.K., Uddandam, A., Krupal, L., Sengupta, J.: A deep learning model that diagnosis skin diseases and recommends medication. In: 1st International Conference on the Paradigm Shifts in Communication, Embedded Systems, Machine Learning and Signal Processing (PCEMS), Nagpur, India, pp. 7–10, (2022).https://doi.org/10.1109/PCEMS55161.2022.9808065
19. Swamy, K. V., Divya, B.: Skin disease classification using machine learning algorithms. In: 2nd International Conference on Communication, Computing and Industry 4.0 (C2I4), Bangalore, India, pp. 1–5 (2021). https://doi.org/10.1109/C2I454156.2021.9689338
20. Jeyalakshmi, J., Poonkuzhali, S.: Study of diabetes mellitus patients for thyroid related co-morbidities using data analytics. Basic Clin. Pharmacol. Toxicol., Wiley Publication **124**(S3), 19 (2019)
21. Haddad, S., Hameed, S.A.: Image analysis model for skin disease detection: framework. In: 7th International Conference on Computer and Communication Engineering (ICCCE), Kuala Lumpur, Malaysia, pp. 1–4, (2018). https://doi.org/10.1109/ICCCE.2018.8539270

22. Rimi, T. A., Sultana, N., Ahmed Foysal, M. F.: Derm-NN: skin diseases detection using convolutional neural network. In: 4th International Conference on Intelligent Computing and Control Systems (ICICCS), Madurai, India, pp. 1205–1209, (2020). https://doi.org/10.1109/ICICCS48265.2020.9120925
23. Melbin, K., Raj, Y.J.V.: An enhanced model for skin disease detection using dragonfly optimization based deep neural network. In: Third International conference on I-SMAC (IoT in Social, Mobile, Analytics and Cloud) (I-SMAC), Palladam, India, pp. 346–351, (2021). https://doi.org/10.1109/I-SMAC47947.2019.9032458
24. Santhiya, M., Shobana, M., Jegatha, R.: A supervised classification techniques to optimize error evaluation and space complexity. In: International Journal of Engineering and Advanced Technology (IJEAT), ISSN: 2278–3075, 8(11S) (2019)

Estimation of Above Ground Biomass Using Machine Learning and Deep Learning Algorithms: A Review

S. Arumai Shiney[(✉)] and R. Geetha

Department of CSE, S.A. Engineering College, Thiruverkadu, India
arumaishiney@saec.ac.in

Abstract. Accurate estimation of above-ground biomass (AGB) plays a crucial role in various ecological and environmental studies. Traditional AGB estimation methods often rely on field measurements and labour-intensive approaches, limiting their scalability and efficiency. In recent years, the emergence of deep learning algorithms has shown promising results in AGB estimation using remote sensing data. These algorithms analyse input data such as remote sensing images or LiDAR data to learn complex patterns and relationships, enabling accurate estimation of AGB without relying on traditional manual methods or field measurements. This research work aims to provide a comprehensive review and analysis of the application of deep learning algorithms for AGB estimation, highlighting their advantages, limitations, and future research directions.

Keywords: Above-ground biomass · Machine Learning · Deep Learning · Remote Sensing

1 Introduction

Accurate AGB estimation is essential for understanding and monitoring ecosystem health, carbon dynamics, biodiversity conservation, and sustainable land use planning. It serves as a fundamental tool for informing policy decisions, supporting environmental assessments, and guiding conservation and management efforts to ensure the long-term sustainability of terrestrial ecosystems.

Traditional AGB estimation methods suffer from limitations related to labor intensiveness, limited spatial coverage, variability, handling non-tree biomass, insensitivity to fine-scale changes, adaptability to remote areas, and cost-effectiveness. These limitations highlight the need for alternative approaches, such as deep learning algorithms, to overcome these challenges and improve the accuracy and efficiency of AGB estimation.

1.1 Objective

The objectives of this research paper on AGB estimation using deep learning algorithms are as follows:

© The Author(s), under exclusive license to Springer Nature Switzerland AG 2025
R. Geetha et al. (Eds.): AAIMB 2023, CCIS 2202, pp. 181–196, 2025.
https://doi.org/10.1007/978-3-031-73065-8_15

i) To provide a comprehensive review of the application of deep learning algorithms for AGB estimation, particularly in the context of remote sensing data.
ii) To analyze the advantages and limitations of deep learning algorithms in comparison to traditional AGB estimation methods.
iii) To explore the different deep learning architectures, such as Convolutional Neural Networks (CNNs), Recurrent Neural Networks (RNNs), Generative Adversarial Networks (GANs), and Transformer Networks, and their suitability for AGB estimation.

The significance of AGB estimation using deep learning lies in its ability to enhance accuracy, scalability, efficiency, integration of heterogeneous data, robustness to variability, and decision support for environmental management. These advantages contribute to a better understanding of carbon dynamics, improved monitoring of ecosystem health, and the promotion of sustainable practices in land use and conservation.

2   Related Work

2.1   Traditional AGB Estimation Methods

Field Measurements: Field measurements involve collecting data directly from sample plots by measuring tree dimensions such as diameter at breast height (DBH) and tree height. These measurements are typically used to derive allometric equations that relate tree dimensions to AGB. Field measurements are considered the most accurate method but are labor-intensive, time-consuming, and limited in spatial coverage.

Allometric Equations: Allometric equations are statistical relationships that estimate AGB based on tree dimensions. They are derived from field measurements and are species-specific or generalized for specific forest types. Allometric equations provide a practical means of estimating AGB but may have limited applicability outside the regions or forest types for which they were developed.

Remote Sensing Approaches: Remote sensing techniques, such as satellite imagery, LiDAR, and Synthetic Aperture Radar (SAR), have been used for AGB estimation. These methods involve the extraction of relevant information, such as vegetation indices, canopy height, or backscattering values, and the use of empirical relationships to estimate AGB. Remote sensing approaches provide broader spatial coverage but are limited by the availability of appropriate data and the need for calibration and validation.

Modeling Approaches: Modeling approaches, such as forest inventory-based models or growth and yield models, utilize a combination of field measurements, environmental variables, and mathematical algorithms to estimate AGB. These models are often based on statistical or mechanistic principles and require calibration and validation with field data. While modeling approaches can provide spatially explicit AGB estimates, they may be limited by the assumptions and simplifications inherent in the models.

Upscaling Techniques: Upscaling techniques aim to extrapolate AGB estimates from small-scale field measurements or sample plots to larger areas or regions. These techniques use statistical or spatial interpolation methods to account for spatial variability and estimate AGB at a broader scale. Upscaling techniques can be useful for obtaining regional or national AGB estimates but may introduce uncertainties and biases due to the assumptions made during the upscaling process.

Limitations of Traditional Methods: Traditional AGB estimation methods have several limitations. They are often labor-intensive and time-consuming, making large-scale or frequent AGB assessments challenging. These methods may also lack spatial representativeness, as they rely on limited sample plots or field measurements. Additionally, traditional methods may struggle to capture the complexity and heterogeneity of forest ecosystems and may be sensitive to the specific forest type or species for which they were developed.

By discussing these traditional AGB estimation methods and their limitations, the literature review provides a foundation for understanding the need for alternative approaches, such as deep learning algorithms, to overcome these challenges and improve AGB estimation accuracy and efficiency.

3 Deep Learning Algorithms in Remote Sensing

The use of deep learning algorithms in remote sensing has gained significant attention in recent years due to their ability to extract complex patterns and features from large-scale and high-dimensional remote sensing data. Deep learning algorithms, such as Convolutional Neural Networks (CNNs), Recurrent Neural Networks (RNNs), Generative Adversarial Networks (GANs), and Transformer Networks, have shown great potential in various remote sensing applications. In this section, we discuss the application of deep learning algorithms in remote sensing and highlight their benefits and challenges.

Convolutional Neural Networks (CNNs): CNNs are widely used for image analysis tasks in remote sensing. They excel at automatically learning hierarchical spatial features from remote sensing imagery, making them well-suited for tasks such as land cover classification, object detection, and change detection. CNN architectures, such as the popular ResNet, DenseNet, and U-Net, have been adapted and optimized for remote sensing applications, leading to improved accuracy and robustness.

Recurrent Neural Networks (RNNs): RNNs are designed to capture temporal dependencies in sequential data. In remote sensing, RNNs are utilized for time series analysis, such as land surface temperature prediction, vegetation phenology monitoring, and rainfall estimation. RNN variants, including Long Short-Term Memory (LSTM) and Gated Recurrent Unit (GRU), can effectively model temporal dynamics and learn patterns over time, enabling improved prediction and understanding of dynamic processes.

Generative Adversarial Networks (GANs): GANs are generative models that consist of a generator network and a discriminator network, which compete against each other during training. GANs have found applications in remote sensing for tasks such as image synthesis, data augmentation, and domain adaptation. GANs enable the generation of realistic synthetic remote sensing images, which can be valuable for training data-scarce scenarios or simulating different environmental conditions.

Transformer Networks: Transformer Networks have revolutionized natural language processing tasks and have recently been applied to remote sensing data analysis. Transformers excel at capturing long-range dependencies and have shown promise in tasks such as image segmentation, object detection, and land cover classification. Their self-attention mechanism allows them to capture both spatial and contextual information, leading to improved performance.

In conclusion, deep learning algorithms offer significant potential for extracting meaningful information from remote sensing data. Their ability to automatically learn relevant features and patterns, scalability, and transferability make them valuable tools in remote sensing applications. By addressing the challenges and considerations associated with deep learning, researchers can leverage these algorithms to advance our understanding of the Earth's surface and improve decision-making in various fields, including environmental monitoring, land cover mapping, and disaster management.

4   AGB Estimation Using Deep Learning

There is a growing body of research focused on AGB estimation using deep learning algorithms in remote sensing. Here, we provide an overview of some existing studies that highlight the application and effectiveness of deep learning in AGB estimation:

1. **Li et al. (2016):** This study employed a CNN-based approach to estimate AGB using airborne LiDAR data. The CNN model was trained to learn the relationship between LiDAR-derived features and AGB measurements from field plots. The results demonstrated the effectiveness of deep learning in accurately estimating AGB at a regional scale.
2. **Latifi et al. (2017):** This comparative study evaluated the performance of various machine learning algorithms, including deep learning methods, for AGB estimation using multi-source remote sensing data. The authors found that deep learning algorithms, such as CNNs and Random Forest with deep features, outperformed traditional machine learning methods, highlighting their potential for accurate AGB estimation.
3. **Jin et al. (2019):** This study explored the use of deep learning, specifically a CNN-based model, for AGB estimation using Sentinel-2 imagery. The CNN model was trained on spectral and texture features extracted from Sentinel-2 bands. The results demonstrated the capability of deep learning to accurately estimate AGB at a fine spatial resolution.
4. **4. Demir et al. (2018):** DeepGlobe is a competition that includes several sub-challenges related to remote sensing analysis, including AGB estimation. The challenge encourages participants to develop deep learning models that leverage satellite imagery to estimate AGB. The competition has provided a platform for researchers to explore innovative deep learning approaches for AGB estimation.
5. **5. Lin et al. (2020):** This study combined LiDAR data and hyperspectral imagery to estimate AGB using a deep learning framework. The authors proposed a fusion-based CNN model that integrated features from both data sources. The results demonstrated the effectiveness of deep learning in capturing complementary information from LiDAR and hyperspectral data for accurate AGB estimation.

These studies highlight the potential of deep learning algorithms, including CNNs, for AGB estimation using various remote sensing data sources, such as LiDAR, multispectral, and hyperspectral imagery. The findings demonstrate the ability of deep learning to overcome the limitations of traditional methods and provide accurate and spatially explicit AGB estimates. Ongoing research continues to explore advanced deep learning architectures, data fusion techniques, and the integration of multi-source data for further improving AGB estimation accuracy and applicability.

5 Deep Learning Architectures for AGB Estimation

Deep learning architectures have been successfully applied to AGB estimation using remote sensing data. These architectures leverage the power of deep neural networks to learn complex relationships and patterns from large-scale and high-dimensional data. Here are some commonly used deep learning architectures for AGB estimation:

Convolutional Neural Networks (CNNs): CNNs are widely used for image analysis tasks and have shown promise in AGB estimation. CNNs consist of multiple layers of convolutional and pooling operations, allowing them to automatically extract spatial features from remote sensing imagery. CNNs have been applied to AGB estimation by inputting spectral bands or derived indices as image inputs and training the network to predict AGB values.

Recurrent Neural Networks (RNNs): RNNs are designed to capture sequential dependencies in data and have been utilized for AGB estimation using time series remote sensing data. By considering the temporal dynamics of vegetation growth, RNNs can capture seasonal variations and long-term trends in AGB. Architectures such as Long Short-Term Memory (LSTM) and Gated Recurrent Unit (GRU) are commonly used in AGB estimation to model the temporal dependencies and predict AGB values.

Auto-encoders: Auto-encoders are unsupervised learning models that aim to reconstruct the input data from a compressed representation. In the context of AGB estimation, auto-encoders can be used to learn compact representations of remote sensing data that capture important features related to AGB. These representations can then be used as inputs for subsequent regression models to estimate AGB.

Generative Adversarial Networks (GANs): GANs are generative models that consist of a generator network and a discriminator network. GANs have been applied to AGB estimation for generating synthetic remote sensing data that closely resemble real AGB patterns. By training the GAN on a combination of real and synthetic data, the discriminator network can learn to distinguish between real and synthetic AGB patterns, leading to improved AGB estimation.

Transformer Networks: Transformer Networks have gained attention for their ability to model long-range dependencies and have shown promise in AGB estimation. Transformer-based architectures, such as the Vision Transformer (ViT), have been adapted for remote sensing data analysis. These models can capture spatial and contextual relationships in remote sensing images, enabling accurate AGB estimation.

It is worth noting that the selection of the appropriate deep learning architecture depends on the characteristics of the data, such as spatial resolution, temporal frequency, and data availability. Additionally, model architecture customization, such as adding attention mechanisms or incorporating multi-scale features, can further improve AGB estimation accuracy. Ongoing research continues to explore and refine deep learning architectures for AGB estimation to enhance the understanding of ecosystem dynamics and support effective environmental management.

6 Data Acquisition and Preprocessing

In the context of AGB estimation using deep learning, several remote sensing data sources have been utilized. These data sources provide valuable information about vegetation structure, spectral characteristics, and environmental conditions, which are essential for accurate AGB estimation. Here are some commonly used remote sensing data sources:

Optical Imagery: Optical sensors, such as those on satellite platforms like Landsat, Sentinel-2, and MODIS, provide spectral information across different wavelengths of the electromagnetic spectrum. Optical imagery captures the reflectance properties of vegetation, allowing for the estimation of vegetation indices (e.g., NDVI, EVI) that are correlated with AGB. These indices serve as inputs to deep learning models for AGB estimation.

LiDAR Data: Light Detection and Ranging (LiDAR) is an active remote sensing technique that uses laser pulses to measure the distance between the sensor and the Earth's surface. LiDAR data provides highly accurate information about the vertical structure of vegetation, including canopy height and canopy density. This data can be used to derive metrics related to AGB, such as biomass profiles or vertical distribution patterns, which can be integrated into deep learning models.

Synthetic Aperture Radar (SAR) Data: SAR sensors, such as those on satellites like Sentinel-1, emit microwave signals and measure the backscattered energy. SAR data is particularly useful in areas with cloud cover or during nighttime when optical sensors may be limited. SAR signals penetrate vegetation and provide information about vegetation structure, biomass, and moisture content. Deep learning models can be trained using SAR data to estimate AGB.

Hyperspectral Imagery: Hyperspectral sensors capture the reflectance of the Earth's surface in hundreds of narrow and contiguous spectral bands. Hyperspectral imagery provides detailed spectral information, allowing for the identification of specific vegetation types and the estimation of biochemical and biophysical properties. Deep learning models can be trained using hyperspectral data to estimate AGB by capturing the unique spectral signatures associated with different AGB levels.

Thermal Imagery: Thermal sensors, such as those on satellites like Landsat and MODIS, measure the thermal radiation emitted by the Earth's surface. Thermal imagery provides information about vegetation water stress, energy balance, and transpiration rates, which are related to AGB. Deep learning models can be trained using thermal data in combination with other remote sensing data sources to improve AGB estimation accuracy.

The integration of multiple data sources, such as combining optical imagery with LiDAR or SAR data, can enhance the accuracy and robustness of AGB estimation using deep learning. These data sources provide complementary information and enable the capture of both spectral and structural characteristics of vegetation, leading to more comprehensive AGB assessments.

7 Data Preprocessing Techniques

Data preprocessing plays a crucial role in preparing remote sensing data for AGB estimation using deep learning algorithms. Preprocessing techniques help to enhance the quality, consistency, and compatibility of the data, thereby improving the performance of the deep learning models. Here are some common data preprocessing techniques used in AGB estimation:

Data Normalization: Normalizing the input data is essential to ensure that features have similar scales and distributions. Common normalization techniques include min-max scaling, z-score standardization, and logarithmic transformations. Normalization prevents features with large values from dominating the learning process and ensures that the model can effectively learn from all features.

Image Resampling: Remote sensing data may have different spatial resolutions, and it is often necessary to resample the data to a consistent resolution. Resampling can be performed using techniques such as nearest-neighbor, bilinear, or cubic interpolation. Resampling ensures that all input images have the same pixel size and aligns the spatial information across different data sources.

Data Augmentation: Data augmentation techniques are used to artificially increase the size of the training dataset by applying transformations to the existing samples. Augmentation can include random rotations, translations, flips, and zooms to generate additional variations of the data. Data augmentation helps in improving the model's generalization ability and reduces overfitting by exposing it to a wider range of training examples.

Feature Extraction: Deep learning models often benefit from input data that is representative of the target variable. In AGB estimation, relevant spectral indices, vegetation metrics, or texture features can be extracted from remote sensing data. These features capture important information related to vegetation structure and composition, which can improve the model's ability to estimate AGB accurately.

Data Fusion: Integration of multiple data sources, such as optical imagery, LiDAR data, and hyperspectral imagery, can provide a more comprehensive representation of the study area. Data fusion techniques aim to combine the strengths of different data sources to improve the estimation accuracy. Fusion can be performed at the pixel level, feature level, or decision level, depending on the characteristics of the data and the specific AGB estimation task.

Quality Control: Remote sensing data may contain artifacts, noise, or missing values. Quality control procedures, such as data filtering, outlier detection, and data gap filling, are applied to remove or correct unreliable data. Ensuring the quality and consistency of the input data is essential for obtaining reliable AGB estimates.

These data preprocessing techniques help to prepare the remote sensing data for deep learning models, enabling accurate and robust AGB estimation. The specific techniques employed depend on the characteristics of the data sources, the availability of ground truth data, and the requirements of the AGB estimation task. Proper data preprocessing ensures that the deep learning model can effectively learn from the data and capture the relevant patterns and relationships.

8   Feature Extraction and Selection

Spectral Information: Spectral information is a fundamental component of remote sensing data and plays a crucial role in AGB estimation using deep learning algorithms. Spectral information refers to the measurements of reflected or emitted energy across different wavelengths of the electromagnetic spectrum. It provides valuable insights into the spectral characteristics of vegetation, which are closely related to AGB.

In AGB estimation, spectral information is typically derived from optical sensors, such as those found on satellite platforms like Landsat, Sentinel-2, or MODIS. These sensors capture electromagnetic radiation in the visible, near-infrared, and shortwave infrared regions. Spectral bands corresponding to specific wavelengths are used to quantify the reflectance properties of vegetation.

The spectral information extracted from remote sensing data can be used in various ways to estimate AGB using deep learning algorithms. Here are some key aspects related to spectral information in AGB estimation:

Vegetation Indices: Vegetation indices are mathematical combinations of spectral bands that capture specific vegetation characteristics. Common vegetation indices used in AGB estimation include the Normalized Difference Vegetation Index (NDVI), Enhanced Vegetation Index (EVI), and Green Chlorophyll Index (GCI). These indices quantify the greenness, vegetation density, and photosynthetic activity of vegetation, which are related to AGB. Deep learning models can be trained using spectral bands or derived vegetation indices as input features to estimate AGB.

Spectral Signatures: Spectral signatures represent the unique reflectance patterns of different land cover types. In AGB estimation, spectral signatures of vegetation can be used to identify and differentiate vegetation classes with varying AGB levels. Deep learning models can learn to recognize and distinguish these spectral signatures to estimate AGB accurately.

Spectral Libraries: Spectral libraries contain spectral signatures of different vegetation species or AGB levels. These libraries provide reference spectra that can be used for comparison and matching with the spectral information derived from remote sensing data. Deep learning models can be trained to identify the best match between the observed spectral information and the spectral library to estimate AGB.

Spectral Unmixing: Spectral unmixing techniques aim to decompose the mixed spectral signals observed in remote sensing data into their constituent endmembers. Endmembers represent pure spectral signatures of different land cover components, including vegetation, soil, and water. Spectral unmixing helps to estimate the fractional abundance of vegetation within a pixel, which can be related to AGB. Deep learning models can learn to perform spectral unmixing or leverage the fractional abundances as input features for AGB estimation.

By leveraging the spectral information captured by remote sensing sensors, deep learning models can effectively learn the relationships between the spectral characteristics of vegetation and AGB. This enables accurate estimation of AGB at different spatial scales and provides valuable insights into ecosystem dynamics, carbon storage, and environmental monitoring.

Texture Analysis: Texture analysis is a technique used in remote sensing and image processing to extract information about the spatial patterns and arrangement of pixels within an image. It complements spectral information by providing additional details about the texture and structure of the land cover, which can be useful for AGB estimation using deep learning algorithms. Texture analysis considers the spatial relationships between neighboring pixels and captures information related to the surface roughness, heterogeneity, and patterns within the image.

In the context of AGB estimation, texture analysis techniques can be applied to remote sensing data to extract textural features that capture important information about vegetation structure and biomass distribution. These textural features can then be used as input variables for deep learning models to improve the accuracy of AGB estimation. Here are some commonly used texture analysis techniques:

Gray Level Co-occurrence Matrix (GLCM): GLCM calculates the frequency of occurrence of pairs of pixel values at specified spatial offsets. It measures the spatial dependencies between pixels and provides information about the texture or pattern within the image. From the GLCM, various statistical measures can be derived, such as contrast, homogeneity, entropy, and correlation, which represent different aspects of the texture. These statistical measures can be used as texture features for AGB estimation.

Local Binary Patterns (LBP): LBP is a simple yet effective texture descriptor that encodes the local variations in pixel intensity. It compares the intensity of a central pixel with its neighboring pixels and assigns a binary code based on whether the neighboring pixels are greater or lesser than the central pixel. By considering the patterns formed by these binary codes, LBP captures the texture variations within the image. The histogram of LBP patterns can be used as a texture feature for AGB estimation.

Gabor Filters: Gabor filters are a set of linear filters that are used to extract texture features at different scales and orientations. These filters mimic the response of human visual system cells and capture texture information at various spatial frequencies. By convolving the remote sensing data with Gabor filters, responses at different scales and orientations can be obtained, which can be used as texture features for AGB estimation.

Haralick Features: Haralick features are a set of texture descriptors derived from the GLCM. They capture different statistical properties of the GLCM, such as angular second moment, entropy, contrast, and correlation. These features describe the texture heterogeneity, smoothness, and patterns within the image and can be used as input features for deep learning models in AGB estimation.

By incorporating texture analysis techniques, deep learning models can capture important spatial information related to vegetation structure and arrangement, which is valuable for AGB estimation. The combination of spectral and textural features provides a comprehensive representation of the remote sensing data, enhancing the ability of deep learning models to estimate AGB accurately and capture fine-scale variations in biomass distribution.

Vegetation Indices: Vegetation indices are mathematical formulas that use the spectral information captured by remote sensing sensors to provide insights into the health, density, and vigor of vegetation. These indices are widely used in AGB estimation and vegetation monitoring studies as they capture the unique reflectance properties of vegetation across different wavelengths of the electromagnetic spectrum. Here are some commonly used vegetation indices:

Normalized Difference Vegetation Index (NDVI): NDVI is one of the most widely used vegetation indices. It quantifies the difference between the reflectance in the near-infrared (NI R) and red spectral bands. The formula for NDVI is: NDVI = (NIR - Red) / (NIR + Red). NDVI values range from − 1 to 1, where higher values indicate healthier and more abundant vegetation. NDVI is sensitive to the presence of chlorophyll and can effectively capture variations in vegetation density and greenness.

Enhanced Vegetation Index (EVI): EVI is an improved version of NDVI that corrects for atmospheric effects and provides a more accurate representation of vegetation conditions. It incorporates additional blue and red-edge bands in addition to the NIR and red bands. The formula for EVI is: EVI = 2.5 * ((NIR - Red) / (NIR + 6 * Red - 7.5 * Blue + 1)). EVI values range from − 1 to 1, and higher values indicate healthier vegetation.

Green Chlorophyll Index (GCI): GCI is specifically designed to capture the chlorophyll content in vegetation. It utilizes the green and red spectral bands. The formula for GCI is: GCI = (Green - Red) / (Green + Red). GCI values range from − 1 to 1, where higher values indicate higher chlorophyll content and healthier vegetation.

Soil Adjusted Vegetation Index (SAVI): SAVI is similar to NDVI but includes a soil background adjustment to account for variations in soil reflectance. It reduces the influence of soil reflectance on the vegetation signal. The formula for SAVI is: SAVI = ((NIR - Red) / (NIR + Red + L)) * (1 + L), where L is a soil adjustment factor. SAVI values range from − 1 to 1, and higher values indicate healthier vegetation.

Normalized Difference Water Index (NDWI): NDWI is used to detect the presence of water bodies within an image. It utilizes the green and NIR spectral bands. The formula for NDWI is: NDWI = (Green - NIR) / (Green + NIR). NDWI values range from − 1 to 1, where higher values indicate the presence of water.

These vegetation indices capture different aspects of vegetation health and density and provide valuable information for AGB estimation using deep learning algorithms. By incorporating these indices as input features, deep learning models can learn the relationships between vegetation spectral characteristics and AGB, enabling accurate estimation of biomass levels across different spatial and temporal scales.

LiDAR and SAR Data: LiDAR (Light Detection and Ranging) and SAR (Synthetic Aperture Radar) data are two remote sensing technologies that provide valuable information for AGB estimation when combined with deep learning algorithms. They offer unique capabilities to assess vegetation structure, biomass, and spatial distribution, complementing the spectral information obtained from optical sensors. Here's an overview of LiDAR and SAR data and their applications in AGB estimation:

LiDAR Data Principle: LiDAR uses laser pulses to measure the distance between the sensor and the Earth's surface, creating highly accurate 3D point cloud data.

Vegetation Information: LiDAR data provides detailed information about vegetation structure, including canopy height, vertical profile, canopy density, and foliage distribution. It captures fine-scale details, such as individual tree crowns and forest understory.

AGB Estimation: LiDAR-derived metrics, such as canopy height model (CHM), canopy cover, leaf area index (LAI), and biomass profiles, are used as input features for deep learning models to estimate AGB. LiDAR data helps capture the vertical structure and biomass distribution, particularly in complex forest environments.

LiDAR is valuable for AGB estimation in forest inventory, carbon sequestration assessment, deforestation monitoring, and ecological modeling.

SAR Data: Principle: SAR sensors emit microwave signals and measure the backscattered energy. SAR operates independently of solar illumination, making it suitable for all-weather and day/night observations.

Vegetation Information: SAR data captures backscatter signals that are influenced by vegetation structure, biomass, and moisture content. It provides information about vegetation density, roughness, and scattering mechanisms.

AGB Estimation: SAR data, combined with ancillary data or ground measurements, can be used to estimate AGB using empirical models or machine learning algorithms. Backscatter coefficients or derived indices from SAR data, such as radar vegetation index (RVI) or biomass indices, can serve as input features for deep learning models.

Applications: SAR data is useful for AGB estimation in areas with frequent cloud cover, dense vegetation cover, or in regions where optical sensors face limitations. It supports applications like forest biomass mapping, deforestation monitoring, and agricultural crop yield assessment.

By integrating LiDAR and SAR data with deep learning algorithms, AGB estimation can benefit from the complementary information they provide. The vertical structure information from LiDAR and the microwave backscatter properties from SAR enhance the understanding of vegetation biomass distribution, improving the accuracy and spatial resolution of AGB estimation models. This fusion of different data sources helps overcome limitations of individual sensors and provides a more comprehensive assessment of AGB at various scales.

9 Training and Evaluation

Training Data Preparation: Preparing training data is a crucial step in AGB estimation using deep learning algorithms. The quality and representativeness of the training data directly impact the accuracy and generalization ability of the model. Here are some key considerations for training data preparation:

Ground Truth Data Collection: Ground truth data refers to field measurements or reliable reference data that provide AGB values at specific locations within the study area. It serves as the basis for training the deep learning model. Ground truth data can be collected through field surveys, biomass harvesting, or destructive sampling. The locations for ground truth collection should be randomly or systematically distributed across the study area to ensure representativeness.

Data Labeling and Annotation: The ground truth data needs to be associated with the corresponding remote sensing data, such as satellite images or LiDAR point clouds. This process involves labeling or annotating the training data by spatially matching the ground truth values with the corresponding pixels or areas in the remote sensing data. The labeling can be done manually or using automated algorithms, depending on the availability of resources and the complexity of the task.

Training Data Selection: From the labeled dataset, a subset is selected as the training data for the deep learning model. It is important to ensure that the training data is representative of the entire study area and captures the variability in AGB levels, vegetation types, and environmental conditions. The selection of training data should consider a balanced representation of different land cover classes and AGB ranges to avoid bias towards specific conditions.

Data Augmentation: Data augmentation techniques can be applied to increase the size and diversity of the training dataset. Augmentation involves applying various transformations, such as random rotations, translations, flips, or changes in brightness and contrast, to the existing training samples. Data augmentation helps the deep learning model generalize better by exposing it to a wider range of training examples and reducing overfitting.

Data Split: The training dataset is divided into training and validation subsets. The training subset is used to train the deep learning model, while the validation subset is used to monitor the model's performance during training and make adjustments if necessary. The data split should be done randomly, ensuring that both subsets have a representative distribution of AGB values and land cover classes.

Data Normalization: The input data, including the remote sensing features and AGB labels, should be normalized to a common scale or distribution. Normalization ensures that all input variables have similar ranges and prevents certain features from dominating the learning process. Common normalization techniques include min-max scaling or z-score standardization.

By carefully preparing the training data, ensuring its quality, diversity, and representativeness, the deep learning model can learn effectively and accurately estimate AGB across the study area. Proper training data preparation sets the foundation for a robust and reliable AGB estimation model.

Loss Functions: Loss Functions: In deep learning, a loss function quantifies the discrepancy between the predicted outputs of the model and the ground truth labels. It serves as a measure of how well the model is performing during training. For AGB estimation using deep learning algorithms, suitable loss functions include:

Mean Squared Error (MSE): MSE is a commonly used loss function for regression tasks, including AGB estimation. It computes the average squared difference between the predicted AGB values and the ground truth labels. MSE penalizes larger errors more heavily and encourages the model to minimize the overall square difference between predictions and labels.

Mean Absolute Error (MAE): MAE calculates the average absolute difference between the predicted AGB values and the ground truth labels. Unlike MSE, MAE does not square the errors and provides a measure of the average magnitude of errors. MAE is less sensitive to outliers compared to MSE and can be used when absolute errors are more meaningful for the problem.

Huber Loss: Huber loss is a combination of MSE and MAE. It behaves like MSE for small errors and like MAE for large errors. Huber loss is more robust to outliers and can handle situations where the training data contains noise or anomalies.

10 Model Evaluation Metrics

Model evaluation metrics are used to assess the performance of a deep learning model for AGB estimation. These metrics provide quantitative measures of how well the model predicts AGB values compared to the ground truth labels. Here are some commonly used evaluation metrics for AGB estimation:

Mean Squared Error (MSE): MSE measures the average squared difference between the predicted AGB values and the ground truth labels. It provides a measure of the overall accuracy of the model's predictions, with higher values indicating larger errors. MSE is widely used for regression tasks, including AGB estimation.

Mean Absolute Error (MAE): MAE calculates the average absolute difference between the predicted AGB values and the ground truth labels. It provides a measure of the average magnitude of errors made by the model. MAE is less sensitive to outliers compared to MSE and can provide a clearer interpretation of the model's performance.

Root Mean Squared Error (RMSE): RMSE is the square root of the MSE and provides a measure of the average magnitude of errors in the same units as the AGB values. RMSE is useful for comparing models and understanding the scale of errors in the predictions.

R-Squared (R2) or Coefficient of Determination: R2 measures the proportion of the variance in the AGB values that is explained by the model. It ranges from 0 to 1, with a higher value indicating a better fit of the model to the data. R2 provides an indication of how well the model captures the variability in AGB and can be used for model comparison.

Relative Root Mean Squared Error (RRMSE): RRMSE is the RMSE normalized by the range of the ground truth AGB values. It provides a relative measure of the error, allowing for comparison across different datasets with varying AGB ranges. RRMSE is useful for assessing the model's performance across different spatial or temporal scales.

Bias: Bias quantifies the systematic deviation of the model's predictions from the ground truth AGB values. It measures whether the model consistently underestimates or overestimates the AGB values. A bias close to zero indicates minimal systematic errors.

Scatterplot and Correlation: Visual inspection of a scatterplot between the predicted AGB values and the ground truth labels can provide insights into the model's performance. Additionally, calculating the correlation coefficient, such as Pearson's correlation coefficient, between the predicted and true AGB values can indicate the strength of the linear relationship between the variables.

It is important to consider multiple evaluation metrics to gain a comprehensive understanding of the model's performance. Some metrics focus on the overall accuracy of the predictions (MSE, MAE, RMSE), while others assess the variability captured by the model (R2). Visualizing the results and inspecting the scatterplot can provide additional insights into the model's strengths and weaknesses.

Limited Training Data: Limited training data is a common challenge in many machine learning tasks, including AGB estimation using deep learning algorithms. When the available training data is insufficient, it can lead to overfitting, where the model fails to generalize well to unseen data.

Data Heterogeneity: Data heterogeneity refers to the presence of variations, inconsistencies, or differences within the training data used for AGB estimation. Heterogeneous data can arise from various sources, such as differences in data sources, sensor characteristics, acquisition dates, spatial resolutions, and environmental conditions. Dealing with data heterogeneity is important to ensure accurate and reliable AGB estimation using deep learning algorithms.

Model Overfitting and Generalization: Overfitting occurs when a machine learning model learns the training data too well, capturing the noise and random variations in the training set instead of the underlying patterns. It happens when the model becomes overly complex and has too many parameters relative to the available training data. As a result, the model performs well on the training data but fails to generalize to new, unseen data.

Interpretability and Explainability: Interpretability: Interpretability refers to the ability to understand and explain how a model arrives at its predictions or decisions. It involves gaining insights into the internal workings of the model, understanding the relationships between input features and the predicted AGB values, and identifying the key factors influencing the model's output. Interpretability helps in building trust in the model's predictions, understanding the underlying processes, and identifying any biases or limitations.

Explainability: Explainability goes a step beyond interpretability by not only understanding the model's internal workings but also providing meaningful explanations for its predictions. It involves presenting the rationale, factors, or evidence that contribute to the model's decision-making process in a way that is understandable to humans. Explainability is especially important when the model's predictions have significant implications or when there are legal, ethical, or regulatory requirements for transparency.

11   Applications of AGB Estimation Using Deep Learning

Forest Monitoring and Management: Forest monitoring and management involve the systematic assessment, tracking, and sustainable utilization of forest resources. It aims to maintain the health, productivity, and biodiversity of forest ecosystems while meeting the socioeconomic needs of society.

Carbon Stock Assessment: Carbon stock assessment is a crucial component of forest monitoring and management. It involves quantifying the amount of carbon stored in forest ecosystems, including above-ground biomass (AGB), below-ground biomass (BGB), and soil organic carbon (SOC).

Climate Change Studies: Climate change studies focus on understanding the causes, impacts, and mitigation of changes in Earth's climate patterns. They encompass a wide range of scientific research, observations, and modeling efforts to examine the complex interactions between the atmosphere, oceans, land surface, and biosphere.

12   Comparative Analysis of Deep Learning Algorithms

Performance Comparison: Define appropriate metrics to assess the performance of the models or strategies being compared. For climate models, common metrics include accuracy in simulating historical climate patterns, ability to reproduce observed trends, and skill in predicting future climate scenarios. For mitigation strategies, metrics may include reductions in greenhouse gas emissions, cost-effectiveness, and long-term sustainability.

Computational Efficiency: Computational efficiency is a critical aspect of any computational system, including those used in climate change studies. It refers to the ability of a system or algorithm to deliver accurate results within a reasonable amount of time and computational resources.

Robustness to Data Variability: Robustness to data variability refers to the ability of a model or algorithm to produce consistent and reliable results despite variations or uncertainties in the input data. In the context of climate change studies, where data can be heterogeneous, noisy, or subject to measurement errors, it is crucial to ensure that models and algorithms are robust to these variations.

Generalization Ability: Generalization ability refers to the capability of a model or algorithm to perform well on unseen or new data that it has not been trained on. In the context of climate change studies, generalization ability is crucial for accurate predictions and reliable assessments.

13 Future Directions

Hybrid Approaches: Hybrid approaches in the context of climate change studies refer to the integration of multiple methods or techniques to address the complexities and challenges associated with climate change modeling, prediction, or mitigation. These approaches combine the strengths of different approaches to enhance accuracy, robustness, or efficiency.

Data Fusion: Data fusion combines multiple sources of data, such as remote sensing imagery, climate models, and ground-based measurements, to improve the accuracy and resolution of climate change assessments. By integrating complementary data sources, data fusion techniques can overcome limitations and uncertainties in individual datasets, providing a more comprehensive understanding of climate variables and their spatiotemporal patterns.

Ensemble Modeling: Ensemble modeling combines the predictions or results from multiple climate models or algorithms to obtain a consensus or weighted average prediction. Ensemble approaches reduce the reliance on a single model and take advantage of the diversity of models to capture a broader range of uncertainties and variability. Ensemble techniques include model averaging, Bayesian model averaging, and model weighting based on performance.

Hybrid Machine Learning Models: Hybrid machine learning models combine different machine learning algorithms or architectures to leverage their individual strengths. For example, a hybrid model can incorporate both deep learning and traditional statistical models to capture complex nonlinear relationships while maintaining interpretability and robustness. Hybrid models can improve the accuracy and generalization ability of climate change predictions.

Transfer Learning and Domain Adaptation: Transfer Learning: Transfer learning aims to transfer knowledge or representations learned from a source domain (where labeled data is abundant) to a target domain (where labeled data is limited). Instead of training a model from scratch on the target domain, transfer learning allows the model to leverage the knowledge gained from the source domain. This is particularly useful when the source and target domains share some underlying patterns or relationships.

Domain Adaptation: Domain adaptation focuses on adapting a model or algorithm from a source domain to a target domain, where the distributions of data may differ. In climate change studies, this can occur when data is collected from different regions, time periods, or using different measurement techniques. The goal of domain adaptation is to mitigate the differences between the source and target domains to improve the model's performance on the target domain.

14 Conclusion

This survey provides a comprehensive review and analysis of the application of deep learning algorithms for AGB estimation, highlighting their advantages, limitations, and future research directions. Future research can focus on developing more advanced deep learning architectures specifically tailored for AGB estimation. This could involve exploring novel network architectures, such as attention mechanisms, graph neural networks, or transformer-based models, that can better capture the complex spatial and spectral relationships associated with AGB. Deep learning models can benefit from the integration of multi-source data, including remote sensing imagery, LiDAR, SAR, and climate data. Future research can investigate effective methodologies for fusing and leveraging diverse data sources to improve AGB estimation accuracy and robustness.

References

1. Li, W., Guo, Q., Li, X., Wang, Z.: Estimating forest aboveground biomass by combining LiDAR data and convolutional neural networks. Remote Sens. **8**(3), 198 (2016)
2. Latifi, H., Galos, B., Fassnacht, F.E., Hartig, F.: Mapping forest aboveground biomass using remote sensing data and machine learning: a comparative study. Int. J. Appl. Earth Observ. Geo-inf. **60**, 40–52 (2017)
3. Jin, X., Liu, D., Yang, Y., Wang, J., Yang, G., & Dai, X.: Estimating Forest Aboveground Biomass using Sentinel-2 Imagery and Deep Learning.: Forests, 10(8), 664.(2019).: Estimating Forest Aboveground Biomass using Sentinel-2 Imagery and Deep Learning by [3]Jin et al. (2019).: DeepGlobe: A Challenge to Parse the Earth through Satellite Images by Demir et al. (2018)
4. Luo, S., et al.: Fusion of airborne LiDAR data and hyperspectral imagery for aboveground and belowground forest biomass estimation. Ecol. Indic. **73**, 378–387 (2017). https://doi.org/10.1016/j.ecolind.2016.10.001
5. Zhang, Y., Chen, S., Zhang, C.: Deep learning for remote sensing data: a technical tutorial on the state of the art. IEEE Geosci. Remote Sens. Mag. **7**(2), 8–23 (2019)
6. Ball, J.E., Woodcock, C.E., Wang, Q.: Data-intensive approaches to mapping forest carbon in Southeast Asia. Carbon Bal. Manage. **14**(1), 1–18 (2019)
7. Li, W., Gong, P.: Deep learning for remote sensing image analysis: a technical tutorial on the state of the art. IEEE Geosci. Remote Sens. Mag. **7**(4), 8–23 (2019)
8. Lu, D., Chen, Q., Wang, G.: A survey of deep learning applications in remote sensing. ISPRS J. Photogram. Remote Sens. **152**, 166–177 (2019)
9. Zhang, S., Li, Y., Yao, X., Zhang, H.: A review of deep learning-based object detection and classification in urban scenes using remote sensing images. Remote Sens. **12**(10), 1652 (2020)

URL Phishing Detection Using Deep Learning and Machine Learning Techniques

R. Jegadeesan[1]($\boxtimes$), Dava Srinivas[1], N. Sankar Ram[2], R. Janakiraman[3], M. Jhansi[1], C. H. Sanjana[1], N. Akshitha[1], and C. H. Saicharan[1]

[1] Department of CSE, Jyothishmathi Institute of Technology and Science, Karimnagar, India
`ramjaganjagan@gmail.com`
[2] Kgisl Institute of Technology, Coimbatore, India
[3] Kummuru Pratap Reddy Institute of Technology, Hyderabad, India

Abstract. Cyber attacks have also increased as smart devices have been used more frequently in recent years. Phishing is a type of fraud in which a person pretends to be someone they can trust by sending emails or using other communication channels to get sensitive information, like login passwords or account information. Here, we contrast AI and profound learning ways to deal with give a framework that is viable at spotting phishing sites through URL investigation, determined to diminish cyber attacks. We show that while tried utilizing URLs from legitimate login pages, existing methodologies had a broad bogus positive rate. Furthermore, via preparing a base model utilizing obsolete datasets and contrasting it with additional ongoing URLs, we exhibit how models lose precision with time utilizing datasets from different years. Phishing Index Login URL (PILU-90K) is a brand-new dataset that consists of 30K phishing URLs and 60K legitimate URLs, including login and index pages. The latest model we propose accomplishes 96.50% exactness on the introduced login URL dataset when Calculated Relapse matched with Term Recurrence - Backwards Record Recurrence (TF-IDF) include extraction.

Keywords: Machine learning · Cyberattack · Term Frequency · Inverse Document Frequency · Logistic Regression model · high false-positive rate

1 Introduction

We all know that the ongoing digital transformation has led to a significant increase in web services usage over the past few years, which has also resulted in numerous cyberattacks. Online administrations like e-banking, web based business, and Programming as a Help have changed how everybody offers their types of assistance. Millions more employees, students, and teachers are now developing their activities remotely as a result of restrictions imposed by the COVID-19 pandemic. This significantly increases the workload for services like email, student platforms, VPNs, and company portals. Thus, more individuals are in danger from becoming focuses of phishing assaults, in which lawbreakers endeavor to imitate dependable sites to take clients' login accreditations or installment data. Phishing, alongside spam messages and sites, is one of the

© The Author(s), under exclusive license to Springer Nature Switzerland AG 2025
R. Geetha et al. (Eds.): AAIMB 2023, CCIS 2202, pp. 197–212, 2025.
https://doi.org/10.1007/978-3-031-73065-8_16

main social designing assaults completed during the Coronavirus scourge, as indicated by ongoing investigations.

The guideline of recognizing phishing sites by their HTTP convention is as of now not material. According to the APWG, less than 25% of phishing websites used the HTTPS protocol in the third quarter of 2017, but this number increased to 83% in the first quarter of 2021. These websites offer secure end-to-end communication, giving customers a false sense of security when making an online purchase. Besides, just between the principal quarters of 2020 and 2021, the Counter Phishing Working Gathering (APWG) kept an obvious expansion in phishing endeavors, going from 165, 772 to 611, 877 sites. This ascent might be brought about by the way that more individuals have gone to, regardless do, internet providers because of the Coronavirus pandemic. We have focused on phishing detection using three main methods in the literature: list-based, programmed, and discovery using Profound Learning and AI calculations.

List-Based Approach:
The rundown based strategy, which is broadly used to distinguish phishing URLs, can be founded on white records or boycotts, contingent upon whether they contain genuine or phishing URLs, individually. All websites that are not on a white list are disabled by a system based on that list that was developed by Jain and Gupta. The boycott based frameworks, then again, are more common since they have a zero bogus positive rate, implying that no legitimate site is recognized as phishing. Phish Net and Google Safe Browse are two examples of these kinds of systems. Nonetheless, assuming that an aggressor changes a URL that is restricted, they could be compromised. They likewise extraordinarily depend on how habitually the framework's records are refreshed. A list-based approach is not a reliable method because there are so many new phishing websites created each day and they only last an average of 21 days.

Machine Learning Methods:
To circumvent the drawbacks of blacklists, researchers have used machine learning algorithms to identify unreported phishing encounters. Based on the input data, these methods can be divided into two groups: 1) URL-based: Jain and Gupta created an anti-phishing system with 14 handcrafted URL descriptors, some of which were obtained from third-party services like WHOIS registers or DNS lookups. They accomplished precision of 76.87% and 91.28% on a 35,491sample private dataset. 2) **Content-based:**
Content-based works for the most part use attributes taken from the source code of sites. However, the majority of the most recent solutions combine these with URLs and other services provided by third parties, such as WHOIS. Saloon, a heuristic framework in view of TF-IDF, was quite possibly of the earliest happy based work. Utilizing TFIDF, Bar takes five words from every site and enters them into the Google search bar. If a domain appeared in the top n results, the page was deemed legitimate; otherwise, it was regarded as phishing.

Deep Learning:

A modified CNN was proposed by Alyan and Ahmadi. The URL protocol was first left out, and subsequently URLs longer than 256 characters were clipped. To create a 128 embedding vector, they employed a 69 character alphabet that included lowercase letters, numbers, and some symbols. A one-dimensional CNN was then used to achieve 95.78% accuracy on a dataset of 2, 307, 800 URLs.A Gated Recurrent NeuralNetwork (GRU) that can recognise patterns and sequences inside URLs was presented by Zhao et al. [42]. They contrasted their strategy with a collection of 21 manually created features and an RF classifier. Results showed that RF outperformed automatic feature extraction when paired with GRUs, with 98.5% and 96.4%, respectively.

Dataset

Phishing Index Login URLS (PILU-90K):

An extended rendition of the Phishing Record Login URL (PILU-60K) dataset, named PILU-90K, is introduced in this work. Three gatherings of 90K URLs each make up PILU-90K: 30K real login URLs, 30K genuine URLs of landing page, and 30K authentic phishing URLs.[6].

2 Related Work

Salvi Siddhi Ravindra, Shah Juhi Sanjay, Shaikh Nisenbaum Ahmed Gulzar-Khodke Pallavi – "Phishing Website Detection Based on URL"-2021.

Due to the increased use of the internet and other online platforms in the modern period, security has received a lot of attention. Every day, there are several cyberattacks, with website phishing being the most prevalent problem. It involves pretending to be a trustworthy website in order to deceive users and obtain their private information. In light of this issue, this will present a potential remedy to prevent such assaults by verifying if the submitted URLs are genuine URLs or phishing URLs. It is a machine learning system, namely supervised learning, where 2000 datasets for authentic and phishing URLs have been provided. Because of its effectiveness and accuracy, they have taken into consideration the Random Forest algorithm. Here all these algorithms work and efficiency is seen.

The new Covid (Coronavirus) pandemic has meaningfully affected society and the economy around the world. It has likewise achieved various network safety gives that should be settled rapidly to safeguard casualties and urgent foundation, alongside other potential challenges across different regions. One of the fundamental systems for causing ruin is to target fundamental framework, like emergency clinics and medical care offices, with social designing based cyberattacks or dangers. Social engineering cyberattacks employ psychological and methodical tactics to influence their targets. The purpose of this study is to investigate the most advanced and widely used social engineering-based attack strategies, platforms, and tactics that are currently used in cyberattacks and threats. Since the beginning of the COVID-19 pandemic, we conduct an organized Multivocal Literature Review (MLR) on the current rise in social engineering-based cyberattacks and threats. This paper simply gives the data pretty much all the proficient calculation working and the contrast between every one of the calculations.

Patel, P, Sarno, D.M., Lewis, J.E., Shoos, M., Neider, M.B., Bohil, C.J.: "Perceptual representation of spam and phishing emails", (2020).

Understanding how computer users focus on the characteristics of potentially harmful emails could reduce costly mistakes. Which characteristics stand out? How consistent is attention distribution across different email features? We made an effort to gauge the mental salience of a number of email elements that are frequently found in spam and/or phishing emails. We prepared two email sets: one containing firm logos and links that could be clicked immediately, and the other without. Participants judged how similar emails in each batch were when compared in pairs. Email psychological similarity was measured using multidimensional scaling (MDS) analysis. A other group evaluated the same emails for the existence of five additional features: significant downloaded content, data collection, account suspension or deletion, advertisements, and huge graphics with clickable content. Regressing feature evaluations onto the MDS coordinates showed that, independent of the presence or absence of firm logos and urgent actionable buttons, similarity judgements were mostly driven by advertisements/large images and gathering personal information.

Governmental, financial, social, and personal privacy can all be severely harmed by phishing websites. Many phishing detection systems are being tested using tiny datasets, which makes them vulnerable to sampling problems like representing legitimate websites by just high-ranking domains, which could reduce the usefulness of their evaluation in actual use. Solutions for detecting phishing that simply require the URL are appealing since they can be utilised with firewalls and other constrained systems. In this study, a convolutional neural network (CNN) model-based method for URL-only phishing detection was put forth. Instead of employing present properties like URL length, the suggested CNN uses the URL as the input. They have gathered over two million URLs in a massive URL phishing detection (MUPD) dataset for training and evaluation. They divided the datasets for MUPD into training, validation, and testing sets and also gives detailed information. Machine learning methods also are important. Sahaja, K., Dr Sujatha, M., Dr. Jegadeesan, R.: "Key Enabled Privacy Scheme Using Edge Computing in Medical Diagnosis", JAC: A Journal Of Composition Theory, Volume xiv, Issue vii, ISSN: 0731–6755, page No. 48–53, (2021), DOI:21.18001.AJCT.2021.V14I7.21.1306.

This paper give an overall clarification of how music can impact a client's temperament and clear up how for pick the fitting music tracks to give a client a much needed boost. The innovation set up is equipped for distinguishing the client's feelings. The system could distinguish between happy, sad, angry, neutral, and shocked feelings. After identifying the user's emotion, the suggested strategy provided the user with a playlist of music matches. When a large dataset is processed, more memory and CPU are used. As a result, development will become more challenging and appealing. The objective is to develop this application on a standard platform at a reasonable cost. Our facial feeling based music proposal framework will make it more straightforward for clients to make and oversee playlists. The ongoing framework doesn't perform well in very terrible light circumstances and unfortunate camera goal in this way gives a chance to add some usefulness as an answer from now on. Pravallika, L., Dr. Jegadeesan, R., Dr. Sujatha, M.: "Identity-Based Encryption Transformation For Flexible Sharing Of Encrypted Data In

Public Cloud", Journal Of Resource Management And Technology, Volume 12, Issue 3, ISSN: 0745–6999, (2021).

This paper give an overall clarification of how music can impact a client's temperament and clear up how for pick the fitting music tracks to give a client a much needed boost. The innovation set up is equipped for distinguishing the client's feelings. The system could distinguish between happy, sad, angry, neutral, and shocked feelings. After identifying the user's emotion, the suggested strategy provided the user with a playlist of music matches. When a large dataset is processed, more memory and CPU are used. As a result, development will become more challenging and appealing. The objective is to develop this application on a standard platform at a reasonable cost. Our facial feeling based music proposal framework will make it more straightforward for clients to make and oversee playlists. The ongoing framework doesn't perform well in very terrible light circumstances and unfortunate camera goal in this way gives a chance to add some usefulness as an answer from now on.

Jegadeesan, R., Beno, A., Manikandan, S.P., Naga Malleswara Rao, D.S., Bharath Kumar Narukullapati,5T. Rajesh Kumar, Batyrkhan Omarov, Areda Batu: "Stable Route Selection for Adaptive Packet Transmission in 5G-Based Mobile Communications", "Wireless Communications and Mobile Computing" Research Article | Open Access Volume 2022 | Article, (2022). ID 8009105 | https://doi.org/10.1155/2022/8009105.

This paper proposed a framework that could recognize a client's feelings and concentrate face tourist spots in view of those enunciations. These facial landmarks would be used to determine the user's particular mood following their classification. Music that is representative of the user's emotion will be played for them after the user has been identified. It's possible that users will be able to select the music they want to listen to to relax. By not having to look up or search for music, the user would save time. Audio recovery, emotion retrieval retrieval, and emotion filtration were the three parts of the suggested design. However, it had a ton of drawbacks, such the way that the proposed approach couldn't as expected catch all temperaments on the grounds that the pre-owned assortment of photos needed an adequate number of pictures. For the processor to deliver right outcomes, the picture it gets should be caught in a sufficiently bright climate. The classifier requires a picture with a base goal of 320p to foresee the client's state of mind precisely. In the outside, carefully assembled qualities ordinarily don't decipher well.

Jegadeesan, R., Beno, A., Manikandan, S.P., Naga Malleswara Rao, D.S., Bharath Kumar Narukullapati,5T. Rajesh Kumar, Batyrkhan Omarov, Areda Batu: "Stable Route Selection for Adaptive Packet Transmission in 5G-Based Mobile Communications", "Wircless Communications and Mobile Computing, (2022).

Stable Course Determination for Versatile Bundle Transmission in 5G-Based Portable Correspondences In this paper different exploration papers are examined. The creator's methodology, execution strategies, merits, negative marks, future degree, finish of each paper is examined. The study's papers are also based on finding people who have disappeared. This paper is fundamentally founded on the continuous examination and toward the finish of this paper a superior method for finding the missing people will be found and the downsides of the past exploration papers will be tended to. The literature review of the previous papers is the subject of this paper.

"Ayush Choudhary, Abhinav Pundir, Baha Ur Rehaan, Jegadeesan, R.: "Trust-based Privacy-Preserving Photo Sharing in Online Social Networks", (2021).

Trust-based Protection Safeguarding Photograph Partaking in Web-based Informal organizations This paper is utilized to distinguish missing individuals. The architecture of our framework is shown in the diagram. The recommended Individual Recognizable proof Framework's engineering. Any revealed missing individual who is seen on a web cam will have their facial highlights matched to the information base and messaged to the police. Our framework separates the picture's facial encodings and analyzes them to the encodings of recently put away photographs in the data set. If a match is found, an alert message will be sent to the concerned police officer.

Jegadeesan, R., Ishank Vasania, Baha Ur Rehaan, Anuj Goyal: "Covid-19 Future Forecasting Using Exponential Smoothing", Strad Research, Volume 8, Issue 8, 2021, ISSN No: 0039–2049, page No. 724–737, (2021), https://doi.org/https://doi.org/10.37896/sr8.8/072.

Covid-19 Future Forecasting Using Exponential Smoothing In this paper, the primary objective is to locate face motion, which further aids the face recognition system, based on the video's location of the faces. Robert edge locator is utilized to distinguish the edge of the countenances later a few number juggling tasks are performed between the closest casing and starting edge. After that, unwanted edges and noises are removed using the Gaussian filtering method. Then initial two result outlines are taken and coherent activity is performed between errorless face frame casing to distinguish the edges which are like the face video. After that, we attempt to use the four corner points to draw a rectangle around the face. Which additionally assists with finding face and make a blueprint of the face alongside each casing? To know whether the position and area of face is changing development of each point is taken after some time.

3   Preliminaries

Researchers have honed in on phishing detection using three key strategies in the literature: automated list-based detection employing deep learning and machine learning methods.

In this, an AI calculation for traffic sign recognition and path location in independent vehicles. Shape models that are able to identify the appropriate shapes for traffic signs and lanes can be trained using the algorithm. The calculation is modified utilizing Python with OpenCV2 and NumPy libraries, and the Hough discovery strategy is utilized to distinguish the proper circles of the traffic lights. The AI calculation is prepared utilizing administered learning strategies, which include giving the calculation named information to gain from. The calculation is prepared to perceive the states of traffic signs and paths, which empowers it to recognize them continuously while the independent vehicle is in activity.

As a standard software architecture for autonomous vehicles, the proposed system makes use of the AUTOSAR architecture. Communication interfaces, device drivers, fundamental software, and a run-time environment for autonomous vehicles are all provided by the AUTOSAR architecture. The autonomous vehicle system's various components will be able to seamlessly communicate with one another thanks to this standardized architecture.

The proposed framework is intended to empower independent vehicles to recognize traffic signs and paths consequently, without the requirement for human mediation. This can assist with working on the security and proficiency of independent vehicles, as they can settle on choices in light of the information they gather from their sensors.

In general, the proposed framework in this is an AI calculation for traffic sign discovery and path location in independent vehicles, which is intended to work with the AUTOSAR design to empower powerful correspondence between equipment parts and applications. The development of a perception algorithm for self-driving cars based solely on vision or camera data is the primary objective of this work. The work is broken up into major sections. In the first section, we create a robust and effective lane detection algorithm that can identify the safe driving area in front of the vehicle. In second part, we foster a start to finish driving model in view of CNN's to gain from the drivers driving information and can drive the vehicle with just the camera information from on-board cameras. Execution of the proposed framework is seen by the execution of the independent vehicle that can have the option to identify and arrange the stop signs and different vehicles.

To give benefits to oppressed ranchers, our drive endeavors to productively deal with the issue of yield cost determining. It uses methods from machine learning to come up with better solutions from a variety of data. With the assistance of prepared information from endorsed datasets, this framework utilizes choice tree relapse strategies to gauge crop values. Productivity may rise as a result of this program. Perceiving and predicting Under shifted ecological settings, cropper incites skin break out. An effective crop price forecasting system can provide customers with choices that can meet their needs in a variety of circumstances. At last, the results are introduced as aweb application so that striving ranchers can rapidly get to them.

The SVM, KNN, Naive Bayes, Random Forest, and Decision Tree regression algorithms were utilized in this paper. Support vector machines (SVM) is set of administered learning techniques utilized for characterization, relapse and exception's revelation. It is a method of classification. Here, we tend to plot each information item to some extent in an n-dimensional house, with the value of each feature being the value of a chosen coordinate (where n is the number of options you have). it's a grouping method. In this algorithmic rule, we tend to plot each information item to some extent in an n-dimensional house, with the value of each feature being the value of a chosen coordinate (where n is the number of options you have). A Help Vector Machine (SVM) is discriminative classifier accurately limited by an isolating hyperplane. In elective words, given marked training data (managed learning), the algorithmic rule yields partner degree best hyperplane that classifies new models. Support vector basic machine (SVM) might be a bunch of managed learning procedures utilized for grouping, relapse and exception's uncovering.

Instead of a single algorithm, Naive Bayes' is a family of algorithmic rules. All gullible Bayes financial classifiers embraces that the worth of a specific element is free of the worth of some other component, given the class variable. A straightforward probabilistic classifier based on the applied theorem (derived from Bayesian statistics) and robust naive independence assumptions could be the Naive Thomas Bayes classifier. it's an order method upheld Bayes' hypothesis with partner degree suspicion of freedom between indicators. In clear terms, a Credulous Thomas Bayes categorified expects that the presence of a particular component in an extremely class is not at all like the presence of the other element. For instance, an organic product could likewise be considered to be partner degree apple in the event that it's red, round, and concerning a couple of creeps in measurement. Regardless of whether these highlights rely upon one another or upon the presence of different elements, a credulous Bayes classifier would think about these properties to freely add to the likelihood that this natural product is an apple. Each training data set's class label is predicted by these Learners.

Using the majority voting method, the class label for the training data set is chosen based on the prediction made by the majority of models. From the ensembled models the guidelines are created. Irregular backwoods square measure partner troupe learning technique for characterization, relapse and various errands, that work by building a wreck of call trees at instructing time and yielding the class that is the method of the classifications or mean expectation of the singular trees. Irregular call backwoods right for bring tree custom of over fitting to their training set. It has been demonstrated that one can use random forests for random objects by employing only pairwise similarities between objects, despite the fact that random forests are naturally designed to figure solely with third-dimensional information. Irregular Backwoods could be a brand name term for partner outfit of call trees. Irregular Woods is collection of call trees (supposed Timberland). To characterize a substitution object upheld credits, each tree offers a categorification and that we say the tree votes in favor of that class.

Regression using a Decision Tree: Trees are used in this method of supervised learning to make decisions. Dissimilar to other administered learning calculations, the choice tree technique can be used to address relapse and order issues. A decision tree can be used to create a training model that can be used to learn fundamental choice rules from previous data (training data) to predict the class or value of the target variable. In Choice Trees, we expect the class name for the record by beginning at the highest point of the tree and working our direction down. A comparison is made between the value of the initial attribute and its cost in the record. We move to the resulting hub by following the branch that is related with that worth in view of the examination. Measures for picking properties It very well might be trying to conclude which credits to use as inside hubs at the root. We teamed up to find replies to the characteristic determination issue. What's more, exhorted utilizing standards, for example, Entropy, Information Gain, the Gini Index, the Gain Ratio, and the Chi-Square These criteria will be used to determine the value of each characteristic. In the case of information gain, the features are arranged in a tree with the most significant feature at the top, and the values are sorted. We take a gander at various leveled characteristics while utilizing data gain as a standard, and we take a gander at nonstop qualities while utilizing the Gini file. We used Choice Tree Relapse on the grounds that the dataset's information is persistent. Because crops are seasonal, their prices fluctuate over time. Precipitation and WPI are utilized in the dataset as the boundaries for crop cost expectation.

A. List-Based

The rundown based system, which is notable for distinguishing phishing URLs, can be founded on whitelists or boycotts, contingent upon whether they contain legitimate or phishing URLs, separately. Jain and Gupta developed a whitelist-based strategy that disables all websites that are not on the list. On the other hand, the blacklist-based systems are more common because they have no false positives, which means that no legal website is flagged as phishing. PhishNet and Google Safe Browse are two examples of these kinds of systems. Nonetheless, assuming that an aggressor changes a URL that is restricted, they could be compromised. They likewise extraordinarily depend on how habitually the framework's records are refreshed. Thusly, a rundown based procedure is certainly not a vigorous arrangement because of the great volume of new phishing sites presented everyday and their short life expectancy, which is assessed to be 21 days by and large.

B. Machine Learning Methods.

o conquer boycott weaknesses, analysts have created AI models to recognize unreported phishing experiences. Contingent upon their input information, these methodologies can be o

1. URL-based

rdered into two classifications: URL-base and content-based.

Buberetal. We have implemented a URL detection system consisting of two sets of functions. The first is a vector of 209 words obtained using his String To Word Vector tool in Weka6.The second is 17NLP (Natural Language Processing) hand-crafted features such as number of subdomains, random words, numbers, special characters, and URL word length measurement. Combining both feature sets, we achieved a high accuracy of 97.20% on a 10% subsample set from the Ebbu 2017 dataset using Wekas RFC (random forest classifier).In the study below, Sahingoz etal. used three different feature sets: Word vectors, NLP, and hybrid sentences that combine both sentences. Using only 38 of his NLP features in the Ebbu 2017 dataset, we achieved an accuracy of 97.98% for Random Forest (RF). In this work, we used his NLP functions from Sahingoz etal. It is used because recent studies have reported state-of-the-art performance.

2. Content-based

One of the main substance based works was Saloon [29], which comprises of a heuristic framework in light of TF-IDF. Using TFIDF, CANTINA introduces five words from each website into the Google search engine. In the event that a space was inside the n first outcomes, the page was thought of as genuine, or phishing in any case. They got a precision of 95% with a limit of n = 30 Google query items. Because of the utilization of outer administrations like WHOIS7 and the high misleading positive rate, creators proposed CANTINA + [30]. Including two filters, their new proposal received a 99.61 percent F1-Score: i) a comparison of known phishing structures with hashed HTML tags; and ii) websites that were discarded without a form.

C. Deep Learning

For deep learning based techniques, Some shaetal. These features are 3URL features based on the number of dots, URL length, and HTTPS, and 6 features such as internal links and images extracted from HTML, broken link rate, and presence or absence of anchors. Link in HTML body. Finally, a third- party numeric function was extracted from Alexa PageRank. These features were extracted from his dataset of 3,526 samples and introduced into an LSTM model, yielding an accuracy of 99.57%. Arjofeyetal. We introduced the RCNN model for classifying phishing URLs. They took a URL as input to at okenizer and used one-hot encoding to represent the URL as a character-level matrix. In the final step, the model input is given a fixed length of 200 characters. If the URL falls below this threshold, the remaining characters are filled with zeros. Otherwise, characters exceeding the limit are truncated. Finally, we used a dataset of 310,642 URLs to feed the RCNN model and achieved 95.02% using the character embedding layer feature described above.

4 Proposed Work

This study utilizes real login sites to gather the URLs from such pages to create a dataset of phishing URLs. We then evaluate machine and deep learning methods in order to more accurately recommend the strategy. We then, at that point, show our speculation on phishing recognition and genuine login URLs by exhibiting how models prepared with authentic landing pages battle to group authentic login URLs. In addition, we demonstrate how accuracy decreases over time when models are evaluated using data from 2020 and trained on 2016 datasets. At long last, we give a rundown of ongoing phishing occurrences while illustrating the methodologies utilized by aggressors. We expanded the size of the PILU-60K (Phishing File Login URL) dataset from 60K to 90K URLs, uniformly split between the three gatherings of phishing, genuine landing pages, and real logins. We created and tried three pipelines for URL phishing recognition utilizing PILU-90K. We likewise prepared four regulated AI classifiers utilizing the handmade component descriptors proposed by Sahingoz et al. as well as programmed include extraction utilizing the Term Recurrence Opposite Report Recurrence (TF-IDF) at character N-gram level joined with the Calculated Relapse (LR) calculation and a CNN at character level. We looked at how reliable the phishing detection was over time. Here, we utilized strategic relapse, nave bayes, sgd (stochastic angle plunge), svm calculations for more exactness.

System Architecture

The categorization receives the phishing dataset and real URLs, which are then pre-processed to put the data in a useful format for analysis. The raw words are then retrieved from the various portions of the URL by dividing the string using a number of symbols, especially '/', '-', '.', '@', '?', '&', ' = ', and '_'. We extracted 38 features suggested by Sahingoz et al. after pre-processing. The number of words, compound words, words that are identical to or similar to well-known brands or a keyword like "secure" or "login," as well as the maximum, minimum, average, and standard deviation of word length, are all retrieved from the raw words. The project's objective is to recognise phishing URLs. Rather of depending on predetermined criteria or signatures, machine learning algorithms can learn to recognise phishing websites based on their attributes. They become stronger as a result and are less likely to produce false positives or false negatives (Fig. 1).

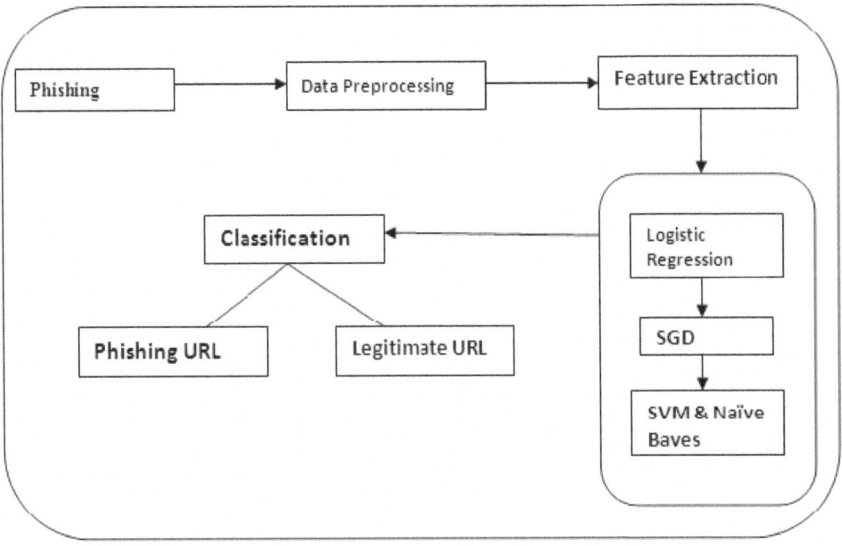

Fig. 1. System Architecture

5 Result and Analysis

Result

For a given information the accompanying picture addresses thus. Based on the proposed algorithms, we classify a URL as Phished, Legitimate, Malware, or Defacement in this

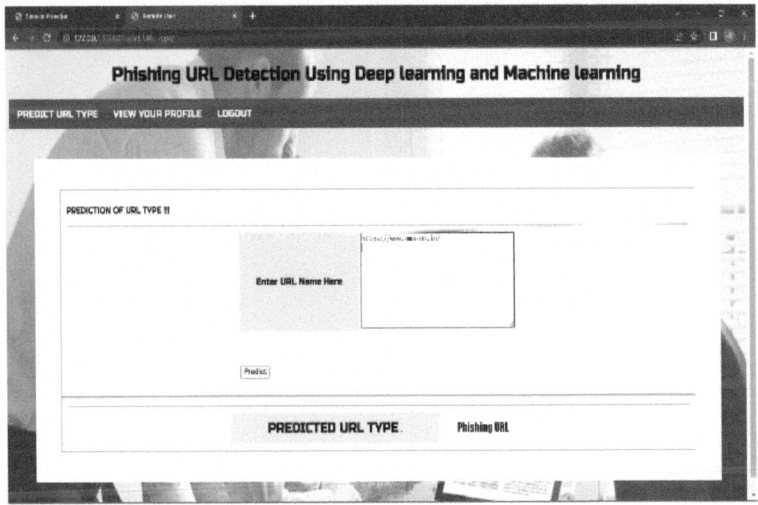

Fig. 2. Output

work. The calculation arranges the information in view of highlights that are proposed (Fig. 2).

Analysis

See Table 1.

Table 1. Algorithm analysis

	Accuracy	Precision	Recall	Fl-score	Support
Naive Bayes	92.780	0.81	0.31	0.45	735
SVM	96.508	0.89	0.55	0.68	735
Logistic Regression	96.150	0.89	0.51	0.65	735
SGD	95.972	0.93	0.44	0.59	735

Graphs

The graphical representations of our work on URL detection can be found in the charts below.

Pie Chart: Using the above approaches, a pie chart may be used to show a collection of various accuracy values for a particular variable (Fig. 3).

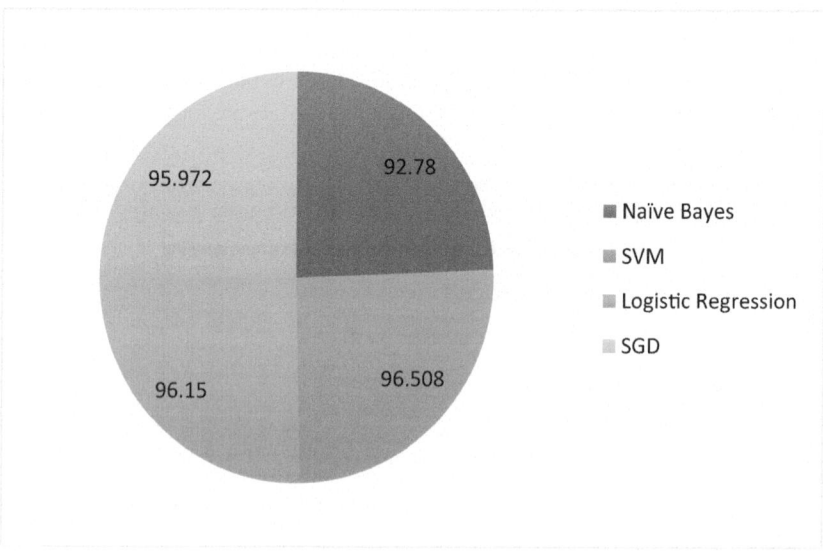

Fig. 3. Pie chart (Accuracy analysis)

Line Graph: A line graph, also known as a line plot, is a type of graph in which lines connect individual data points (Fig. 4).

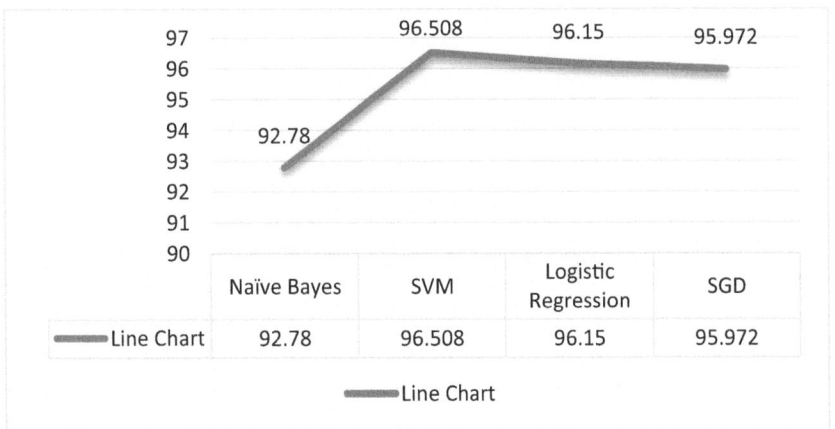

Fig. 4. Line chart (Accuracy analysis)

Bar Graph: A bar graph uses rectangles (or bars) with equal widths and varied heights to depict numerical data. Every bar should have an equal spacing between them (Fig. 5).

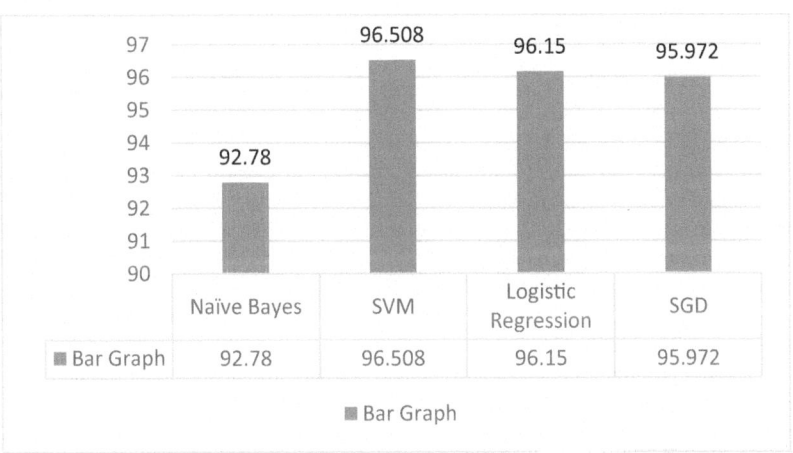

Fig. 5. Bar Chart (Accuracy Analysis)

6   Conclusion

In order to protect users from bogus login forms, the phishing detection mechanism aims to advance the blacklist methods currently in use. The PILU-90K dataset has been updated so that researchers can train and test their methods. This dataset contains true login URLs, which are the most dependable portrayal of genuine world phishing assaults.

Utilizing profound learning and AI instruments prepared on phishing and genuine home URLs, we explored various URL-based identification models. The low pace of bogus up-sides while recognizing this kind of URL is the critical advantage of our strategy. Out of all the models examined, TFIDF, the N-gram, and the LR algorithm produced the best results with an accuracy ratio of 96:50%. Contrasting our method with the cutting edge, as talked about in Segment II, uncovers three key advantages are no reliance on outside administrations, Login site discovery, Refreshed and genuine world dataset. At last, we showed that AI models utilizing hand tailored URL highlights diminished their exhibition more than time, up to 10:42% exactness on account of the Light GBM calculation from the year 2016 to 2020. To avoid significant ageing from its release date, machine learning methods should therefore be trained with recent URLs.

References

1. Ravindra, S.S., Sanjay, S.J., Ahmed Gulzar, S.N., Pallavi, D.: Phishing website detection based on URL (2021)
2. Jegadeesan, R., Srinivas, D., Umapathi, N., Karthick, G., Venkateswaran, N.: Personal health-care chatbot for medical suggestions using artificial intelligence and machine learning. Eur. Chem. Bull. **12** (S3), 6004–6012 (2023). https://doi.org/10.31838/ecb/2023.12.s3.670. (Scopus)
3. Hijji, M., Alam, G.: A multivocal literature review on growing social engineering based cyber-attacks/threats during the covid-19 pandemic: Challenges and prospective solutions. IEEE Access **9**, 7152–7169 (2021)
4. Patel, P., Sarno, D.M., Lewis, J.E., Shoos, M., Neider, M.B., Bohil, C.J.: Perceptual representation of spam and phishing emails (2020)
5. Al-Alyan, A., Al-Ahmadi, S.: Robust URL phishing detection based ondeep learning (2020)
6. Jegadeesan, R., Vasania, I., Ur Rehaan, B., Goyal, A.: Covid-19 future forecasting using exponential smoothing. Strad Res. **8**(8), 724–737 (2021), ISSN No: 0039-2049. https://doi.org/10.37896/sr8.8/072 (UGC Care Group II Journal and Web Of Science Group)
7. De', R., Pandey, N., Pal, A.: Impact of digital surge during covid-19 pandemic: a viewpoint on research and practice. Int. J. Inf. Manag. **55** (2020)
8. Chaudhry, J.A., Chaudhry, S.A., Rittenhouse, R.G.: Phishing attacks and defenses. Int. J. Secur. Appl. **10**(1), 247–256 (2016)
9. Umapathi, N., Karthick, G., Venkateswaran, N., Jegadeesan, R., Srinivas, D.: Desktop's virtual assistant using python. Eur. Chem. Bull. **12** (S3), 5975–5984 (2023). https://doi.org/10.31838/ecb/2023.12.s3.667. (Scopus)
10. Moghimi, M., Varjani, A.Y.: New rule-based phishing detection method. Expert Syst. Appl. **53**, 231–242 (2016)
11. Jegadeesan, R., Vijayakrishna Rapaka, E., Himabindu, K., Ranjan Behera, N., Shukla, A.K., Dangi, A.K.: Grey wolf optimizer with deep learning based short term traffic forecasting in smart city environment. In: IEEE paper was Published in: 2023 5th International Conference on Smart Systems and Inventive Technology (ICSSIT), Date of Conference: 23–25 January 2023, Electronic ISBN: 978-1-6654-7467-2 (2023). https://doi.org/10.1109/ICSSIT55814.2023.10061127, Publisher: IEEE , Link https://ieeexplore.ieee.org/document/10061127
12. Khonji, M., Iraqi, Y., Jones, A.: Phishing detection: a literature survey. Commun. Surv. Tutor. IEEE **15**(4), 2091–2121 (2013)
13. Mouton, F., Leenen, L., Venter, H.S.: 'Social engineering attack examples, templates and scenarios.' Comput. Secur. **59**, 186–209 (2016)

14. Ferreira, A., Teles, S.: Persuasion: how phishing emails can influence users and bypass security measures. Int. J. Human Comput. Stud. 19–31 (2019). https://doi.org/10.1016/j.ijhcs.2018.12.004

15. Jegadeesan, R., et al.: Stable route selection for adaptive packet transmission in 5G-based mobile communications. Wirel. Commun. Mobile Comput. Open Access (2022) Article ID 8009105. https://doi.org/10.1155/2022/8009105

16. Srinivas, D., Umapathi, N., Karthick, G., Venkateswaran, N., Jegadeesan, R.: Rnn-Dea: cyber-bullying detection in social mediaplatform (Twitter). Eur. Chem. Bull. **12**(S3), 6013–6021 (2023). https://doi.org/10.31838/ecb/2023.12.s3.671. (Scopus)

17. Karthick, G., Venkateswaran, N., Jegadeesan, R., Srinivas, D., Umapathi, N.: Forecast the autism spectrum disorder using various machine learning techniques. Eur. Chem. Bull, **12**(S3), 6030–6037 (2023). https://doi.org/10.31838/ecb/2023.12.s3.673. (Scopus)

18. Venkateswaran, N., Jegadeesan, R., Srinivas, D., Umapathi, N., Karthick, G.: Utilizing ensemble learners help prevent unauthorized access into Iot networks. Eur. Chem. Bull. **12**(S3), 5994–6003 (2023). https://doi.org/10.31838/ecb/2023.12.s3.669. (Scopus)

19. Jegadeesan, R., Srinidhi, A., Pranitha, P.: Sales analysis and prediction in big mart. J. xi'an Univ. Archit. Technol. **XV**(7), 186–191 (2023), ISSN No : 1006-7930. https://doi.org/10.37896/JXAT15.7/32216 (Scopus)

20. Sravanthi, P., Pranitha, P., Sujatha, M., Jegadeesan, R.: Authentication and key agreement based on anonymous identity for Peerto-Peer cloud. J. Resour. Manag. Technol. **12**(3, 173–180 (2021)), ISSN: 0745-6999, 12.15433.JRMAT.2021.V12I3, 2021.29 (UGC Care Group II Journal)

21. Anusha, K., Jyothi Prabha, A., Sujatha, M., Jegadeesan, R.: Discovery of ranking fraud for mobile apps. JAC : J. Compos. Theory **xiv**(vii), 42–47 (2021), ISSN: 0731-6755 (UGC Care Group II Journal), (2021)

22. Jegadeesan, R., Vasania, I., Ur Rehaan, B., Goyal, A.: Implications of machine learning for autonomic network operation and management. J. Xidian Univ. (Science and Technology Edition) **15**(8) , 307–315 (2021), ISSN No:1001-2400. https://doi.org/10.37896/jxu15.8/031 (UGC Care Group II Journal and Scopus)

23. Jegadeesan, R., Vasania, I., Ur Rehaan, B., Goyal, A.: Modelling and predicting cyber hacking breaches. J. Chengdu Univ. Technol. **26**(7) (2021). ISSN No: 1671-9727 (UGC Care Group II Scopus Active)

24. Jegadeesan, R., Vasania, I., Ur Rehaan, B., Goyal, A.: Covid-19 future forecasting using exponential smoothing. Strad Res. **8**(8), 724–737 (2021). ISSN No: 0039-2049. https://doi.org/10.37896/sr8.8/072 (UGC Care Group II Journal and Web Of Science Group)

25. Ur Rehaan, B., Pundir, A., Chaudhary, P., Choudhary, A., Jegadeesan, R.: Finding psychological instability using machine learning. Strad Res. **8**(9), 302–314 (2021), ISSN No: 0039-2049. https://doi.org/10.37896/sr8.9/037 (UGC Care Group II Journal and Web Of Science Group)

26. Choudhary, A., Pundir, A., Ur Rehaan, B., Jegadeesan, R.: Trust-based privacy-preserving photo sharing in online social networks. JAC : J. Compos. Theory **XIV**(X), 30–40 (2021). ISSN : 0731-6755, 21.18001.AJCT.2021.V14I10.21.1604 (UGC Care Group II Journal) Impact Factor 5.7

27. Jeyabharathi, J., Seedha Devi, Krishnan, B., Samuel, R., Imran Anees, M., Jegadeesan, R.: Human ear identification system using shape and structural feature based on SIFT and ANN classifier. In: International Conference on Communication, Computing and Internet of Things, IC3IoT) I 978-1-6654-7995-0/22/$31.00 ©2022 IEEE (2022). https://doi.org/10.1109/IC3IOT53935.2022.9767893

28. Jegadeesan, R., et al.: Stable route selection for adaptive packet transmission in 5G-based mobile communications. Wirel. Commun. Mobile Comput. Res. Article Open Access (2022) Article ID 8009105. https://doi.org/10.1155/2022/8009105

29. Pranitha, P., Manjula, A., Jegadeesan, R., saipriya, P.: Driver drowsiness monitoring system using visual behaviour and machine learning. Strad Res. **8**(7), 149–156 (2021). ISSN: 0039-2049. https://doi.org/10.37896/sr8.7/017 (UGC Care Group II Journal)

30. Ur Rehaan, B., Jegadeesan, R., Maheen, A.: Modelling and predicting cyber hacking breaches. J. Chengdu Univ. Technol. **26**(7) (2021). ISSN: 1671-9727. UGC Care Group II Journal and Scopus)

31. Pravallika, L., Jegadeesan, R., Sujatha, M.: Identity-based encryption transformation for flexible sharing of encrypted data in public cloud. J. Resour. Manage. Technol. **12**(3), 157–164 (2021). ISSN: 0745-6999 (UGC Care Group II Journal)

32. Divya, M., Venkateshwaran, N., Sujatha, M., Jegadeesan, R.: Secure data transfer and deletion from counting bloom filter in cloud computing. JAC : J. Compos. Theory **xiv**(vii), 36–41 (2021). ISSN: 0731-6755, 21.18001.AJCT.2021.V14I7.21.1304. (UGC Care Group II Journal)

33. Sahaja, K., Sujatha, M., Jegadeesan, R.: Key enabled privacy scheme using edge computing in medical diagnosis. JAC : J. Compos. Theory **xiv**(vii), 48–53 (2021). ISSN: 0731-6755, 21.18001.AJCT.2021.V14I7.21.1306. (UGC Care Group II Journal)

Enhanced Disease Recognition and Classification in Black Gram Plant Leaves Using Deep Learning

K. Prasanth[1]([✉]), P. Kabilamani[2], G. Sangar[2], V. Kaliraj[1], and V. Rajasekar[2]

[1] SA Engineering College, Chennai, India
{prasanth,kallis}@saec.ac.in
[2] SRMIST Vadapalani, Chennai, India
{kabilamp,rajasekv2}@srmist.edu.in

Abstract. While urad bean is an important grain crop in several locations, it is frequently plagued by a number illness that has a devastating effect on harvest yields and quality. Using Convolutional Neural Net- works (CNNs) with 50 Layers, Local Binary Pattern and Support Vector Machines (SVMs), we offer a deep learning-based method for accurately recognising and categorising disorders of leaves in black gram plants. To begin, we amassed a database of pictures of black gram leaves affected by diseases like leaf blight, leaf spot, and yellow mosaic virus. We did some preliminary processing to clean up the photographs and make them look better. Then, we educated a convolutional neural network model to use to reduce the number of parameters and the LBP used to extract disease classification features from the plant images. Our model's accuracy was enhanced by employing SVM as a classifier on the CNN50 with LBP model's outputs features. We were able to decrease the amount of incorrect classifications by using the SVM model to boost classification precision. Using criteria like accuracy, precision, recall, and F1-score, we analysed how well our method performed. Our experiments demonstrate indicates the suggested method executes better than earlier state-of-the- art methods with a degree of 98.69%. In conclusion, we show that deep learning methods, specifically CNN and SVM, may be used to precisely detect and categorise black gram plant leaf diseases. This method may help producers better detect and combat plant illnesses, which would ultimately boost food production and quality.

Keywords: Convolutional Neural Networks · Support Vector Machines · Local Binary Pattern · Precision · Black gram plant leaf diseases · Accuracy · Fusion Model

1 Introduction

To diagnose a crop illness, one must first determine whether or not an illness is present in a crop and what kind of disease it is. As early detection and treatment of diseases can help prevent significant yield loss and the dissemination of disease to adjacent crops,

R. Geetha et al. (Eds.): AAIMB 2023, CCIS 2202, pp. 213–224, 2025.
https://doi.org/10.1007/978-3-031-73065-8_17

this is a vital component of the farming process. The following procedures can aid in the detection of agricultural diseases: Careful observation of the crop is the first stage in diagnosing a crop disease. Check the plant's leaves, stems, and fruit for signs of damage, discoloration, and anomalies. Take careful inventory of the side effect's geographical spread. Once you have a good look at the crop, you can try to pinpoint the exact signs of the disease. Leaves, stems, and fruit may wilt, become yellow or brown, or develop spots or various discoloration as symptoms. Lesions, cankers, and growths are other possible manifestations of whatever ails the plant. Check the symptoms you've noticed against a list of diseases we already know about. A plant scientist or agricultural expert can be consulted, as can crop disease identification guides or online tools. Once a potential disease has been identified, further testing can be done to confirm the diagnosis. As part of this process, it may be necessary to collect plant tissue samples for later investigation in a clinical setting. If a disease is discovered, it must be treated immediately to stop its spread and save as many crops as possible from destruction. Fungicides, insecticides, and other treatments advised by a crop specialist may be used for treatment. Overall, efficient crop disease control relies on prompt diagnosis and identification. Farmers may preserve their crops and reduce losses by being vigilant in spotting the early warning signs of crop diseases.

There are many illnesses that can affect the leaves of black gram vegetation, all of which have the potential to drastically lower the crop's output and quality. Black gram plants often fall victim to the following leaf diseases:

Leaf spot: The fungus Alternaria alternata causes leaf spot, which manifests as tiny, irregularly shaped brown or yellow dots on the leaves. These blemishes can range in color from brown to black, with a yellowish aura. The leaves may wither and fall off in extreme circumstances.

Powdery mildew: The fungal infection Erysiphe polygoni causes powdery mildew, which appears as a white powder on the leaves. Reduced efficiency of photosynthesis and hence decreased productivity may result from this disease.

Rust: Small, dark brown blisters appear on the undersides of leaves and are the tell-tale sign of rust, which is triggered by a mold called Uromyces vignae. It's also possible for the leaves to turn yellow and fall off early.

Anthracnose: Anthracnose: A fungus called anthracnose that develops depressed, round sores on the leaves. It is caused by the fungus Colletotrichum lindemuthianum. These spots could be black or brown and surrounded by a yellowish halo. Yellowing and eventual leaf drop may occur in extreme situations.

Good crop management measures, such as crop rotation, correct irrigation, and the adoption of resistant to illness cultivars, are essential for preventing and controlling these pests. Diseases can also be managed with the help of fungicides, but their usage must be managed carefully to prevent the evolution of resistant varieties.

2   Dataset

Datasets about black gram plant leaf disease (BPLD) are assemblages of information about the ailment that causes leaf discoloration and eventual death in black gram vegetation. Images, information, and comments are only some of the data kinds included in the databases. Techniques for computerised identification and evaluation of diseases using artificial intelligence for BPLD are predominantly trained and tested using these datasets (Table 1).

Table 1. Amount of photos are included in every illness group within the BPLD Dataset.

Type of leaf disease	No. of Pictures
Anthracnose	230
Healthy	221
Leaf Crinkle	152
Powdery Mildew	180
Yellow Mosaic	224
Total	1007

They can also be used to learn more about the disease's epidemiology, examine the environmental factors that play a role in its spread, and assess the success of current disease control practices. Photos of BPLD-affected black gram plants are included in this data set.

More than a thousand pictures are included in the dataset, which can be used for study. Researchers and agricultural professionals working on BPLD management can benefit greatly from these datasets. They can aid in the improvement of disease detection, the creation of better control plans, and the decrease of agricultural losses due to BPLD.

3   Neural Network Model

Convolutional Neural Networks (CNNs) are a classic example of deep neural network technology that can be taught to extract features from photographs. By examining data obtained from the Urad bean crop leaf illness (BPLD) dataset, a convolutional neural network (CNN) may learn to recognise and extract characteristics that distinguish healthy leaves from those that are ill. How to take features out of the BPLD dataset using a convolutional neural network (CNN).

Data Preparation: The first order of process is to get the BPLD dataset ready to be used for CNN50 training. Preprocessing the photos includes reducing them to a consistent size, standardizing the pixel standards, and using data enhancement methods to enhance the size of the training set, and then splitting the information into training, validation, and test sets (Fig. 1).

Building the CNN Model: CNN50 are known for capturing hierarchical structures in data. With increasing depth, the network extracts more abstract features, encompassing both low-level details and high-level semantic patterns. LBP captures local texture patterns and micro-structures in images. CNN with 50 convolutional layers and not using any pooling the spatial dimensions of the output will depend on the padding used in the convolutional operations. Using a 3×3 filter with "valid" padding will reduce each

(a) Anthracnose   (b) Healthy   (c) Leaf Crinkle

(d) Powdery Mildew   (e) Yellow Mosaic

Fig. 1. Classes of Disease images in BPLD Dataset

spatial dimension by 2 after each convolution. After 50 layers, the reduction in size would be $50 \times 2 = 100$. So, for an input of 256×256, the output dimensions would be $(256–100) \times (256–100) = 156 \times 156$. Combining a deep CNN50 with the Local Binary Pattern (LBP) can generates new features by combining CNN and LBP, can leverage both the abstracted high-level features and the localized texture patterns. Many real-world datasets contain complex variability, including texture, shape, and semantic patterns. The fusion of CNN50 and LBP can enhance the recognition capabilities by ensuring both texture and hierarchical patterns are considered. The tasks where either the texture or other intricate patterns play a crucial role, the fusion can provide more robust performance. Fusing handcrafted features (like LBP) with deep learned features (from CNNs) can offer a better generalization in scenarios with limited data. Handcrafted features can act as a regularizer, potentially preventing overfitting. Multiple methods can be used to fuse CNN and LBP features, such as simple concatenation, fusion at the feature level using methods like Canonical Correlation Analysis (CCA), or even at decision level where predictions based on both features are combined. This flexibility allows for tailored solutions depending on the task (Fig. 2).

Convolutional neural networks (CNN50) are an effective method for extracting features from the urad bean crop leaf disease dataset. Scientists can use this dataset for Deep learning to find pertinent characteristics that differentiate healthy leaves from diseased ones, which can aid in identifying and taking care of plant illnesses. This is accomplished by developing and teaching an CNN50 with LBP algorithm on the dataset.

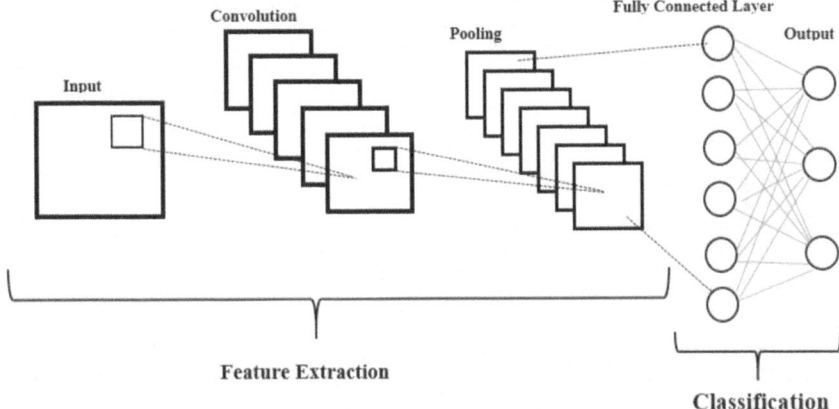

Fig. 2. Feature Extraction and Classification by using CNN

4 Support Vector Machine

Classification problems, such as determining whether or not an image contains Black gram plant leaf disease (BPLD), might benefit from the application of Support Vector Machines and other guided ML techniques. Here are the measures required to predict BPLD in the dataset using an SVM classifier:

Preparation of Data: In order to train the SVM classifier, the BPLD dataset must first be prepared. Preprocessing the photos entails scaling them to the same dimensions, normalizing the pixel values, and flattening them into a vector before the dataset is split into training and testing sets. In order to train the SVM classifier, it is necessary to extract features from the preprocessed images. Images can have features extracted from them with the use of feature extraction methods like Deep learning-based discovery of features using an already trained CNN50, Histogram of Oriented Gradients (HOG), and Local Binary Patterns (LBP). After feature extraction is complete, the SVM classifier can be trained on the extracted features and the training dataset. The classifier learns, through training, to distinguish between feature patterns that indicate either healthy or diseased leaves. The SVM method determines the optimal hyperplane for classifying data, increasing the gap between the two groups as much as possible.

The accuracy of the SVM classifier in predicting BPLD can be evaluated after using the test dataset. The machine learning algorithm determines whether a picture is healthy or unhealthy based on its characteristics vectors. Adjusting the SVM Classifier: The precision of the SVM classifier on the BPLD dataset can sometimes be enhanced by fine-tuning it. The regularization parameter (C) and the kernel's type are two examples of SVM parameters that can be tweaked to enhance the method's efficiency on the dataset. Black gram plant leaf disease detection in images is greatly aided by SVM classifiers. Using the BPLD dataset for training an SVM classifier, researchers can create a reliable model that can aid in the diagnosis and treatment of plant diseases by extracting features from the preprocessed images (Fig. 3).

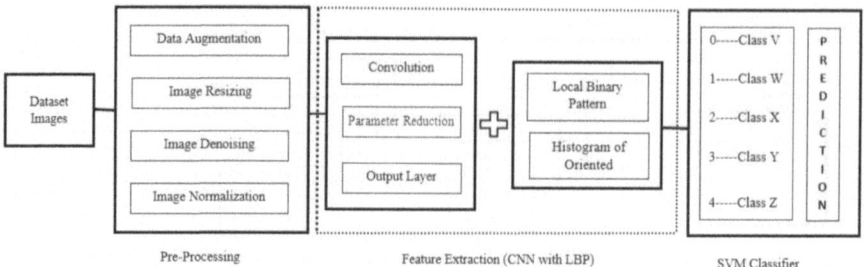

Fig. 3. Proposed Architecture Model

5 Study of Literature

Srinivas Talasila, Kirti Rawal, Gaurav Sethi, Sanjay MSS, Surya Prakash Reddy M [1] Black gram, or Vigna mungo in its scientific form, is more commonly known as Urad in India. The writer presents the Urad Bean Plant Leaf Disease dataset. The black gram crop has been severely impacted by illnesses such as a.anthracnose, b.leaf crinkle, c.powdery mildew, and d.yellow mosaic, costing farmers a lot of money. In recent years, a combination of image processing and computerizing algorithms has seen widespread use in applications aimed at diagnosing and classifying diseases that manifest themselves in plant leaves. A Urad Bean Plant Leaf Disease database was constructed and quickly explained in this paper in order to identify and group together plant leaf diseases that damage evaluating the beginning stages of a plant's health using visual computing techniques. A total of 1000 photos, representing 5 categories (4 disorders + 1 healthy class) are included in the dataset. Under actual cultivation fields in Nagayalanka, Krishna, Andhra Pradesh, cameras and smartphones gathered the photographs included in the offered dataset. Experts in agriculture helped sort and analyse the photographs after they were captured.M. Dhasarathan, S. Geetha, A. Karthikeyan, D. Sassikumar, N. Meenakshi- ganesan, [2] Using gamma rays to induce substantial interpopulation variation and the potential for selecting elite variants to improve output is supported by the findings of this investigation. Grain yield was found to be positively correlated with cluster density, pod density, and pod density per group. High-yielding blackgram variants were found thanks to G E evaluation, and gamma-ray annealing led to the discovery of operational mutants including G13, G7, and G34 that show promise in environments E1, E2, and E3, respectively. In order to get appropriate segregants for long-term blackgram growth through fusion breeding, it is necessary to conduct further evaluations of high-yielding mutants for specific settings in bigger plots. X.E. Pantazi, D. Moshou, A.A. Tamouridou [3] The approach shown here employs Local Binary Patterns (LBPs) for character extraction and Single Class Classifier for sorting, automating the process of diagnosing agricultural illnesses in pictures of leaf samples from various crops. A separate One Class Classifier is used in the proposed technique to discriminate between various stages of a crop's health, such as healthy, downy mildew, powdery mildew, and black rot. The algorithms based on grape skins were tried on a variety of crops with great results. When knowledge instances may be classified into many categories, a revolutionary method that resolves conflicts amongst One Class Classifiers produces the best accurate classification. The total yield rate for the 46 plant-condition combinations studied was 95%.Talasila, K. Rawal, G. Sethi, [4] Crop administration duties including recognizing species, finding diseases, and categorization all rely on the ability to determine leaf regions from photographs of plant leaves. In order to accomplish the procedure of separating a leafy area from its backdrop, several methods were created. While most of these techniques have been utilized on photos captured in controlled lab settings or with a uniform background, it is critical that they also be used on images captured in real time from cultivation fields, which often feature varied backdrops. For black gram leaves photos specifically, a good technique that mechanically divides leaf region from the complicated environment is still being created as of yet. N.K. Gautam, K. Kumar, M. Prasad [5] Severe puckering, rugosity, and creasing of leaflets caused by urdbean leaf crinkle disease (ULCD) causes substantial seasonal yield decreases in the world's leading urdbean-producing regions. The urdbean leaf crinkle virus (ULCV) is responsible for this illness.

When compared to other kinds of waves, urdbean (Vigna mungo L. Hepper) is especially vulnerable to leaf crinkle disease. The urdbean, a staple crop grown all over South and Southeast Asia, thrives in the dry and humid environments. The disease has been transmitted via aphids, insects, and whiteflies. Sap injection, transplantation, and seed can potentially spread the virus. Depending on genetic makeup, location, and period of infection, urdbean crops might lose anywhere from 35% to 81% of their seed production due to ULCD. The virus spread because of contaminated materials and warm temperatures. Diseased plants undergo structural and metabolic alterations. Reports of genetic variants in screened germoplasm highlight the need for ongoing screening of both existing and new germoplasm for the purpose of discovering novel features (genes) and potential new reservoirs of disease protection. Rarely do we hear about efforts to breed ULCD-resistant varieties and then release them to the public. Blackgram identification has mostly made use of RAPD (random amplified polymorphic DNA) and divide-simple sequence repeat (ISSR) genetic markers, relatively few studies on indicators having sequence-tagged micro sites (STMS). Many RNA viruses have evolved strategies to combat the mutagenesis process, often by accumulating amino acids known as suppressors. However, no information has been released for ULCV, despite it being the causal agent of urdbean leaf crinkle disease, that specifies which defense route functions for its susceptibility in the plants or the fact that the This virus employs the similar silenced suppression technique. Antiviral principles (AVP) identified in plant leaf samples from a range of species can inhibit numerous viruses from entering a host. There are a lot of substances that have been documented to prevent viral proliferation in plants. Increasing the height of the wall plants is also a useful strategy for preventing the propagation of the disease.

6   Experimental Model

6.1   Preprocessing

Due to the wide range of gadgets, the initial RGB photos were of varying proportions. The squarish aspect ratio of each image in the collection was then measured. Pictures that weren't originally squares were resized using software to produce a square version of every leaf area. When analysing was complete, all images with square dimensions larger than 512 were reduced down to 512.

6.2   Feature Extraction and Classification

Identification and categorization of features To correctly recognise and group together plant leaf diseases, feature mining and classification may be accomplished using machine learning or deep learning models. Obtaining features is a crucial step in machine learning, as it is the process by which the right data that defines each class is sought out. Gathering the most relevant features from raw photos is the goal of the extraction of features. A learning algorithm is used to perform state-of-the-art machine learning approaches for feature extraction avoiding the need for manual intervention from a domain expert. Computations are made for a number of characteristics of the images, for example, their association, entropy, deviation, uniformity the contrary, energy, and mean. The classifiers use these picture attributes to make associations between the input and the desired output. In deep learning, Convolutional Neural Networks (CNNs) carry out both feature extraction and identification. Since CNNs may gather features automatically, there is no need to hand-extract them. We built to demonstrate the use of neural network models

Fig. 4. Data augmented images

in identifying black gramme plant leaf diseases addressed in this paper, the algorithms were trained using verification and testing its level of efficacy on unknown information, and the dataset was increased to 15000 images after the preparation processes were given. We partitioned the data set by setting K = 5, creating five subsets. So, there are 3,000 pictures of various diseases on each fold. All of the adapted models were trained using the same set of hyperparameters, including an SVM solver, an initial learning rate of 0.001, a small batch size of 32, and 30 training iterations. Results from studies with validation were given as mean values in Table (Fig. 4).

6.3 Evaluation Metrics

Accuracy: The ratio of correctly classified samples to the total no. of samples in the dataset. It is given by:

$$Accuracy = (T.P + T.N)/(T.P + F.P + T.N + F.N) \tag{1}$$

where TP stands for true positives, TN stands for actual negatives, FP stands for false positives, and FN stands for false negatives. Precision is defined as the ratio of genuine wins to the entire amount of favourable samples expected. It is provided by:

$$Precision = T.P/(T.P + F.P) \tag{2}$$

Recall: The ratio of true positives to the total number of actual positive samples. It is given by:

$$Recall = T.P/(T.P + F.N) \tag{3}$$

F1-Score: The harmonic mean of precision and recall. It is given by:

$$F1 - Score = 2 * (Precision * Recall)/(Precision + Recall) \tag{4}$$

Based on the results of the study, the support vector machine (SVM) depth performed better than the other trained models, with an average classification ac- curacy of 98.69%, consistency of 98.65%, recalls of 98.58%, F1_Score of 98.60%, and accuracy of 98.69%.

6.4 Result Discussion

The proposed method was tested experimentally on a more extensive dataset. Table 2 displays the experimental outcomes. In the data table, the higher the obtained value for every assessed measure is shown in dark text. Higher values represent more favorable outcomes. Almost every metric we looked at showed that the proposed approach was superior. Therefore, when contrasted to rival approaches, our proposed approach significantly improved accuracy.

Table 2. Performance Comparison.

Model	Precision	Recall	F1-Score	Accuracy
Automated CNN	95.52%	95.45%	95.50%	95.60%
CNN Based	96.35%	96.30%	96.32%	96.37%
Deep CNN	95.28%	95.22%	95.27%	95.29%
DCNN vs Transfer learning	97.76%	98.02%	97.89%	97.89%
Transfer Learning	98.30%	98.10%	98.20%	98.20%
Proposed Model	98.65%	98.58%	98.60%	98.69%

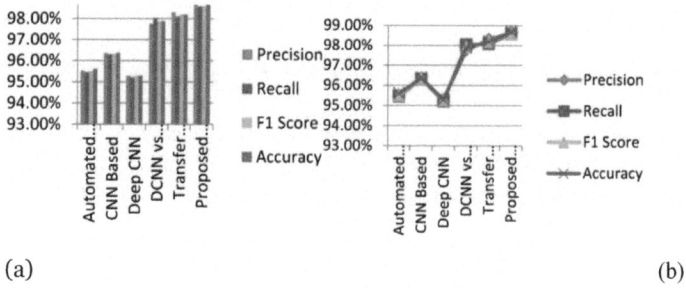

(a)   (b)

Fig. 5. Result Analysis by using bar and line chart

7   Conclusion

In this research, we describe a deep-learning-based method that is both effective and harmless for the detection of viruses in the Vigna Mungo plant. The suggested approach estimates a plant's leaf condition and divides leaves into fit, mildly contracted, and badly affected groups. Infected Vigna Mungo crop fields were used to build CNN50 model for to reduce the parameters and To determine the characteristics of Leaf pictures, the Local Binary Pattern (LBP) was applied. We reduced the amount of misclassifications by using the SVM model to increase the precision of classification. We assessed the effectiveness of our method using common measures such as accuracy, precision, recall, and F1-score. Our results show that the proposed strategy has a higher success rate (98.69%) than the prior state-of-the-art methods. Here, we demonstrate how CNN50 with LBP and SVM, two deep learning techniques, may be used to accurately identify and categorize leaf illnesses in black gram plants. Deep learning models will increasingly be considered as black boxes in the future, with an increased emphasis on models which are comprehensible and explainable. Understanding why a model produced a certain forecast might be critical for faith and future actions in agricultural (Fig. 5).

References

1. Talasila, S., Rawal, K., Sethi, G., MSS, S., M.S.P.: Black Gram Plant Leaf Disease (BPLD) dataset for recognition and classification of diseases using computer-vision algorithms. Data Brief **45**, 108725 (2022). https://doi.org/10.1016/j.dib.2022.108725
2. Dhasarathan, M., Geetha, S., Karthikeyan, A., Sassikumar, D., Meenakshi Ganesan, N.: Development of novel blackgram (Vigna Mungo (L.) hepper) mutants and deciphering genotype × environment interaction for yield-related traits of mutants. Agronomy **11**(7), 1287 (2021). https://doi.org/10.3390/agronomy11071287
3. Pantazi, X.E., Moshou, D., Tamouridou, A.A.: Automated Leaf disease detection in different crop species through image features analysis and one class classifiers. Comput. Electron. Agri. **156**, 96–104 (2019). https://doi.org/10.1016/j.compag.2018.11.005
4. Talasila, S., Rawal, K., Sethi, G.: PLRSNet: a semantic segmentation network for segmenting plant leaf region under complex back-ground. Int. J. Intell. Unmanned Syst. **11**(1), 132–150 (2021). https://doi.org/10.1108/ijius-08-2021-0100
5. Gautam, N,K., Kumar, K., Prasad, M.: Leaf crinkle disease in Urdbean (Vigna Mungo L. Hepper): an overview on causal agent, vector and host. Protoplasma **253**(3), 729–746 (2016). https://doi.org/10.1007/s00709-015-0933-z
6. Vishalakshi, B., et al.: RAPD assisted selection of Black Gram (Vigna Mungo L. Hepper) towards the development of multiple disease resistant germplasm. Biotech **7**(1) (2017). https://doi.org/10.1007/s13205-016-0582-8
7. Joshi, R.C., Kaushik, M., Dutta, M.K., Srivastava, A., Choudhary, N.: VirLeafNet: automatic analysis and viral disease diagnosis using deep-learning in Vigna Mungo Plant. Ecol. Inform. **61**, 101197 (2021). https://doi.org/10.1016/j.ecoinf.2020.101197
8. Özacar, T., Öztürk, Ö., Güngör Savaş, N.: Hermos: an annotated image dataset for visual detection of grape leaf diseases. J. Inf. Sci. 016555152210918 (2022). https://doi.org/10.1177/01655515221091892
9. Alessandrini, M., Calero Fuentes Rivera, R., Falaschetti, L., Pau, D., Tomaselli, V., Turchetti, C.: A grapevine leaves dataset for early detection and classification of Esca Disease in vineyards through machine learning. Data Brief **35**, 106809 (2021). https://doi.org/10.1016/j.dib.2021.106809
10. Jepkoech, J., Mugo, D.M., Kenduiywo, B.K., Too, E.C.: Arabica coffee leaf images dataset for coffee leaf disease detection and classification. Data Brief **36**, 107142 (2021). https://doi.org/10.1016/j.dib.2021.107142
11. Talasila, S., Rawal, K., Sethi, G.: Conventional data augmentation techniques for plant disease detection and classification systems. Intell. Syst. Sustain. Comput. 279–287 (2022). https://doi.org/10.1007/978-981-19-0011-2_26
12. Amanullah, Hatam, M.: Yield potential of Blackbean (Vigna Mungo (L.) hepper) germplasm. Pakistan J. Biol. Sci. 3(10), 1571–1573 (2000). https://doi.org/10.3923/pjbs.2000.1571.1573
13. Ghafoor, A., Sharif, A., Ahmad, Z., Zahid, M.A., Rabbani, M.A.: Genetic diversity in black-gram (Vigna Mungo L. Hepper). Field Crops Res. **69**(2), 183–190 (2001). https://doi.org/10.1016/s0378-4290(00)00141-6
14. Knauer, U., Matros, A., Petrovic, T., Zanker, T., Scott, E.S., Seiffert, U.: Improved classification accuracy of powdery mildew infection levels of wine grapes by spatial-spectral analysis of hyperspectral images. Plant Methods **13**(1) (2017). https://doi.org/10.1186/s13007-017-0198-y
15. KC, K., Yin, Z., Wu, M., Wu, Z.: Depthwise separable convolution architectures for plant disease classification. Comput. Electron. Agri. **165**, 104948 (2019). https://doi.org/10.1016/j.compag.2019.104948

16. Arsenovic, M., Karanovic, M., Sladojevic, S., Anderla, A., Stefanovic, D.: Solving current limitations of deep learning based approaches for plant disease detection. Symmetry **11**(7), 939 (2019). https://doi.org/10.3390/sym11070939

17. Vamsee Kongara, R.K., Siva Charan Somasila, V., Revanth, N., Polagani, R.D.: Classification and comparison study of rice plant diseases using pretrained CNN Models. In: International Conference on Inventive Computation Technologies (ICICT) (2021) https://doi.org/10.1109/icict54344.2022.9850784

18. Li, L., Zhang, S., Wang, B.: Plant disease detection and classification by deep learning—a review. IEEE Access **9**, 56683–56698 (2021). https://doi.org/10.1109/access.2021.3069646

19. Dai, Q., Cheng, X., Qiao, Y., Zhang, Y.: Crop leaf disease image super-resolution and identification with dual attention and topology fusion generative Adversarial Network. IEEE Access **8**, 55724–55735 (2021). https://doi.org/10.1109/access.2020.2982055

20. Pantazi, X.E., Moshou, D, Tamouridou, A.A.: Automated Leaf disease detection in different crop species through image features analysis and one class classifiers. Comput. Electron. Agri. **156**, 96–104 (20219). https://doi.org/10.1016/j.compag.2018.11.005

21. Daub, M.E., Hangarter, R.P.: Light-induced production of singlet oxygen and superoxide by the fungal toxin, cercosporin. Plant Physiol. **73**(3), 855–857 (1983). https://doi.org/10.1104/pp.73.3.855

22. Sanida, T., Sanida, M.V., Sideris, A., Dasygenis, M.: A lightweight CNN model for tomato crop diseases on heterogeneous embedded system. In: 12th International Conference on Modern Circuits and Systems Technologies (MO- CAST) (2023). https://doi.org/10.1109/mocast57943.2023.10176582

23. Ramachandra, A.C., Rajesh, N., Aishwarya, K., Madugonda, N., Akash, R.A., Namratha, G.: Classification of crop diseases through remote-sensed data using multi-class SVM. In: IEEE 2nd Mysore Sub Section International Conference (MysuruCon) (2022). https://doi.org/10.1109/mysurucon55714.2022.9972622

24. Ahmed, I., Habib, G., Yadav, P.K.: An approach to identify and classify agricultural crop diseases using machine learning and Deep Learning Techniques. In: International Conference on Emerging Smart Computing and Informatics (ESCI) (2023). https://doi.org/10.1109/esci56872.2023.10099552

25. Tiwari, S., Kumar, S., Tyagi, S., Poonia, M.: Crop recommendation using machine learning and plant disease identification using CNN and Transfer-Learning Approach. In: IEEE Conference on Interdisciplinary Approaches in Technology and Management for Social Innovation (IATMSI) (2022). https://doi.org/10.1109/iatmsi56455.2022.10119276

Ensemble Deep Learning Approach for Identification of DDOS Attack

C. Balakrishnan[(⊠)] and V. S. Prassana kumar

Department of Computer Science and Engineering, S.A.Engineering College, Chennai, India
balakrishnan@saec.ac.in

Abstract. Numerous IoT web apps have been hindered by multiple safety hazards and network intrusions as a result of the home control dataset's on going enhancement. Automating your home has always been centred on safety and the recognition of DDOS Threats. It is essentially possible to make use of numerous IoT web-based resources through input of a set of data for the home's automation or by tapping on a hyperlink in the application itself. The cloud manages the aforementioned difficulties in the concept of the Edge of Things. An attacker can create various web strikes. By embedding executable instructions or injecting illicit software inside a DDOS threat, many simultaneous intricate models are employed to improve system reliability and the ease of maintaining data disclosure. By accurately recognising malicious automated homes information set, it is crucial to enhance the steadfast reliability and integrity of IOT web apps. The site of harmful home automation information set detection according to character-centered material categorization emphasises will be investigated using an artificial learning algorithm system in this work. The fact that hazardous buzz phrases are exceptional home automation information set, it is possible to divide the equipment in an ordinary intelligent home setting into four classes: Class 1 is very high traffic reliability, Class 2 is high traffic certainty, Class 3 is medium traffic accuracy, and Class 4 is low traffic stability. According on the test outcomes, our suggested neural network identification algorithm is actually appropriate for high-accuracy tasks such as categorization. When compared to different categorization designs, the model's accuracy rate is over %. Considerable speculative and empirical advantages for web safety studies can be derived from the use of deep learning to aggregate home automation information to identify Web visitor's goals, which opens up novel possibilities for clever safety studies.

Keywords: DDOS · we attack · Malware detection · Machine learning · deep learning

1 Introduction

Internet safety experts have been looking at new protection systems that utilise deep learning since attacks via the internet constantly increase in complexity and variation. The emergence of deep learning offers unique answers to safety concerns in these types of systems, but classic online attack identification methods exhibit vulnerabilities in big

R. Geetha et al. (Eds.): AAIMB 2023, CCIS 2202, pp. 225–233, 2025.
https://doi.org/10.1007/978-3-031-73065-8_18

data situations. Applications for deep learning that use big data analysis to evaluate large amounts of internet flow demonstrate a better ability to identify hostility. Security for IoT networks has improved and been made easier to create due to these deep learning techniques. In this study, we suggest an entirely new WAD targeting Internet of Things (IoT) networks that is based on collective deep learning. To evaluate URL responses in internet traffic and spot abnormal queries with web threat packages included, the suggested EDL-WADS focuses on the use of deep learning algorithms. Three models based on deep learning are used in our strategy to uncover the related aspects that are concealed in the questions.

- The system under consideration offers common vulnerability architecture for various code injection-based app attacks.
- In order to develop defensive strategies, we classify and evaluate suggested countermeasures by an established list of parameters.
- We discover a technique to depict all DDOS Attack types. We quickly substitute each word or character with ones from vocabularies that have been already established in the context of various attacks.
- Developed novel artificial neural network architecture for network-based information for intrusion detection that utilises gated recurrent modules and support vector machines (SVM).
- In response to choices made regarding prejudice, measures are planned. If a DDOS Attack is identified as normal, operations will proceed normally.

Dataset collection:

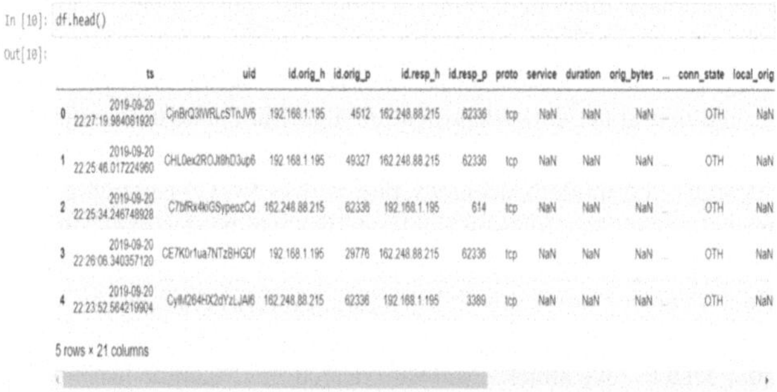

```
In [10]: df.head()
Out[10]:
```

	ts	uid	id.orig_h	id.orig_p	id.resp_h	id.resp_p	proto	service	duration	orig_bytes	...	conn_state	local_orig
0	2019-09-20 22:27:19.984081920	CjnBrQ3tWRLcSTnJV6	192.168.1.195	4512	162.248.88.215	62336	tcp	NaN	NaN	NaN	...	OTH	NaN
1	2019-09-20 22:25:46.017224960	CHL0ex2ROJt8hD3up6	192.168.1.195	49327	162.248.88.215	62336	tcp	NaN	NaN	NaN	...	OTH	NaN
2	2019-09-20 22:25:34.246748928	C7bfRx4kiGSypeozCd	162.248.88.215	62336	192.168.1.195	614	tcp	NaN	NaN	NaN	...	OTH	NaN
3	2019-09-20 22:26:06.340357120	CE7K0r1ua7NTzBHGDf	192.168.1.195	29776	162.248.88.215	62336	tcp	NaN	NaN	NaN	...	OTH	NaN
4	2019-09-20 22:23:52.564219904	CyIM264HX2dYzLJAi6	162.248.88.215	62336	192.168.1.195	3389	tcp	NaN	NaN	NaN	...	OTH	NaN

5 rows × 21 columns

Fig. 1. Dataset under test

Fig. 1 shows the sample view of collected dataset on smart city under test. It holds various activity in the web portals as recorded data.

The remainder of the essay is structured as a thorough literature review in the second portion. In the third part, the choice of system tools and issue identifications are covered. In the fourth chapter, the system architecture and specific system design stages are covered. Potential improvement is discussed as the end result of the paper.

2   Background Study

Z.Tian et al. (2020): The technology is implemented on the edge computers and is intended to identify attacks through the web. In the context of the Edge of Things (EoT) concept, the cloud addresses the aforementioned issues. To increase system reliability and upgrade simplicity, several parallel deep models are deployed. Utilising various datasets, we ran trials on the system using two parallel deep models and contrasted it with alternatives. The testing results show that the system is effective in identifying online attacks with a precision rating of 99.410%, TPR at 98.91%, and DRN about 99.55%.

The technology uses edge machine URLs to identify web threats using decentralised deep learning. By voluntarily acquiring attributes, the system can differentiate between unusual and typical queries. Several simultaneous algorithms are used in our system to improve the reliability of the recognising mechanism.

M.H.Amouei et al. (2021): A complete payload set of malware instances are routinely analysed by RAT to identify characteristics and group together instances that share those tendencies. In order to effectively explore the clusters and find nearly all circumventing assault structures, it then applies an evolutionary reinforcement learning approach in conjunction with an innovative adaptive search method. Our tests demonstrate that RAT beats modern techniques in the real world when it comes to locating attack payloads that avoid defences in a sizable payload area.

In order to simulate complex assault behaviours and improve efficiency, we use the n-gram feature acquisition technique. We analyse the clustering's outcomes and provide a technique for combining comparable threat payloads. In order to improve the effectiveness of our black box testing strategy, we employ the -greedy method and an innovative adaptive search methodology.

Researchers in artificial intelligence (AI) frequently use AI to offer more effective approaches for a variety of issues, particularly safety evaluation. Because of its instability and the local minima issue, the decision tree algorithm is less effective; the issue must be resolved.

R.J. Olanrewaju et al. (2021): With the help of an analyser and the authorization operation, the system evaluates these information and creates an account profile for the user. According to the research's findings, the system that was suggested outperforms alternative forms of authentication for a web application that is already in use. A comparative study of the suggested approach and conventional methods of authentication for high-end internet-based applications reveals that it offers roughly a 10% latency decrease, 7% quicker reaction time, and 11% reduced usage of memory.

In accordance with a description of the user's conduct, an algorithm is presented that generates registration alternatives, enhancing the user interface. In order to model frictionless and secure user authentication (FSUA) in online applications, a statistical approach is used. After a statistical component that lets visitors choose a verification technique, a logic-based activity monitoring is built.

This demonstrates conclusively that safety issues relating to internet-based apps remain a persistent issue. In their work, Chang and Choi's authors talked about control over access and authentication for users while also attempting to address a critical issue and its accompanying study obstacles.

D. Cole et al. (2021): An intruder might quickly mimic the target in the current face authentication platforms using an innovative data intoxication technique which doesn't need any understanding of the server level but rather requires a few fraudulent picture uploads. Then, using deep learning methods for automatically recognising these kinds of assaults, we suggest an innovative method of defence termed DEFEAT. The results of our thorough studies, which were done on actual information sets, demonstrate that our defensive technique yields over ninety per cent precision for detection.

We investigate a novel technique for data poisoning approach on face identification that enables the perpetrator to quickly assume the true nature of the target. To stop the aforementioned imitation threat, we suggest new identifiers. Our tests using genuine data sets show that our detector delivers extremely precise detection in a variety of situations.

There is a possibility that software for managing passwords might be able to solve the problem of the password inflation issue. The man-in-the-middle attack, that continues to be an important safety concern on numerous home routers, might be used to carry out this attack.

3 System Design

The application of letters as text categorization characteristics in a technique based on machine learning for the identification of distributed denial of service (DDOS) attacks. In this section, we examine different online code injection attack defence strategies. We provide an attack utilising code injection victimisation paradigm that shows how machine learning principles are used for the majority of the processes required to carry out various code injection attempts.

Both the server and the client endpoint may be involved in that avoidance of attacks. Usually, a proxy is positioned in the path of the server in order to review its replies shortly before they are sent to a dataset for home automation. The answers are examined for possible hacking using an altered browser or an application which has been properly fetched from the server. On the contrary, there exists a concern that such systems employing only one model have the potential to be utilised by hackers because just one discriminative framework had been selected to make categorization in the prior studies. The majority of methods, yet, neglected to consider how to upgrade the systems they had. In the actual world, with no revisions, systems might not be able to identify emerging threats.

The proposed approach considers Fast text model, Long short-term memory model, and multiple regression network (MRN) model. The hybrid approach considers the benefit of one or more machine learning models considered together (Fig. 2).

4   Methodology

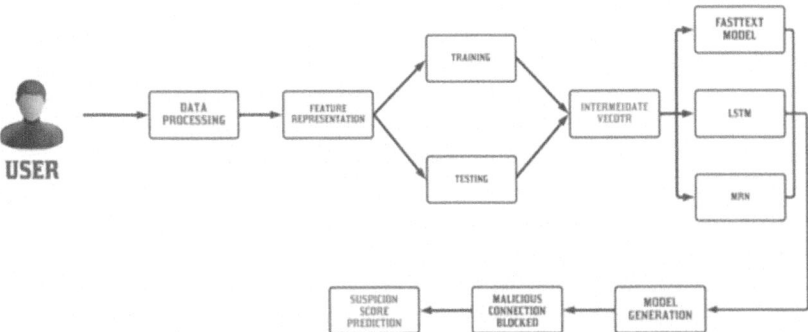

Fig. 2. System architecture

Particularly now that people are beginning to shift their apps as well as private information to the cloud, web applications are crucial to everyday life for individuals. Web apps are widely used and retain a lot of sensitive user data, which makes them ideal targets for attackers. Therefore, it is essential to safeguard online applications from invasions. The majority of all vulnerabilities fall under the DDOS category, which can be attacked through the delivery of carefully crafted queries. The first and third primary online application safety issues, respectively, are mentioned as injections, such as SQL injection and cross-site scripting. As indicated above, there are often two methods to identify attacks. In the first, requests are examined to identify particular types of attacks using a signature-based approach. In the second, requests are examined using an anomaly-based approach, which involves creating standard request characteristics in order to distinguish between regular and unusual ones. Because the signature-based method typically has reduced instances of false alarm and delivers a greater degree of precision, it is more commonly utilised than the anomaly-based methodology.

A Web Application Firewall (WAF) has an enormous amount of criteria that may identify SQL Injection, Cross-Site Scripting. The rule-based approach continues to be troublesome despite how useful it is. First of all, it cannot detect assaults that are not included in its signature dataset because it can only detect attacks that fall within the scope of the criteria set. Furthermore, by simply changing the phrases used in already unwanted requests or repeatedly encoding them, WAF can be readily defeated. Thirdly, completing template comparing requires an immense amount of computational power when dealing with exceptionally big threat pattern sets or responses with extended durations.

Feature Normalization Module

Basic characteristics from internet traffic entering the network within the organization where the secured servers are located are collected in this section and utilized to create traffic logs for a predetermined amount of time. By focusing exclusively on pertinent incoming traffic, tracking and evaluating at the final point in the network minimize the overhead of identifying harmful actions. Additionally, since the genuine traffic profiles utilized by the sensors are created for a fewer amount of internet solutions; this permits the sensor to offer security that is most appropriate for the intended network's internals.

Multivariate Correlation Analysis

The "Triangle Area Map Generation" module is used in this multidimensional correlation study for obtaining connections among two separate characteristics inside every traffic document from the initial phase or the traffic record that has been adjusted by the "Feature Normalization" module in this phase. These correlations vary when network breaches happen, and these alterations can be utilized as markers to spot the invasive movement. Triangle areas kept in Triangle Area Maps (TAMs), which contain all the derived associations, are then utilized to convey traffic data instead of the fundamental characteristics that were initially there or the adjusted data.

Decision Making Module

The process of making choices in this section uses an anomaly-based identification technique. Despite the need for attack-specific expertise, it makes it easier to identify any DoS attempts. Additionally, in this instance of misuse-based identification, the costly threat assessment and constant revision of the attack profile records are averted. Since hackers must create attacks that fit the typical traffic patterns created by a particular identification approach, the technique increases the resilience of the suggested sensors thereby rendering it more difficult for them to be avoided.

MRN algorithm:

1. Initialise the dataset and assign equal weight to each of the data point.
2. Provide this as input to the model and identify the wrongly classified data points.
3. Increase the weight of the wrongly classified data points.
4. if (got required results)

   Goto step 5
   else
   Goto step 2

5. End

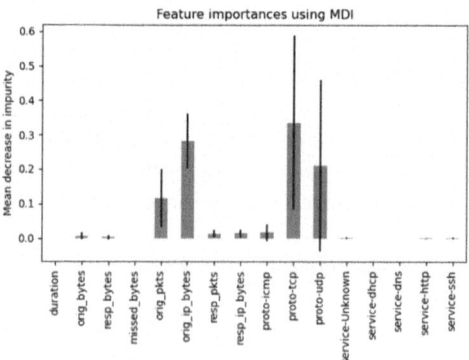

Fig. 3. Feature extraction

5   Results and Discussions

Fig. 3 shows the feature extraction using MDI

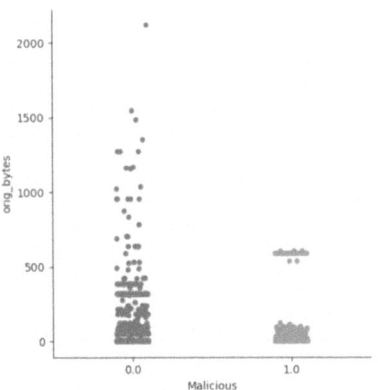

Fig. 4. Detection of malicious activity

Fig. 4 shows the regression plot on malicious activity

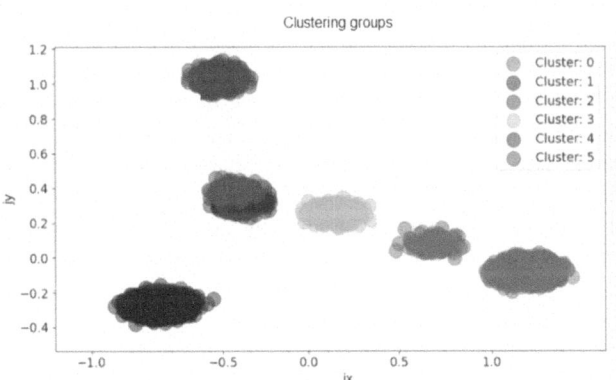

Fig. 5. Clustering groups

Fig. 5 shows the clustering groups associated within the dataset.

Fig. 6 shows the detection of malicious node in the network, Cyber physical systems are highly impacted by the external attacks. The novel system extract the unique attributes using multiple modality of feature extraction technique. The challenges in the unstructured dataset need to be overcome with the hyper-parameter tuning procedures. The hyper-parameters are optimized with respect to the given process. in future ensemble approach need to be adopted.

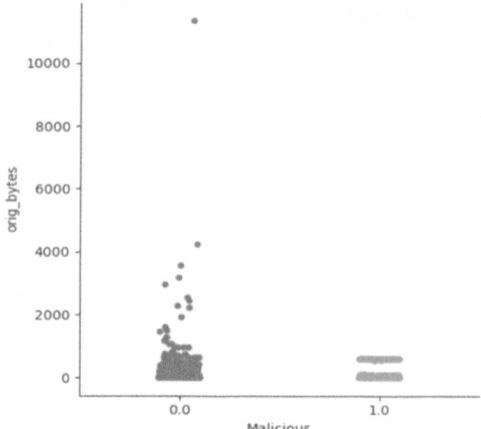

Fig. 6. Detection of Malicious node

6 Conclusion

In order for a security tool to be used over time, it must provide the user with some sort of benefit. The value should, in particular, be greater than the cost of its use. The cost may be incurred due to the time needed to use the tool, any inconvenience it may create, any erroneous alerts that might develop, etc., even though it isn't truly a financial expense. Poor testing, too much overhead, the lack of readily accessible designs, transmitting challenges, and compromised protection are some of the problems we have been exploring here. The value of an exploration fill in as a useful tool would increase if any one of these viewpoints had been enhanced, and so would the fill in's value for the actual research. The evaluation of alternative approaches would be aided by accurate identification specification. Unrealistic goals and key interest activity elsewhere can be discovered by broad implementation predictions. The ability of obtaining the actual code improves fundamental logical tasks like affirmation and copying. Testing becomes simpler when things are laid out simply. The foundation for developing procedures with greater participation can be defined by safe methods.

References

1. Tekerek Bay, O.: Design and implementation of an artificial intelligence-based web application firewall model. Neural Netw. World **29**(4), 189–206 (2019)
2. Vartouni, A.M., Teshnehlab, M., Kashi, S.S.: Lever-aging deep neural networks for anomaly-based web application firewall. IET Inf. Secur. **13**(4), 352–361 (2019)
3. Hao, S., Long, J., Yang, Y.: Bl-ids: Detecting webattacks using bi-lstm model based on deep learning. In: International Conference on Security and Privacy in New Computing Environments. Springer (2019)
4. Zhang, L., Zhang, D., Wang, C., Zhao, J., Zhang, Z.: Art4sqli : the art of sql injection vulnerability discovery. IEEE Trans. Reliab. **68**(4), 1470–1489 (2019)
5. Lv, C., Zhang, L., Zeng, F., Zhang, J.: Adaptive random testing for xss vulnerability. In: 26th Asia-Pacific Software Engineering Conference (APSEC) (2019), IEEE., pp. 63–69 (2019)

6. Osborne, C.: Backdoor malware is being spread through fake security certificate alerts. ZDNet: Security (2020)
7. Cimpanu, C.: Let's Encrypt to revoke 3 million certificates on March 4 due to software bug. ZDNet: Security (2020)
8. Yu, Z., Xue, D., Fan, J., Guo, C.: DNSTSM: DNS cache resources trusted sharing model based on consortium Blockchain. IEEE Access **8**, 13640–13650 (2020)
9. Lashkari, B., Musilek, P.: A comprehensive review of Blockchain consensus mechanisms. IEEE Access. **9**, 43620–43652 (2021)
10. Mlitz, K.: CIO COVID survey current and future trends in remote workworldwide from 2020 to 2021. Statista (2021). Accessed on: (2022)

ROCLT: Enhanced Text Classifier for Sentiment on Imbalanced Multiclass Tweet Data Using Hybrid Deep Learning Techniques

M. Rameshraja and J. Arunadevi[✉]

PG and Research Department of Computer Science, Raja Doraisingam Government Arts College, Sivaganga, India
arunardm2015@gmail.com

Abstract. In this study, we propose a text classification model that addresses the challenge of imbalanced text data using a combination of random oversampling, Convolutional Neural Network (CNN) with Long Short-Term Memory (LSTM), and a self-attention mechanism using Transformer. The model aims to achieve high accuracy in classifying imbalanced text data by leveraging various techniques. To handle the class imbalance issue, random oversampling is applied to generate synthetic samples from the minority class. This technique balances the class distribution and ensures that the model learns from both majority and minority classes effectively. The model architecture combines CNN and LSTM layers to capture local and global dependencies in the text. The CNN layer applies filters of different sizes to extract local features and detect patterns at various scales. The LSTM layer captures long-term dependencies and sequential information in the text, enabling the model to understand the context and structure of the input. Furthermore, the model incorporates a self- attention mechanism using the Transformer architecture. This mechanism allows the model to focus on important words or phrases within the text, enhancing its ability to recognize crucial information and make accurate predictions. Experimental results on a benchmark dataset demonstrate the effectiveness of the proposed model. It achieves an impressive accuracy of 98.08%, outperforming previous approaches for imbalanced text classification. The model's high accuracy can be attributed to the combination of random oversampling, CNN-LSTM layers, and the self-attention mechanism, which collectively address the challenges posed by imbalanced text data.

Keywords: Imbalanced Text classification · Random Over sampling · CNN · LSTM · BiLSTM · Transformers · etc,.

1 Introduction

In the field of natural language processing, text classification plays a crucial role in various applications such as sentiment analysis, spam detection, and fake news identification. However, one common challenge in text classification tasks is dealing with imbalanced datasets, where the number of instances in different classes is significantly unequal.

R. Geetha et al. (Eds.): AAIMB 2023, CCIS 2202, pp. 234–245, 2025.
https://doi.org/10.1007/978-3-031-73065-8_19

Imbalanced datasets can lead to biased models that perform poorly on minority classes. To address this challenge, we have developed a text classifier that combines various techniques to handle imbalanced text data. The model incorporates random oversampling, Convolutional Neural Network (CNN) with Long Short-Term Memory (LSTM), and a self-attention mechanism using Transformer.

Firstly, random oversampling is employed to address the class imbalance issue. Random oversampling generates synthetic samples by randomly replicating instances from the minority class, thereby balancing the distribution of classes in the dataset. This technique helps the model to learn from the minority class and improve its ability to make accurate predictions.

Secondly, the model utilizes a combination of CNN and LSTM layers to capture both local and global dependencies in the text. The CNN layer extracts local features by applying filters of different sizes to capture important patterns at different scales. On the other hand, the LSTM layer captures long-term dependencies and sequential information in the text. This combination allows the model to effectively learn meaningful representations from the input text.

Lastly, the model incorporates a self-attention mechanism using the Transformer architecture. Self-attention mechanisms enable the model to focus on important words or phrases within the text, allowing it to understand the context and importance of different parts of the input. This attention mechanism enhances the model's ability to recognize crucial information and make accurate predictions, especially in complex and lengthy texts.

By combining these techniques, our text classifier aims to address the challenge of imbalanced text data and improve the performance of the model on minority classes. The random oversampling technique helps to balance the dataset, while the CNN-LSTM layers and self-attention mechanism enable the model to capture both local and global dependencies, as well as important contextual information.

Through this approach, we aim to build a robust and accurate text classifier that can handle imbalanced text data effectively, making it applicable in real-world scenarios where imbalanced datasets are common.

2 Background Study

Imbalanced text data is a common challenge in natural language processing (NLP) tasks, including text classification. Imbalanced text data refers to datasets where the number of instances in different classes is significantly unequal. This issue is prevalent in various real-world applications, such as sentiment analysis, spam detection, and fake news identification. The class imbalance problem can lead to biased models that favor the majority class and perform poorly on minority classes.

To address imbalanced text data, researchers have explored different approaches. One common technique is data-level approaches, which aim to rebalance the class distribution by oversampling the minority class, undersampling the majority class, or generating synthetic samples. Another approach is algorithm-level methods, which modify the learning algorithm to handle the class imbalance, such as cost-sensitive learning or ensemble methods.

Random oversampling is a data-level technique commonly used to address the imbalanced class distribution. This technique involves randomly replicating instances from the minority class to increase its representation in the dataset. By increasing the number of minority class samples, random oversampling aims to balance the class distribution, allowing the model to learn from both classes effectively.

Several studies have explored the effectiveness of random oversampling in imbalanced text classification. Chawla et al. [1] found that random oversampling can improve the performance of classifiers by providing more training samples for the minority class. They observed that random oversampling reduced the bias towards the majority class and improved the classification accuracy for imbalanced datasets.

Japkowicz and Stephen [2] highlighted the importance of random oversampling in handling imbalanced text data. They emphasized that random oversampling can help address the skewed class distribution and improve the performance of classifiers. The authors also mentioned the potential issues of overfitting and the need for careful evaluation when using random oversampling.

Convolutional Neural Networks (CNNs) have been widely used in the field of text classification due to their ability to capture local patterns and hierarchies in textual data. In this section, we review relevant literature on the application of CNNs for text classification.

Kim [3] introduced one of the seminal works on using CNNs for text classification. The author proposed a simple yet effective CNN architecture that operates on top of word embeddings. The model uses multiple parallel convolutional layers with different filter sizes to capture local features at various scales.

Max- pooling is then applied to extract the most salient features from each feature map. This approach achieved competitive performance on various text classification tasks, demonstrating the effectiveness of CNNs for text classification.

Zhang et al. [4] extended the application of CNNs by incorporating dynamic k- max pooling, which allows the model to capture multiple high-level features. The authors demonstrated the effectiveness of their model on sentiment analysis and topic categorization tasks. The dynamic pooling mechanism provided better modeling of variable-length texts and improved the discriminative power of the model.

Similarly, Gehring et al. [5] proposed a CNN-based model for text classification that utilized dilated convolutions. Dilated convolutions enable the model to have a larger receptive field without increasing the number of parameters, facilitating the extraction of contextual information from the text. The authors showed that their model achieved state-of-the-art performance on several benchmark datasets.

Apart from traditional CNN architectures, some studies have explored the combination of CNNs with other neural network layers. For instance, Johnson and Zhang [6] proposed a model that combined CNNs with Recurrent Neural Networks (RNNs) to capture both local and global dependencies in text. The CNN layer captures local features, while the RNN layer captures sequential information. This hybrid architecture achieved improved performance on text classification tasks.

Moreover, researchers have investigated the use of pre-trained word embeddings in CNN-based text classification models. Howard and Ruder [7] introduced Universal Language Model Fine-tuning (ULMFiT), which utilized pre-trained language models and transfer learning to improve text classification performance. The authors fine- tuned the pre-trained model using CNNs and achieved state-of-the-art results on various text classification tasks.

CNNs have demonstrated their effectiveness in text classification tasks by capturing local patterns and hierarchies in textual data.

Long Short-Term Memory (LSTM) networks have gained significant popularity in text classification tasks due to their ability to capture sequential dependencies and long-term context in textual data. In this section, we review relevant literature on the application of LSTMs for text classification.

Hochreiter and Schmidhuber [8] introduced LSTM as a specialized recurrent neural network (RNN) architecture that addresses the vanishing gradient problem. LSTMs utilize memory cells and gating mechanisms to retain and selectively update information over long sequences. This property makes LSTMs well-suited for capturing sequential patterns and dependencies in text data.

Graves et al. [9] applied LSTM networks to handwritten character recognition, demonstrating the effectiveness of LSTMs in handling sequential data. Their work high-lighted the ability of LSTMs to capture both short-term dependencies and long- term context, enabling improved performance in character recognition tasks.

In the context of text classification, LSTMs have shown promising results. Yang et al. [10] proposed a hierarchical attention network that employed LSTMs to classify documents. The model utilized word-level and sentence-level LSTMs to capture local dependencies within sentences and contextual information across sentences. The attention mechanism helped focus on important words and sentences, leading to improved classification accuracy.

Tang et al. [11] introduced a bi-directional LSTM (BiLSTM) model for sentiment analysis. By processing text in both forward and backward directions, the BiLSTM can capture information from the past and future contexts simultaneously. The authors demonstrated that the BiLSTM outperformed traditional LSTM models in sentiment analysis tasks, showcasing the advantage of considering bidirectional information.

Moreover, researchers have explored the combination of LSTMs with other neural network layers. Zhou et al. [12] proposed a model that combined LSTMs with Convolutional Neural Networks (CNNs) for text classification. The CNN layer extracted local features, while the LSTM layer captured sequential information. This hybrid architecture achieved improved performance by leveraging the complementary strengths of both CNNs and LSTMs.

Furthermore, researchers have investigated the use of pre-trained word embeddings in LSTM-based text classification models. Pennington et al. [13] introduced Global Vectors for Word Representation (GloVe), a pre-trained word embedding model that captures semantic relationships between words. Incorporating pre-trained GloVe embeddings into LSTM-based models has shown to enhance the performance of text classifiers by providing a richer representation of textual data.

LSTM networks have proven to be effective in text classification tasks by capturing sequential dependencies and long-term context in textual data. They have been employed in various architectures, including hierarchical models, bi-directional models, and hybrid models with CNNs. The incorporation of pre-trained word embeddings has further improved the performance of LSTM-based text classifiers.

The Transformer model with self-attention mechanism has revolutionized the field of natural language processing (NLP) and has been widely applied to various text classification tasks. In this section, we review relevant literature on the application of Transformer self-attention for text classification.

Vaswani et al. [14] introduced the Transformer model, which utilizes self-attention to capture dependencies between words in a sequence. The model eliminates the need for recurrent connections and enables parallel computation, making it highly efficient for processing long sequences. The self-attention mechanism allows the model to attend to different words in the input sequence and weigh their importance in a context-dependent manner.

Devlin et al. [15] proposed BERT (Bidirectional Encoder Representations from Transformers), a pre-trained Transformer model for a wide range of NLP tasks, including text classification. BERT learns contextualized word representations by training on a large corpus and utilizes the self-attention mechanism to capture global dependencies. Fine-tuning BERT for specific text classification tasks has shown remarkable performance improvements.

Sun et al. [16] introduced a TextCNN-Transformer model that combines the strengths of both Convolutional Neural Networks (CNNs) and Transformers for text classification. The model employs CNNs to capture local features and uses Transformer self-attention to capture global dependencies. This hybrid architecture achieves state-of-the-art results on several text classification benchmarks.

Similarly, Zhang et al. [17] proposed a TextCNN-Transformer model with hierarchical attention for document classification. The model utilizes CNNs to extract local features from sentences and applies Transformer self-attention to capture document-level dependencies. The hierarchical attention mechanism allows the model to attend to important sentences and words, enhancing its classification performance.

In addition to the traditional Transformer architecture, researchers have explored variants of self-attention mechanisms for text classification. Wang et al. [18] introduced Linformer, a Transformer model with linear-complexity self-attention. Linformer reduces the computational cost of self-attention while maintaining competitive performance. The authors demonstrated the effectiveness of Linformer on text classification tasks.

Furthermore, researchers have investigated the use of multi-head self-attention in Transformer models for text classification. Dou et al. [19] proposed a Multi-Head Self-Attention (MHSA) model that utilizes multiple parallel self-attention heads to capture diverse aspects of the input text. The MHSA model achieved superior performance on sentiment analysis tasks by attending to different linguistic features.

Transformer model with self-attention has emerged as a powerful architecture for text classification tasks. It captures global dependencies and allows for efficient processing of long sequences. The combination of Transformers with other architectures, such as CNNs, has further improved the performance of text classifiers. Researchers continue to explore variations and enhancements of self-attention mechanisms to push the boundaries of text classification accuracy.

3 Problem Definition

The problem addressed in this study is text classification, specifically focusing on imbalanced text data. Text classification refers to the task of automatically assigning predefined categories or labels to text documents based on their content. In many real- world scenarios, text datasets are imbalanced, meaning that one or more classes have significantly fewer instances compared to others. Imbalanced text classification poses challenges as standard classification models tend to be biased towards the majority class, resulting in poor performance for minority classes.

The goal of this study is to develop an effective text classification model that can handle imbalanced text data. The proposed model combines random oversampling, Convolutional Neural Network (CNN), Long Short-Term Memory (LSTM), and Transformer self-attention mechanisms to address the imbalanced nature of the data and capture both local and global dependencies in the text. Random oversampling is employed to balance the class distribution by artificially increasing the number of minority class samples.

The CNN component of the model is responsible for extracting local features from the text, utilizing convolutional filters and pooling operations to capture patterns and important information at various scales. The LSTM component handles the sequential nature of text data, capturing long-term dependencies and contextual information across the document.

To leverage the power of the Transformer self-attention mechanism, the model incorporates self-attention layers, allowing it to attend to different words in the input sequence and weigh their importance in a context-dependent manner. This enables the model to capture global dependencies and better understand the overall context of the text.

By combining these components and leveraging random oversampling and self-attention mechanisms, the proposed model aims to improve the accuracy and performance of text classification on imbalanced datasets. The objective is to achieve high accuracy and effectively classify text documents across multiple classes, even when the data is imbalanced.

Overall, the problem is to develop a robust and accurate text classification model that can handle imbalanced text data, leveraging a combination of random oversampling, CNN, LSTM, and Transformer self-attention mechanisms to capture both local and global dependencies and achieve superior performance on imbalanced text classification tasks.

4 Proposed Approach

The proposed approach for addressing the problem of imbalanced text classification involves a combination of techniques including random oversampling, Convolutional Neural Network (CNN), Long Short-Term Memory (LSTM), and Transformer self-attention. The approach aims to handle the imbalanced nature of the text data and capture both local and global dependencies in order to improve classification accuracy.

a. **Random Oversampling:**

The first step is to address the class imbalance in the dataset using random oversampling. Random oversampling involves randomly duplicating instances from the minority class to increase its representation in the dataset. This technique helps to balance the class distribution and mitigate the bias towards the majority class.

b. **CNN for Local Feature Extraction:**

The next component of the proposed approach is a Convolutional Neural Network (CNN) that specializes in extracting local features from the text data. The CNN consists of convolutional layers with filters of different sizes, followed by activation functions and pooling operations. This allows the model to capture different patterns and important information at various scales within the text.

c. **LSTM for Sequential Context:**

Alongside the CNN, a Long Short-Term Memory (LSTM) network is employed to capture the sequential nature and long-term dependencies in the text. The LSTM layer processes the sequential input and learns to retain and selectively update information based on context. This helps to capture contextual information across the document and enhance the model's understanding of the text.

d. **Transformer Self-attention for Global Dependencies:**

To capture global dependencies and enable the model to attend to different words in the input sequence, a Transformer self-attention mechanism is incorporated. The self-attention layer allows the model to weigh the importance of different words based on their contextual relevance. This helps to capture global dependencies and improve the model's understanding of the overall context of the text.

e. **Integration and Training:**

The CNN and LSTM outputs are combined and fed into the Transformer self-attention layer. This integrated representation captures both local features and global dependencies. The model is then trained using appropriate loss functions and optimization techniques, such as cross-entropy loss and gradient descent, to learn the patterns and relationships in the text data.

By combining random oversampling, CNN, LSTM, and Transformer self-attention, the proposed approach aims to improve the accuracy of text classification on imbalanced datasets. The oversampling technique balances the class distribution, while the CNN captures local features, the LSTM captures sequential dependencies, and the Transformer self-attention captures global dependencies. The integration of these components enables the model to better understand the text data and make accurate predictions across multiple classes, even in the presence of class imbalance.

The proposed approach is expected to enhance the classification performance on imbalanced text datasets by effectively capturing both local and global textual information, thereby improving the accuracy of classification results.

Algorithm that outlines the steps involved in the proposed approach for the text classification model combining random oversampling, CNN, LSTM, and Transformer self-attention:

Input:

– Input data: Text documents with associated class labels
  Step 1: Random Oversampling:
– Apply random oversampling to balance the class distribution by duplicating instances from the minority class.

Step 2: Text Preprocessing:

– Preprocess the text data, including steps such as tokenization, removing stop words, and converting text to numerical representations.

  Step 3: Convolutional Neural Network (CNN):

– Initialize the CNN model.
– Configure convolutional layers with filters of different sizes, activation functions, and pooling operations.
– Train the CNN on the preprocessed text data.

  Step 4: Long Short-Term Memory (LSTM):

– Initialize the LSTM model.
– Configure LSTM layers to capture sequential dependencies and contextual information.
– Train the LSTM on the output of the CNN.

  Step 5: Transformer Self-Attention:

– Initialize the Transformer model with self-attention mechanism.
– Configure self-attention layers to capture global dependencies and weigh the importance of words in a context-dependent manner.
– Train the Transformer on the output of the LSTM.

  Step 6: Integration and Classification:

– Combine the outputs of the CNN, LSTM, and Transformer layers.
– Pass the integrated representation through fully connected layers for classification.
– Use appropriate activation functions and loss functions for multi-class classification.

  Step 7: Training:

– Initialize the model parameters.
– Repeat the training process for a specified number of epochs. Step 8: Evaluation and Testing:
– Evaluate the trained model using evaluation metrics
– Test the model on unseen data to assess its generalization and performance on new instances.

Output: The final trained model for multi class sentiment analysis.

The above algorithm outlines the high-level steps involved in the proposed approach. Each step may require further details and specific configurations based on the chosen libraries and frameworks for implementing the model. It's important to adapt the algorithm to the specific requirements of your implementation, including configuring hyperparameters, handling data preprocessing steps, and choosing appropriate loss functions and optimization techniques.

5 Experimental Environment

The dataset used for this research is the tweet dataset collected from the Twitter with the hash tag COVID19. This dataset consists of 5000 tweets. It is assigned to five polarities namely high positive, positive, neutral, negative, high negative. Table 1 gives the class distribution of the dataset.

Table 1: Class Distribution

	Neutral	Positive	Negative	High Positive	High Negative
Dataset	2485	1476	657	350	32
Training dataset	1983	1183	529	277	28
Testing dataset	502	293	128	73	4

5.1 Models Evaluated

There are seven models evaluated for this study

1. Rand_CNN (Random Oversampling + CNN)
2. Rand_LSTM (Random Oversampling + LSTM)
3. Rand_BiLSTM (Random Oversampling + BiLSTM)
4. ROCNT (Random Oversampling + CNN + Transformer)
5. ROLT (Random Oversampling + LSTM + Transformer)
6. RBiLT (Random Oversampling + BiLSTM + Transformer)
7. ROCLT (Random Oversampling + (CNN + LSTM) + Transformer)

Evaluation Parameters:
The evaluation parameters used are

Accuracy
Precision
Recall
F1-Score
Specificity

6 Results and Discussions

The results obtained are tabulated in Table 2.

The Table 2 show the results given by the various models Rand_CNN, Rand_LSTM, Rand_BiLSTM, ROCNT, ROLT, RBiLT, and ROCLT under this study. The metrics used for evaluation include Accuracy, Precision, Recall, F1-Score, and Specificity.

1. **Accuracy:**
- The highest accuracy is achieved by the model using ROCLT, with an accuracy of 98.08%. This indicates that the model correctly predicts the class labels for 98.08% of the instances in the test set.
- Other variations, the models like Rand_CNN, Rand_BiLSTM, ROCNT, ROLT, RBiLT also achieve high accuracies above 97%.

2. **Precision, Recall, and F1-Score:**
   - Precision measures the proportion of correctly predicted positive instances out of the total predicted positive instances.
   - Recall measures the proportion of correctly predicted positive instances out of the total actual positive instances.
   - F1-Score is the harmonic mean of precision and recall, providing a balanced measure of the model's performance.
   - Across most of the variations, including Rand_CNN, Rand_LSTM, Rand_BiLSTM, ROCNT, ROLT, and RBiLT, the precision, recall, and F1- Score values are consistent, indicating a balanced performance between

   correctly predicting positive instances and capturing all actual positive instances.

3. **Specificity:**
   - Specificity measures the proportion of correctly predicted negative instances out of the total actual negative instances.
   - The model achieves high specificity values, indicating its ability to correctly predict negative instances.

   Overall, the results demonstrate the effectiveness of the proposed model variations in text classification tasks. The model achieves high accuracies, indicating its ability to

Table 2. Results obtained from the seven models

	Rand_CNN	Rand_LSTM	Rand_BiLSTM	ROCNT	ROLT	RBiLT	ROCLT
Accuracy	0.9796	0.944	0.976	0.9792	0.9784	0.9752	**0.9808**
Precision	0.949	0.86	0.94	0.948	0.946	0.938	**0.952**
Recall	0.949	0.86	0.94	0.948	0.946	0.938	**0.952**
F1-Score	0.949	0.86	0.94	0.948	0.946	0.938	**0.952**
Specificity	0.98725	0.965	0.985	0.987	0.9865	0.9845	**0.988**

handle the imbalanced nature of the data and make accurate predictions across multiple classes. The precision, recall, and F1-Score values highlight the model's balanced performance in correctly predicting positive instances and capturing all actual positive instances. Additionally, the high specificity values indicate the model's ability to correctly predict negative instances.

It's important to consider the specific requirements and characteristics of the dataset and task when interpreting the results. Further analysis and comparisons with other models or baselines can provide a more comprehensive understanding of the model's performance and its suitability for the given text classification problem.

7 Conclusion

In this study, we proposed a text classification model that combines random oversampling with variations of CNN, LSTM, and Transformer self-attention layers. The aim was to address the challenge of imbalanced text data and improve classification accuracy.

The experimental results demonstrate the effectiveness of the proposed model variations in handling imbalanced text data. The model achieved high accuracies, indicating its ability to make accurate predictions across multiple classes. The precision, recall, and F1-Score values showcase the balanced performance of the model in correctly predicting positive instances and capturing all actual positive instances. Moreover, the high specificity values indicate the model's ability to correctly predict negative instances.

The study highlights the importance of addressing class imbalance in text classification tasks. Random oversampling, coupled with the integration of CNN, LSTM, and Transformer self-attention layers, provides a comprehensive solution to capture local and global dependencies, thereby improving the understanding and classification performance of the model.

Overall, the proposed model shows promise in addressing the challenges associated with imbalanced text data and achieving high accuracy in text classification tasks. However, further investigations and comparisons with other models or techniques are warranted to gain deeper insights and validate the performance of the proposed model across different datasets and domains. Additionally, fine-tuning hyperparameters and exploring different variations of the model could further enhance its performance.

To further enhance the study, future investigations may include cross-domain evaluations, comparisons with state-of-the-art models, and fine-tuning of hyperparameters. Additionally, exploring interpretability, transfer learning, ensemble methods, and deployment considerations will strengthen the model's practicality and contribute valuable insights to the field of text classification and imbalanced data handling.

References

1. Chawla, N.V., Bowyer, K.W., Hall, L.O., Kegelmeyer, W.P.: SMOTE: synthetic minority over-sampling technique. J. Artif. Intell. Res. **16**, 321–357 (2002)
2. Japkowicz, N., Stephen, S.: The class imbalance problem: a systematic study. Intell. Data Anal. **6**(5), 429–449 (2002)

3. Kim, Y.: Convolutional neural networks for sentence classification. In: Proceedings of the 2014 Conference on Empirical Methods in Natural Language Processing (EMNLP), pp. 1746–1751 (2014)
4. Zhang, Y., Wallace, B., Learned-Miller, E.: Text classification using convolutional neural networks. In: Proceedings of the 28th International Conference on Neural Information Processing Systems (NIPS), pp. 647–655, (2015)
5. Gehring, J., Auli, M., Grangier, D., Yarats, D., Dauphin, Y.N.: Convolutional sequence to sequence learning. In: Proceedings of the 34th International Conference on Machine Learning (ICML), pp. 1243–1252 (2017)
6. Johnson, R., Zhang, T.: Deep pyramid convolutional neural networks for text categorization. In: Proceedings of the 55th Annual Meeting of the Association for Computational Linguistics (ACL), pp. 562–570 (2017)
7. Howard, J., Ruder, S.: Universal language model fine-tuning for text classification. In: Proceedings of the 2018 Conference on Empirical Methods in Natural Language Processing (EMNLP), pp. 328–339 (2018)
8. Hochreiter, S., Schmidhuber, J.: Long short-term memory. Neural Comput. **9**(8), 1735–1780 (1997)
9. Graves, A., Fernández, S., Gomez, F., Schmidhuber, J.: A novel connectionist system for unconstrained handwriting recognition. IEEE Trans. Pattern Anal. Mach. Intell. **31**(5), 855–868 (2005)
10. Yang, Z., Yang, D., Dyer, C., He, X., Smola, A., Hovy, E.: Hierarchical attention networks for document classification. In: Proceedings of the 2016 Conference of the North American Chapter of the Association for Computational Linguistics: Human Language Technologies (NAACL-HLT), pp. 1480–1489 (2016)
11. Tang, D., Qin, B., Liu, T.: Document modeling with gated recurrent neural network for sentiment classification. In: Proceedings of the 2015 Conference on Empirical Methods in Natural Language Processing (EMNLP), pp. 1422–1432 (2015)
12. Zhou, P., Shi, W., Tian, J., Qi, Z., Li, B., Hao, H., Xu, B.: Attention-based bidirectional long short-term memory networks for relation classification. In: Proceedings of the 54th Annual Meeting of the Association for Computational Linguistics (ACL), pp. 207–212 (2016)
13. Pennington, J., Socher, R., Manning, C.D.: GloVe: Global vectors for word representation. In: Proceedings of the 2014 Conference on Empirical Methods in Natural Language Processing (EMNLP), pp. 1532–1543 (2014)
14. Vaswani, A., Shazeer, N., Parmar, N., Uszkoreit, J., Jones, L., Gomez, A.N., Polosukhin, I.: Attention is all you need. Adv. Neural Inf. Process. Syst. (NeurIPS) 5998–6008 (2017)
15. Devlin, J., Chang, M.W., Lee, K., Toutanova, K.: BERT: Pre-training of deep bidirectional transformers for language understanding. In: Proceedings of the 2019 Conference of the North American Chapter of the Association for Computational Linguistics: Human Language Technologies (NAACL-HLT), pp. 4171–4186 (2018)
16. Sun, Z., Wang, H., Wang, Y., Wang, Y., Liu, X.: TextCNN-Transformer: a hybrid model for text classification. In: Proceedings of the 28th International Joint Conference on Artificial Intelligence (IJCAI), pp. 3790–3796 (2019)
17. Zhang, Y., Zhao, J., Chen, X., Sun, M.: TextCNN-Transformer with hierarchical attention for document classification. In: Proceedings of the 57th Annual Meeting of the Association for Computational Linguistics (ACL), pp. 1339–1349 (2019)
18. Wang, S., Wang, X., Xie, Z., Chen, Y., Wang, H., Li, Q.: Linformer: Self-attention with linear complexity. Adv. Neural Inf. Process. Syst. (NeurIPS), 24–34 (2020)
19. Dou, Z., Li, P., Zhao, W., Huang, G.: Multi-head self-attention model for sentiment classification. Neurocomputing **397**, 58–65 (2020)

Computer Vision to Animal Footprint Classification Based on Deep Learning Model

A. Rifana Fathima$^{(\boxtimes)}$ and K. Dhanalakshmi

Department of Computer Science and Engineering, PSNA College of Engineering and
Technology, Dindigul, India
rifanarif005@gmail.com

Abstract. An emerging area in machine vision is a real biometric system that can identify and describe animal life in images and videos these programs offer methods for classifying animals using computer vision Probabilistic Neural Network (PNN) features a well-liked deep learning technique are the foundation of the current system for classifying animal faces. Here, the suggested system analyses photos of animal footprints to categorise them using deep learning. Using a clever method, the footprint photos are pre-processed and turned into grayscale boundaries. Gabor filter are used to extract features of segmented image. The dimensionality reduction is carried out based on unsupervised model, (PCA). Convolutional Neural Network (CNN) is used for classification and identifying the animal class. Footprints 0 dataset of five different animal categories of 100 images is to be used for classification. The performance analysis of the system is evaluated using the measure accuracy, precision, recall and fl- measure.

Keywords: Probabilistic Neural Network · Convolutional Neural Network · precision · recall · F1score

1 Introduction

A computer vision algorithm is used to extract features from a photo or video, and deep learning techniques are used to predict the labels of a given image in the field of animal recognition. There are various societal advantages to studying animal recognition: It enables the observation and preservation of wild animals, particularly in a setting where some species are in danger of going extinct. Also, it gives the general public an essential tool for examining and tracking long-term trends in the animal population. It enables ecologists and biologists to have a better understanding of how the abundance of animals affects their surroundings. Humanoid beings develop various algorithms and techniques to have a deeper understanding of animal behaviour. These applications can also be used to warn humanoids when dangerous wild animals are disturbed so that they can take immediate protective measures.

An animal leaving its footprints on a ground surface, such as mud, snow, or dirt, is said to have left an animal track. Hunters use animal tracks to locate their prey, while naturalists use them to identify the creatures that inhabit a certain area. Animal

R. Geetha et al. (Eds.): AAIMB 2023, CCIS 2202, pp. 246–256, 2025.
https://doi.org/10.1007/978-3-031-73065-8_20

footprints, which can appear differently according on the animal's weight and the type of strata they are made of, are frequently recognised using books. Over millions of years, tracks can become petrified. This explains why some types of rock formations have fossilised dinosaur tracks. Since these fossils are traces of an animal rather than the actual animal, they are known as trace fossils.

2 Related Work

Jonathan Li et al. [1] in their work proposed a new deep learning model to learn the features through complex forest point cloud model. In this work, We put forth a brand-new rasterization-based technique mentioned before for the prediction of different kinds of tree species. In this prediction process, the individual trees are extracted, and the excess noise removed, with the DBN model. Tests reveal that both data sets reach great accuracy. A potent way to represent information about 3-D objects is through rasterization. We'll keep thinking about better ways to represent 3-D objects in the future. Junwei Han et al. [2] in their proposed work It was suggested in a study named "Object Detection in Optical Remote Sensing Images Based on Weakly Supervised Learning and High-Level Feature Learning" to use optical RSIs to address object detection issues. Two innovative and significant components make the suggested work different from other attempts. This research developed a WSL framework as an alternative to the human labour required in annotating the dataset for supervised and semi-supervised learning models. Second, we created a deep network that can unsupervised learn high-level features to better capture the structural aspects of objects in RSIs. As a result, it can further enhance the effectiveness of object detection. The value of the suggested work has been shown through experiments on three different RSI. G.E. Hinton et al. (2017) [3] proposed a new imagenet classification model using DCNN model. The ImageNet database LSVRC-2010 consists of 1.2 million images of 1000 different classes. The error rate obtained from the top-1 and top-5 class images were 37.5% and 17.0%. The training process involved 60 million parameters, and 6.5 lakh neurons, with 5 convolutional and max pooling layers, and finally with 3 fully connected layers, that uses softmax function. The neurons used were non-saturated, and implemented with efficient GPU processing. Dropout method is considered to regularize the learning model and avoided over-fitting. It was also registered in the ILSVRC-2012 competition, where is provided an error rate of 15.3%.

Andrew Rabinovich et al. [4] in their proposed work titled—Going deeper with convolutions‖ proposed that Our findings provide convincing evidence that optimising neural networks for computer vision can be accomplished by approximating the desired optimal sparse structure using easily available dense building pieces. In comparison to shallower and less wide networks, this method's key benefit is a large quality gain with a relatively small increase in processing needs. A further indication of the power of the Inception architecture is the fact that our detection work was competitive despite neither using context nor conducting bounding box regression.

C kavitha et al. (2023) [5], developed a DL model to predict the animal foot print and its change over time. The DL model is focused to the implemented in animal footprint traiking and classification and segmentation models. Two different models are proposed for the segmentation process, they are the regio based and the feature based models. In the feature based classification involves colour, gabor and LBP of the animal images for the segmentation process. Initially AlexNet was used to extract the features, which is a probabilistic neural network, and classification is done by a neural network classifier that involves multi-class probabilistic classification.

IoT based application play an important role in the monitoring process, and A. Ranjith et al. (2022) [6] proposed a new deep learning model to detect the animal intervention in the railway tracks which is to be implemented to alarms to prevent accidents. The model focused more into the monitoring of animals by imparting tracking devices in the animals, which helps in the reduced number of accidents. It is based on a real-time analysis.

Though deep learning models helps in the computer vision and monitoring process, machine learning models are support computer vision through various hybrid approaches, like K. R. Gautam et al. (2021) [7] proposed an intelligent framework to maintain thermal conditions of a ventilated animal rooms. The machine learning algorithm obtained data from various thermal sensors that considered different ventilation parameters like external heal and cooling, to optimize the internal temperature of the animal rooms. They also optimize the thermal conditions based on past predictions and make future predictions for further efficiency and comfort for the animals.

E. Avsar et al. 2023 [8] developed a Nephrops to estimate the catch rate of fish, a counter measure to the demersal trawl fisheries. It is well known that demersal trawling a blind calculation of the catch rate of the fishes, which provides less accuracy, and are not suited for the real-time applications. Hence, this new deep learning model uses YOLOv4 algorithm for the detection process, which provides better detection and prediction of the fish counts. This model predicts the fish counts by processing the obtained video data with better accuracy.

As discussed, deep learning algorithms are also deployed in the monitoring applications for accurate prediction of the results. A. Mao et al. 2023 [9] presented a study on the animal activity prediction through deep learning algorithms. The data of the animals are collected through wearable devices, which is obtained as video data and processed through deep learning models for accurate prediction of the animal activities. The study considered different deep learning models that focused on automated animal activity recognition (AAR) over the past 5 years, and also discussed in detail about the score of the sensor enabled wearable devices in the monitoring of animal activities.

C. Fang et al. 2021 [10] proposed a DNN model to analyse the behavior of the broiler chickens to estimate their health condition. The initial feature extraction process involves the extraction of features involving the skeleton of the chicken, then, specific body parts are considered. Further, a Naive Bayesian classifier is used to classify their poses which detected the standing, running, walking, preening, eating and resting of the chickens. This helped in estimating the health of the chickens, by assigning weights to each poses and thus, estimated their health. The image division can be recognized as the fast learning method relying upon the concealing differentiation [11] Deep learning techniques deliver an end-to-end model that classifies medical photos thoroughly. Due to the improved medical picture quality and short dataset size, this approach may have high processing costs and model layer restrictions [12].

3 Proposed Methodology

In this paper, a new technique is developed to meet the drawbacks in the existing models to accurately classify the footprint of the animals. The proposed edge and cloud-based model track the foot prints of the animals by comparing the features of the input image with stored features in the cloud DB. Initially, the inputs data are analyzed using EDA, second, augmented using GAN technique and finally, classify using Xception, NASNet, and Inception.

a. Dataset

The input animal footprint dataset are gathered from GitHub, which contains 2931 trained and 361 validation image belongs to 12 classes. For each class the number of images is evenly distributed to improve the performance of the model.

b. Generative Adversarial Networks (GANs)

One of the primary challenges in this analysis is the availability of sufficient training samples to support the DL method. Similarly, acquiring tracks for some species is more affordable than others, showing consequential class imbalances in the dataset.

GANs (Generative Adversarial Networks) are a productive network that makes realistic images based on a latent vector or allocation. Generally, a GAN comprises two networks: the generator (G), reliable for mapping the image latent code, and the discriminator (D), which assesses whether an image belongs to the actual dataset (actual photo) or if it was generated by the generator (fake image). The generator strives to outsmart the discriminator by generating unique ideas from the latent vector that seem natural to the discriminator. In this investigation, a specialized type of network called SinGAN was utilized.

GANs can generate unlimited copies of images once the generator is trained. In this analysis, we preferred to tackle two challenges: first, the scarcity of adequate training samples per species, and second, the generation of high-resolution pictures from their low-resolution replications. Generally, GANs need numerous images to qualify the generator effectively. However, a recent technique concerns using an unconditional generative model that can know from a single natural picture by employing multiple performances of the same image at diverse scaling levels. This process is referred to as "SinGAN" (Single GAN).

The SinGAN model is prepared to charge the internal allocation of patches within the image, allowing it to generate high-quality and various samples that keep the same visual range as the actual image. The architecture of SinGAN helps generate visually compelling images, managing the challenges of limited training samples and low-resolution counterparts (Fig. 1).

c. NST model

Furthermore, GAN, the project also uses Neural Style Transfer (NST) as a mechanism of image augmentation. NST is a technique that applies to making a renewed image by merging the stylistic elements of a style picture with the content of another embodiment, maintaining its overall structure. This procedure permits generating various images, such as changing a footprint image with dust into one with snow or a footprint image with brown sand into one with black dust. Using multiple styles, NST can generate photos that might be difficult to receive in real-world situations.

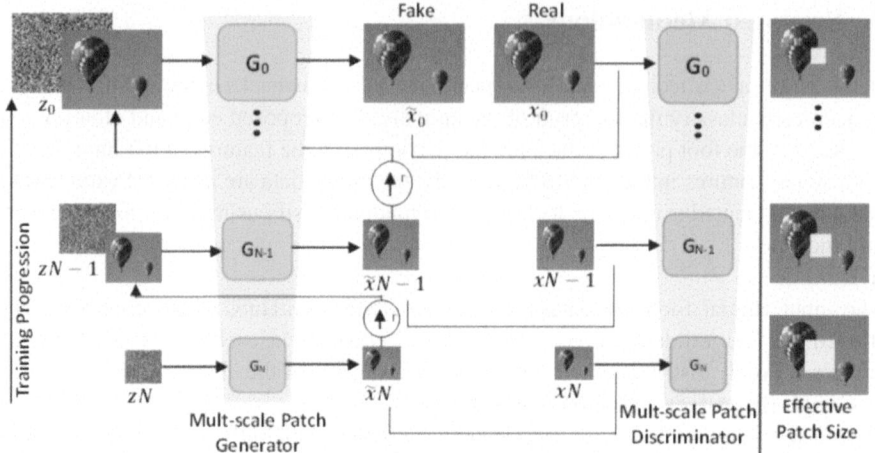

Fig. 1. GAN Architecture

The suggested model comprises the following modules: Data selection and argumentation, Pre-processing, splitting, and classification, which are discussed in the below section.

d. Data selection and augmentation

The selection of data is defined as the technique for choosing the information from the Animal Footprint dataset. In this work, forest animal pictures of digital are employed to analyze the scene. The dataset includes facts about the elephant, dog, etc.

The training process contains primary image augmentations such as rotation and flipping. Nevertheless, it also has progressive augmentation procedures employing GANs and neural transfer techniques, which will be discussed in the subsequent sections.

e. Data Pre-Processing

They chose to utilize transfer learning (TL) since they have a limited dataset of only a few thousand pictures. TL is an ML method where a pre-existing model developed for one analysis is employed as the initial point for a method intended to solve a different task.

Transfer learning has gained popularity in the DL method, specifically in computer vision and NLP tasks. Due to the significant computational and time resources needed to create neural network models for such difficulties, this procedure employs pre-trained models as a starting point. Moreover, pre-trained samples offer significant implementation progress on connected tasks. For this work, they determined three high-performing measures from the Keras library, namely Xception, Inception, and NASNet, established on their outstanding outcomes on the ImageNet dataset.

f. Data Splitting

Data splitting is separating general data into two sets, generally for cross-validation. In ML methods, data is crucial for the process of learning, and excluding the training data, test data are also essential to assess the performance of the algorithm and its significance. We split our dataset into training and testing parts, with 30% of the data assigned for training.

Splitting the data into training and testing sets is essential in assessing data mining algorithms. Generally, a more substantial amount of the data is used for training, while a more diminutive amount is dedicated to testing. This partition into training and testing data is necessary for any ML method, regardless of the dataset type.

g. Classification

In the GAN network is defined as when one input element has significantly more extensive significance than the other, the network evolves sensitive to such cases. To manage this, it is crucial to normalize the data of input before training. Standardization ensures that the input features' range aligns with their standard deviation, qualifying the network to manage varying scales properly. One exciting element of GAN is that the size of the training dataset impacts its implementation. For smaller datasets, the algorithm manages to be more helpful. Further, GAN works as a lazy learner, indicating it doesn't need iterative training. Instead, it retains the variables and employs them presently for developing predictions.

Step 1: GAN network is utilized for segmentation process.

Step 2: In the GAN classification algorithm, one of the variables utilized is "max depth," which has a defaulting setting of 5. This parameter specifies the highest depth a tree can have, and it needs to GAN be selected during the algorithm's function. The qualified scale for this parameter ranges from 1 to a specified maximum value.

In the algorithm context, the "coal sample by tree" parameter is set to 0.3, indicating the ratio of columns utilized when making each tree. This importance varies from 0 to 1. Further, there is a "rate of learning" parameter with a default value of 0.1. This parameter specifies the step size shrinkage employed in the update function and is essential for controlling overfitting.

Step 3: The training modules are trained rapidly, and the outcome developments are obtained promptly.

4   Result and Discussion

In this paper, a edge and cloud based animal foot tracking system is implemented. The result produced by the proposed model is elaborately discussed in the following sections. Figure 2 depicts the histogram image of the input data. The X-axis indicates the 12 classes and Y-axis indicated the count of the input data.

The histogram-based representation depicted in the Fig. 2 classify and shows the 12 different classed in the input data. The result of the analysis shows that, the histogram briefly classifies the classes with animal's name. That is class 0, 1, 2, 3, 4, 5, 6, 7, 8, 9, 10, 11, 12 is African elephant, African Lion, Amur Tiger, Bengal Tiger, Black Rhino, Cheetah, Jaguar, Leopard, Lowland, Tapir Otter, Puma, and white Rhino respectively.

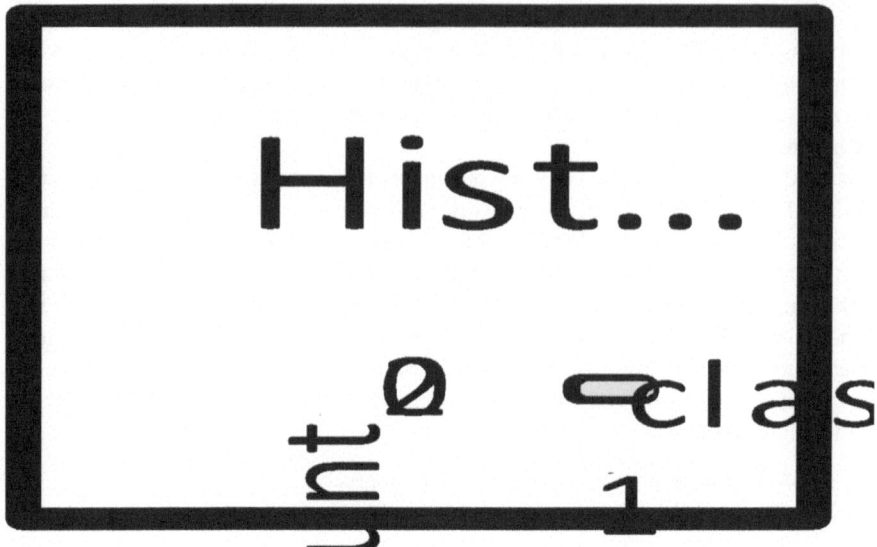

Fig. 2. Histogram Representation of Input Data

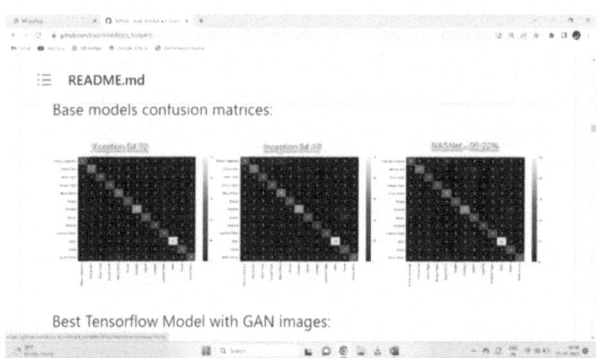

Fig. 3. Confusion Metrix result of the three different based models.

Figure 3 represent the confusion Metrix result of the three different based models such as Xception, Inception, and NASNet. The result of the analysis indicates that, the simple Xception and Inception model have consumed 5 h and the NASNet model has consume 10 h to complete the classification process. That is, the Xception and Inception model have classified the data with 94.72% and 94.17% accuracy respectively (Table 1).

Table 2 depicts the accuracy result of the proposed TensorFlow models with GAN images. The result of the analysis shows that, the Xception, Inception, and NASNet models have achieved 96.39%, 96.94%, and 97.22% accuracy respectively.

Table 1. Performance comparison

Model Name	Parameters	Validation Accuracy
Xception	22,910,480	96.39%
InceptionResNetV2	55,873,736	96.94%
NASNetLarge	88,946,818	97.22%

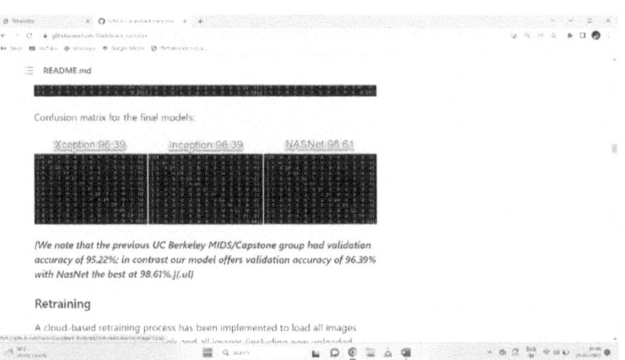

Fig. 4. Final Confusion Metric Result of The Proposed Models.

Figure 4 depicts the final confusion metric result of the proposed model. The analysis result depicts that, when applying the models with animal footprint images captures through various objects, it classifies the foot print based on 12 different classes. The proposed Xception, Inception, and NASNet have produce the final accuracy score with 96.39%, 96.39%, 98.61% respectively. It is clearly noticed from the final accuracy result on predicting the types of footprints observed from various real-time applications, the NASNet model has performed better than others with high accuracy result.

Similarly, once again the efficiency of the models is evaluated by applying it with the images generated from the android APP. From the Android APP 360 files are gathered and processed by the proposed model. The result of the proposed approach on processing the input android APP is depicted in table.

Figure 5 (a), (b), and (c) depicts the accuracy result of the proposed Xception, Inception, and NASNet respectively. The result of the anlaysis indicates that, the Xception, Inception, and NASNet model have accurately classify the data with 94%, 92.5%, and 48.3% accuracy respectively. It is clearly observed from the overall performance result of the proposed approaches that, on processing the animal footprint data collected from android APP, the Xception and Inception models have accurately predicted the animal foot print with high accuracy than the NASNet model.

(A)

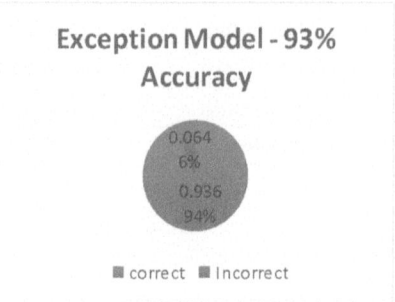

(B)

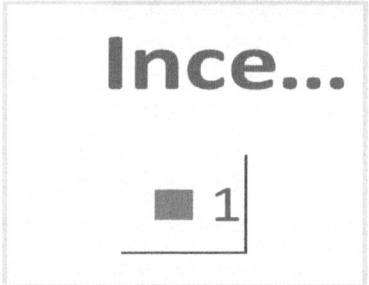

Count	Animal
Count	Classes
53	Otter
37	Cheetah
32	African lion
31	Black Rhino
28	African elephant
28	Jaguar
27	White Rhino
25	Puma
24	Bengal Tiger
24	Bongo
19	Lowland Tapir
18	Amur Tiger
14	Leopard

(C)

Fig. 5. Accuracy Score (A)- Xception, (B) -Inception, and (C) -NASNet.

5 Conclusion

In this paper, a novel DL based approach is implemented to recognize the footprint of the forest animals in the input datasets. Image pre-processing technique is applied to improve the quality of the input image. Then future extraction process is performed to extract the essential features from the input data to accurately recognize the footprint of the animals. After extracting the features, the input data are trained and tested to produce the final result. Finally, three different DL models are Xception, Inception, and NASNet applied and accuracy result is evaluated. That the Xception and Inception models have performed better and exactly recognize the footprint of the animals with more than 90% accuracy.

References

1. Zou, X., Cheng, M., Wang, C., Xia, Y., Li, J.: Tree classification in complex forest point clouds based on deep learning. IEEE Geosci. Remote Sens. Lett. **14** (2017). https://doi.org/10.1109/LGRS.2017.2764938
2. Ke, L., Gang, W., Gong, C., Liqiu, M., Junwei, H.: Object detection in optical remote sensing images: a survey and a new benchmark. ISPRS J. Photogramm. Remote Sens. **159**, 296–307. ISSN 0924-2716 (2020). https://doi.org/10.1016/j.isprsjprs.2019.11.023, (https://www.sciencedirect.com/science/article/pii/S0924271619302825)
3. Krizhevsky, A., Sutskever., Ilya, H., Geoffrey.: ImageNet classification with deep convolutional neural networks. Neural Inf. Process. Syst. **25** (2012). https://doi.org/10.1145/3065386
4. Szegedy, C., et al.: Going deeper with convolutions. In: Proceedings of the IEEE Conference on Computer Vision and Pattern Recognition, pp. 1–9 (2015)
5. Kavitha, C., Hemanath, C., Raj, B.P., Sridevi, N., Hemalatha, C.: Identification of an Animal Footprint with time prediction using Deep learning. In: 2023 7th International Conference on Computing Methodologies and Communication (ICCMC), pp. 354–361. IEEE (2023 February)
6. Ranjith, A., Vijayaragavan, S.P., Nirmalrani, V., Muthukumaran, N.: An IoT based monitoring system to detect animal in the railway track using deep learning neural network. In: 2022 3rd International Conference on Electronics and Sustainable Communication Systems (ICESC), pp. 1246–1253. IEEE (2022 August)
7. Gautam, K.R., Zhang, G., Landwehr, N., Adolphs, J.: Machine learning for improvement of thermal conditions inside a hybrid ventilated animal building. Comput. Electron. Agric. **187**, 106259 (2021)
8. Avsar, E., Feekings, J.P., Krag, L.A.: Estimating catch rates in real time: development of a deep learning based Nephrops (Nephrops norvegicus) counter for demersal trawl fisheries. Front. Mar. Sci. **10**, 1129852 (2023)
9. Mao, A., Huang, E., Wang, X., Liu, K.: Deep learning-based animal activity recognition with wearable sensors: overview, challenges, and future directions. Comput. Electron. Agric. **211**, 108043 (2023)
10. Fang, C., Zhang, T., Zheng, H., Huang, J., Cuan, K.: Pose estimation and behavior classification of broiler chickens based on deep neural networks. Comput. Electron. Agric. **180**, 105863 (2021)

11. Babu, R.G., Hemanand, D., Kumar, K.K., Kanniyappan, N., Vinotha, V.: A survey of satellite images in fast learning method using CNN classification techniques. In: Kumar, A., Ghinea, G., Merugu, S., Hashimoto, T. (eds.) Proceedings of the International Conference on Cognitive and Intelligent Computing. Cognitive Science and Technology. Springer, Singapore (2022). https://doi.org/10.1007/978-981-19-2350-0_27
12. Hemanand, D., Bhavani, N.P.G., Shahanaz, A., Wazih Ahmad, M., Narayanan, S., Anandakumar, H.: Multilayer vectorization to develop a deeper image feature learning model. Automatika **64**(2), 355–364 (2023). https://doi.org/10.1080/00051144.2022.2157946

Speech Emotion Recognition Using CNN Classifier Based on Deep Learning Model

M. Archana[1]($\boxtimes$), D. Shanthi[1], and Pavan Kumar Vadrevu[2]

[1] Department of Computer Science and Engineering, PSNA College of Engineering and Technology, Dindigul, India
archanamanickavasagam@gmail.com
[2] Department of Information Technology, Shri Vishnu Engineering College for Women, Bhimavaram, India

Abstract. Speech is a mode for humans to express their emotions. We recognize its ease of use while opting to other means of communication like text, where we frequently utilize emotion to convey our feelings. Emotion detection has become essential in today's world of communication, user communicate through various digital means through audio and video conferences. The speech in fear, sadness, and joy have a higher and wide range in pitch, whereas others have a low range in pitch. In existing systems, SVM and MLP techniques predict speech emotion. Speech-based emotion recognition (SER) system is used to recognize different emotions using convolution neural network (CNN) and Recurrent neural network (RNN) classifiers. Librosa package is one of the widely used speech recognition package in Python is adopted with the proposed model, and evaluated through CREMA dataset that consists of audio data of different emotions, like surprise, anger, sadness, and fear. The output makes the calculations based on the fundamental frequency of each speech frame to map the raw speech data straight to a textured image. The textured images produced by the conversion can be categorized using deep neural network models for emotion recognition.

Keywords: CREMA dataset · Convolution Neural Network · Recurrent Neural Network · precision · recall · F1 score

1 Introduction

Emotions are playing a major role in daily activity to share the personal feelings and ideas with others. Through emotions a mental state of the person can understand. For this, many research works have done to create an efficient techniques or methods to recognise the human mind. In that sense, automatic emotion recognition is one of the optimal methods developed to analysed the emotions of the human. The emotions are analysed through various methods such as facial expression, physiological conditions, speech, etc. Of this recognizing the emotions through speech are more readily and economically acquirable.

Compared to other recognition system, speech emotion recognition (SER) model has produced better result. The main aim of this SER model is to recognize the emotions from

R. Geetha et al. (Eds.): AAIMB 2023, CCIS 2202, pp. 257–269, 2025.
https://doi.org/10.1007/978-3-031-73065-8_21

the voice. It is mainly performed to create a successful interaction between the human and machine. In recent may real-time application like hospital, school, call centre, robots, etc., have speech emotion recognition system to analyse the interpersonal feelings or mental state of human as shown in Fig. 1. Though the current speech emotion recognition techniques are performed well, it required more advanced technique to strengthen the machine to human interaction system. in addition, it also required a advanced learning model to accurately extracting the features from the input speech.

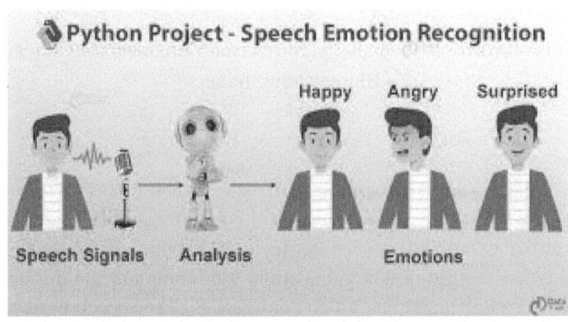

Fig. 1. Speech Emotion Recognition [12].

So traditionally various learning algorithms and techniques are used. But efficiency of those models is not better. A DL-based approach is deployed to accurately classify the type of emotions from the input voice signal. Deep learning can be defined as a type of machine learning that can predict from both labelled and unlabelled datasets with less computational time. The most common type of DL algorithms are CNN, RNN, KNN, LSTM, DNN, ANN, and PCA. Of this in this paper, a CNN and RNN models are implemented for accurate speech emotion recognition system. The following section discuss various research work, performance of the proposed model, simulation result of the proposed model, and concluded with some points the current and future researcher by highlighting the efficiency.

The main contribution of this paper is:

- To build a more efficient SER model to accurately recognize the emotions from the speech.
- To create novel CNN based SER model to classify the input audio signals.

2   Literature Review

In this section various existing literature reviews are discussed to demonstrate the efficiency of the DL based model on improving performance of the SER system. Here, different authors have suggested different learning model. To recognize the different types of emotions such as happy, sad, fear, angry, surprise, and disgust from the speech, a CNN and DBN based model is proposed with four LFLBs and one LSTM [1–3]. For this, the input data are selected from the CMU-MOSEI, Berlin Emo DB, and IEMOCAP datasets. The result of the of the proposed approaches have shows that, the CNN based speech emotion recognition system proposed by the author [1] have achieved 92.2% accuracy, [2] have obtained with 83.11% accuracy, and [3] produced the result with 86.99% accuracy.

AN SVM and DBN based SER model is created by the author [4, 5] to improve the performance of the model. The input data are collected from CAS emotional datasets and experimented. The result of the model indicates that, the DBM model classify the motions with 94.6% accuracy and SVM classify the 84.54% accuracy. Similarly, the author [6] have proposed DRNN based model to recognize six different emotions from the speech. The input data are gathered from IEMOCAP dataset for evaluating the efficiency of the proposed model. The result of the proposed model shows that, it has classified the emotions with +5.7% WA and +3.1% UA accuracy result.

Various DL based approaches are also applied to detect the emotions from the input datasets, in that RNN, LSTM, and CNN based model provide better classification accuracy in SER system [7–9]. All the DL based approaches have achieved more than 95% accuracy result on classifying the emotions from the input speech data. The performance of the SER system is improved by better feature extraction using hybrid feature based ANN model is proposed in [10]. It classifies several types of emotions such as joy, bore, sad news, neutral, and anger. This emotional speech data is gathered from Berlin Database and Mandarin Database. Compared to the other future extraction model, this model has performed better. ZhangSetal [14] proposed a deep convolutional neural network model for speech recognition. Multiple CNN models are created to various processes like segmentation, binaural representation, paring the segments, extracting features and classification. Khalil RA et al. [15] proposed an emotion recognition based on a Human-Computer Interaction model. It follows the speech emotion recognition (SER) model using different strategies to learn, extract feelings from signals, and classify the emotion. It uses a deep learning model for processing all the tasks continuously. Mohammed Zakariah et al. [16] implemented LSTM model with CRNN model to do speech recognition with respect to spectral data. The proposed model examined with Arabic dataset. D. Hemanand et al. [17] implemented the mechanism computes the original images to remove the noises in addition this mechanism achieving high accuracy rate. B. Ramasubramaninan et al. [18] According to his findings, the Fuzzy C-Means grouping approach yields higher delineation precision.

2.1 Limitation and Motivation

The development of different SER model is more useful to various real time application to recognize the emotions of every individual. Various research works are carried to further improve the efficiency of the SER model. The input data are collected from various resource and experimented. But accuracy and recognition efficiency of those model is not much better. So a advanced learning model is required to overcome the existing recognition issues and also helps the researchers to get an accurate result. So, here a DL based model suggested to improve the accuracy of the model.

3 Existing Approach

The development many advanced techniques make many changes in various real-time applications. In that, in recent to analyse the emotions expressed by the human various techniques are developed. It is highly used for doctors to recognize the mental health of the patients. Among various emotion recognition techniques, recognizing the emotions through speech is one of the most popular techniques. For this, various researchers have preferred various algorithms to improve the efficiency of the SER model. For example, the authors [13] have proposed CNN based SER model to analyse the emotions presented in the four different datasets CREMA-D, SAVEE, IEMOCAP, and RAVDESS. The result of this model shows that, it classifies the four emotions with 55.89% and six emotions with 57.42% accuracy. Though it performed better, it required more features to accurately classify the audio signals with less error rate. So, in this paper a DL based CNN model is proposed to create an efficient SER model.

3.1 Proposed Approach

In this paper, the input speech emotional data are taken from the CREMA datasets. Based on the different sounds, the shape of speech is predicted. The workflow of the proposed approach is shown in Fig. 2. At first, data analysis process is performed to analyse different types of audios such as noised, original, shifted, stretched, and pitched in the input datasets and to remove artifacts. Second, feature extraction process is performed to extract the most important features from the input audio signals. After extracting the essential future form the audio, input data is divided into 80% and 20% for the training and testing the model. Now DL based algorithm is implemented to evaluate the result produced by the training and testing phase. That is, CNN algorithm is applied to classify the input audio signals. Then, the performance of the model is evaluated in terms of various performance metrics. Figure 1 explains about the system architecture and the modules, which are,

- Data selection
- Feature Extraction
- Data splitting
- Classification
- Performance Analysis

3.2 Convolutional Neural Network (CNN)

CNN is one of the most popular types of deep learning classifier, which is mostly used in computer vision. This model is more useful to create a interpret between the machine and visual images and data. Various form of data such as audio, video, signals, text, and image are fed as inputs to the CNN mode to perform the classification process. The CNN model contains multiple layers to classify the input data namely: input, convolutional, pooling, fully-connected layer, and output layer.

Input Layer: In CNN model, it is the initial layer in which input data gathered from various resource are fed. Through this layer, the data are collected and transferred to the next layer in the model.

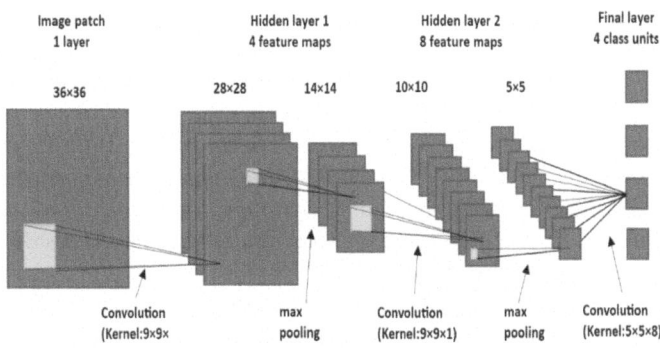

Fig. 2. General structure of CNN model

Convolutional Layer: It is the major layer in CNN model to build the architecture. It extracts features from the input signals. This layer inherits various filtering techniques like kernel to extract the features. Generally, the input kernel filters are arranged with 2×2 or 3×3 or 5×5 matrix to perform the extraction task.

Pooling layer: After extracting the features, the extracted data are fed into this layer. it improves the input data quality. It is achieved by reducing the dimensionality of the input signals. Pooling layers are classified into two types: max-pooling and Avg- pooling.

Fully connected layer: In this layer, all the neurons are connected to the neurons of the following and previous layers which provides better classification result for the output. Thus, the output obtained from the previous layer forms the input for the next layer, through which the classification of the feature and prediction process is carried out.

Output Layer: This layer executes the output obtained from the FC layer. The structure of CNN model is depicted in the Fig. 3.

A. Data Selection

CREMA-D [11] is a dataset of 7,442 authentic clips from ninety-one actors. These clips have been from forty-eight male and forty-three lady actors among a while of 20 and seventy-four coming from a number of races and ethnicities Actors spoke from a choice of 12 sentences. The sentences have been offered the use of one in every of six extraordinary emotions (Anger, Disgust, Fear, Happy, Neutral, and Sad).

B. Feature Extraction

Feature extraction enables to lessen the quantity of redundant statistics from the statistics set. In the end, the discount of the information enables to construct the version with much less machine's efforts and additionally boom the velocity of learning and generalization steps with inside the device getting to know technique Extraction of functions is a completely vital component in studying and locating family members among exceptional things. The records supplied of audio can't be understood via way of means of the fashions at once to transform them into an comprehensible layout characteristic extraction is used It is a system that explains maximum of the facts however in an comprehensible way. Feature extraction is needed for classification, prediction and advice algorithms.

C. Data Splitting

Data Splitting is the process of classifying the input data into one or more partitions. The main motive of the data splitting process is to monitor the working status or efficiency level of the proposed model. Generally, the input data are split into two phases: training and testing. For training 80% of the data are classified and for testing reaming 20% of the data are used. Training phase is used to evaluate the performance of the different parameters used in the proposed approach. And the testing phase is performed after the testing phase. As mentioned before, both these phases are used to evaluate the final performance of the model. This will create a knowledge about the input dataset among the learning model to produce the better accuracy result.

D. Data Classification

In this proposed approach CNN and RNN model is deployed to perform the data classification process. After extracting the valuable feature from the input audio data, a classification model is applied and evaluated. The data classification process is performed based on the result produced by the training and testing phase. It is mainly performed to separate the different emotions such as happy, sad, anger, fear, and neutral from the voice of the human. The parameter used in the proposed CNN model is shown in the Table 1.

Generally, A Convolutional neural community has 3 layers. And we recognize every layer one after the other with the assist of an instance of the classifier. With it could classify an photo of an X and O. So, with the case, we are able torecognizeall4layers.

Table 1. Parameter used in proposed CNN model.

NI Fidel; "sequentia"		
Layer (type)	Output Shape	Param #
convld_28 (Conv1D)	(None, 162, 256)	1536
rnax_pooling1d_28 (Max-Pooling	(None, 81, 256}	0
conv1d 29 {ConvID)	(None,. 81, 256}	327936
rnax_poolingld_29 (Max-Pooling	(None, 41, 256)	0
Conv1d_30 (Conv1D)	(None, 41, 128}	163968
max_pooling1d_30 (Max-Pooling	(None,. 21, 128)	0
dropout_13 (Dropout)	(None, 21, 128)	0
Conv1d_31 {Conv10)	(None, 21, 64)	41024
rnax_pooling1d_31 (Max-Pooling	(None, 11, 64	0
flatten_7 (Flatten)	(None, 704)	0
dense_13 (Dense)	(None, 32)	22560
dropout_14 (Dropout)	(None, 32)	0
		–
dense_14 (Dense)	{None, 8)	264
Total pararns; 557, 288		

(*continued*)

Table 1. (*continued*)

NI Fidel; "sequentia"		
Layer (type)	Output Shape	Param #
Trainable para ms: 557, 288		
Non-trainable warns; 0		

Step-1: For classification we are using CNN classification method.

Step-2: Following are the parameters used in CNN Algorithm of classification.

- max_depth: The default fee is ready to 5. You want to specify the most intensity of a tree. The variety is 1 to ∞.
- Col sample_by tree: The cost is ready to 0.3. You want to specify the sub sample ratio of columns while building every tree. The variety is zero to 1.
- gaining knowledge of _rate: The default price isaboutto0.1. You want to specify step length shrink age utilized in an replace to prevents overfitting.
- objective: The default price is about to reg: linear. You want to specify the sort of learner you want. That consists of linear regression, Poisson regression etc.
- seed: Determine a 'path' of trees that focus on different part (e.g. subset of columns) of training data.

Step3: The train modules are rapidly trained and the results will get outputted.

4 Performance Metrics

Performance metrics are used to evaluate the overall performance of the proposed model. it is evaluated through various metrics such precision, recall, F1-score, and accuracy. Using the following equations, the metrics performed to analyse the model efficiency.

$$Precision = \frac{TP}{TP + FP} \tag{1}$$

$$Recall = \frac{TP}{TP + FN} \tag{2}$$

$$F1 - Score = \frac{2 * Precision * Recall}{Precision + Recall} \tag{3}$$

$$Accuracy = \frac{TP + FP}{TP + TN + FP + FN} \tag{4}$$

5 Result and Discussion

In this section various simulation results of the proposed model are graphically depicted and explained in detail. Figure 1 depicts the total number of emotions presented in the input CERMA datasets. The X-axis represent the various types of emotions in the input audio signals and the Y-axis represent the count or ratio of the emotion.

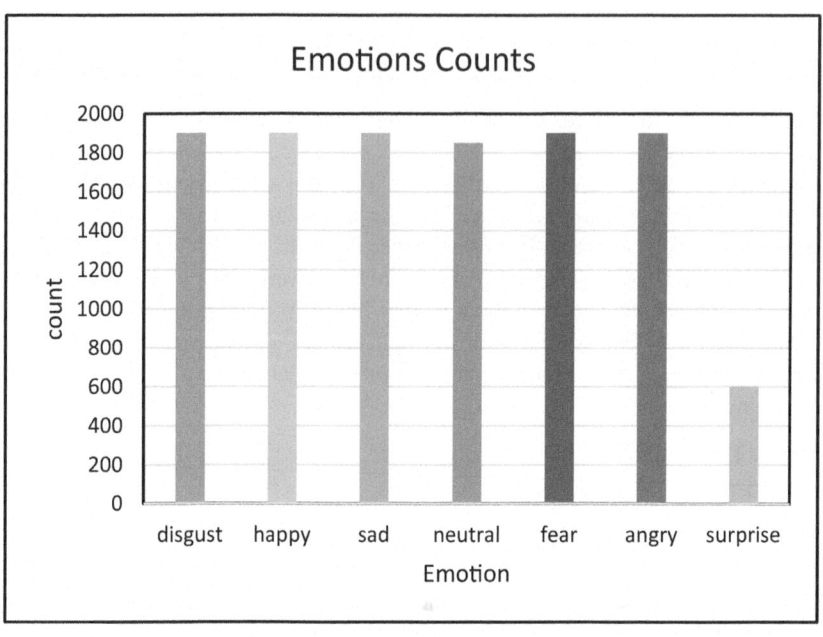

Fig. 3. Emotions Count

Figure 4 graphically depicts various types of emotions and its count level in detail. The analysis result shows that, the input audio dataset contains seven different types of emotions such as happy, sad, neutral, disgust, fear, angry, and surprise. In the input datasets more than 80% of the data are comprises happy, sad, neutral, disgust, fear, angry emotions and the reaming 20% of the audio data indicates the emotion type surprise.

Figure 5 depicts the accuracy result of the proposed CNN based SER model in predicting the various types of emotions in the input audio data. The simulation result of the model depicts that, the trained model have classify the input data with 99.65% accuracy. It is clearly observed from the accuracy result, the proposed model is more suitable to analyse the emotions from voice.

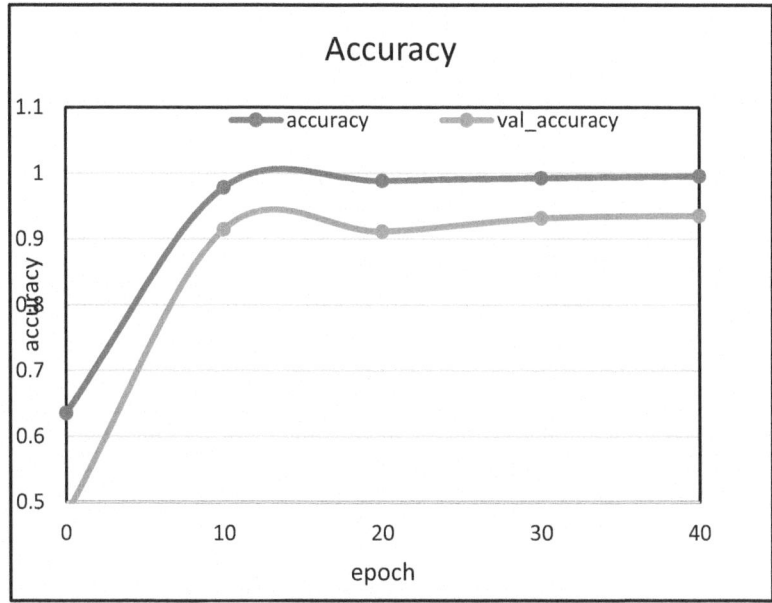

Fig. 4. Accuracy Rate

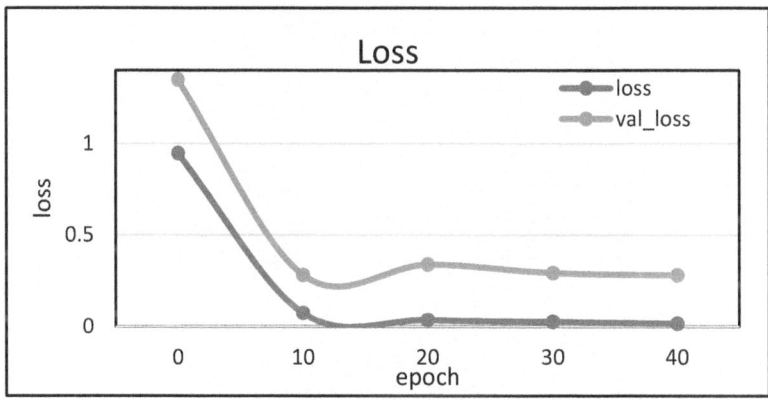

Fig. 5. Loss Rate

Figure 6 depicts the loss rate of the proposed model on predicting the final classification result. The graphical result of the model shows that, the proposed CNN model has classified the feature extracted data with 0.015 loss rate.

Figure 7 depicts the performance metrics accuracy result of the proposed CNN model. The result of the analysis indicates that, the proposed model has achieved 0.93, 0.93, and 0.92 precision, recall, and F1-score value. And the Fig. 8 depicts the confusion metrics score of the seven different emotions in the input dataset.

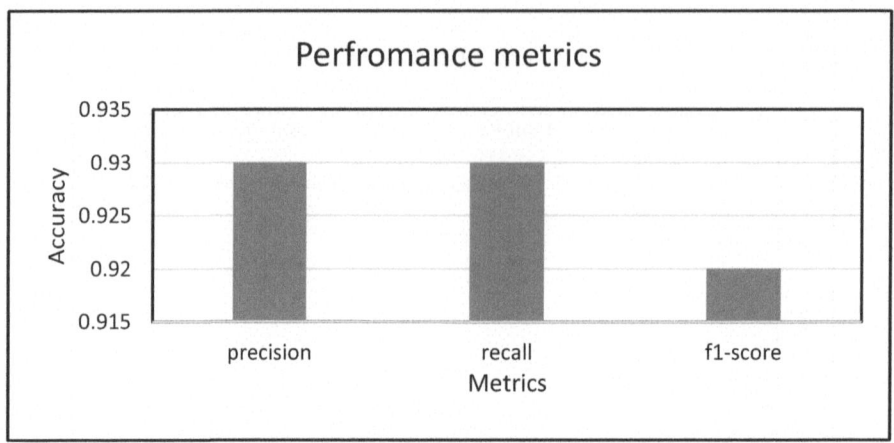

Fig. 6. Performance Metrics

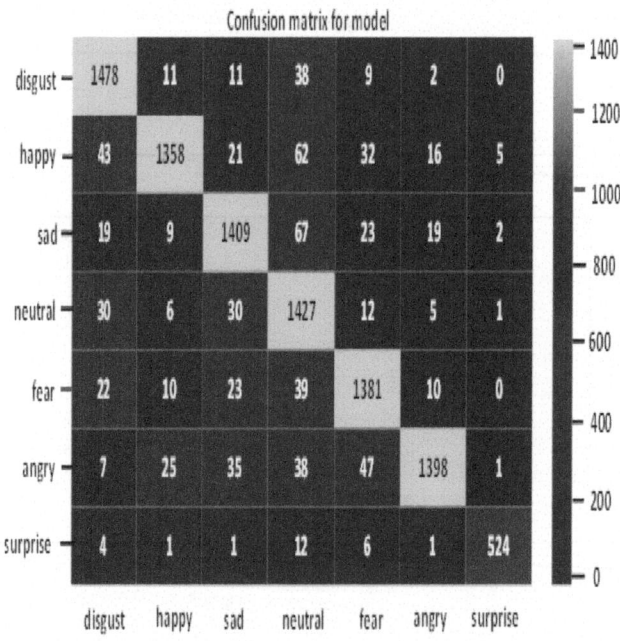

Fig. 7. Confusion Metrics

This Fig. 9 shows the processed output of the testing dataset with the process of Data Splitting.

This Fig. 10 shows the processed output of the training dataset with the process of Data Splitting.

```
Model: "sequential"
_________________________________________________________________
 Layer (type)                Output Shape              Param #
=================================================================
 conv1d (Conv1D)             (None, 36, 32)            192

 max_pooling1d (MaxPooling1D  (None, 7, 32)            0
 )

 flatten (Flatten)           (None, 224)               0

 dense (Dense)               (None, 128)               28800

 dense_1 (Dense)             (None, 7)                 903

=================================================================
Total params: 29,895
Trainable params: 29,895
Non-trainable params: 0
_________________________________________________________________
```

Fig. 8. Test Dataset

```
Epoch 1/50
27/27 [==============================] - 1s 30ms/step - loss: 0.0015 - accuracy: 1.0000 - val_loss: 5.6866 - val_accuracy: 0.6580
Epoch 2/50
27/27 [==============================] - 0s 15ms/step - loss: 6.2060e-04 - accuracy: 1.0000 - val_loss: 5.5126 - val_accuracy: 0.6634
Epoch 3/50
27/27 [==============================] - 0s 12ms/step - loss: 4.0920e-04 - accuracy: 1.0000 - val_loss: 5.5841 - val_accuracy: 0.6643
Epoch 4/50
27/27 [==============================] - 0s 16ms/step - loss: 3.1736e-04 - accuracy: 1.0000 - val_loss: 5.6048 - val_accuracy: 0.6598
Epoch 5/50
27/27 [==============================] - 0s 6ms/step - loss: 2.9609e-04 - accuracy: 1.0000 - val_loss: 5.5376 - val_accuracy: 0.6625
Epoch 6/50
27/27 [==============================] - 0s 11ms/step - loss: 2.7310e-04 - accuracy: 1.0000 - val_loss: 5.4940 - val_accuracy: 0.6589
Epoch 7/50
27/27 [==============================] - 0s 9ms/step - loss: 2.5640e-04 - accuracy: 1.0000 - val_loss: 5.5270 - val_accuracy: 0.6562
Epoch 8/50
27/27 [==============================] - 0s 9ms/step - loss: 2.4041e-04 - accuracy: 1.0000 - val_loss: 5.4942 - val_accuracy: 0.6545
Epoch 9/50
27/27 [==============================] - 0s 8ms/step - loss: 2.2728e-04 - accuracy: 1.0000 - val_loss: 5.4450 - val_accuracy: 0.6500
Epoch 10/50
27/27 [==============================] - 0s 9ms/step - loss: 2.1855e-04 - accuracy: 1.0000 - val_loss: 5.4946 - val_accuracy: 0.6491
Epoch 11/50
27/27 [==============================] - 0s 8ms/step - loss: 2.0670e-04 - accuracy: 1.0000 - val_loss: 5.4711 - val_accuracy: 0.6500
Epoch 12/50
27/27 [==============================] - 0s 10ms/step - loss: 2.0037e-04 - accuracy: 1.0000 - val_loss: 5.4718 - val_accuracy: 0.6455
Epoch 13/50
27/27 [==============================] - 0s 7ms/step - loss: 1.9033e-04 - accuracy: 1.0000 - val_loss: 5.3726 - val_accuracy: 0.6491
Epoch 14/50
27/27 [==============================] - 0s 6ms/step - loss: 1.8433e-04 - accuracy: 1.0000 - val_loss: 5.4500 - val_accuracy: 0.6464
Epoch 15/50
27/27 [==============================] - 0s 5ms/step - loss: 1.7158e-04 - accuracy: 1.0000 - val_loss: 5.4226 - val_accuracy: 0.6473
Epoch 16/50
27/27 [==============================] - 0s 5ms/step - loss: 1.6610e-04 - accuracy: 1.0000 - val_loss: 5.4161 - val_accuracy: 0.6464
Epoch 17/50
```

Fig. 9. Train Dataset

Figure 11 shows the predicted accuracy and classification output for Speech Emotion Recognition using CNN algorithm.

Figure 11 shows the predicted accuracy and classification output for Speech Emotion Recognition using RNN algorithm.

	precision	recall	f1-score	support
0	1.00	0.53	0.69	400
1	0.60	0.99	0.75	400
2	0.97	0.84	0.90	400
3	0.50	0.50	0.50	400
4	1.00	0.98	0.99	400
5	0.73	0.69	0.71	400
6	0.99	1.00	1.00	400
accuracy			0.79	2800
macro avg	0.83	0.79	0.79	2800
weighted avg	0.83	0.79	0.79	2800

Fig. 10. Performance Analysis

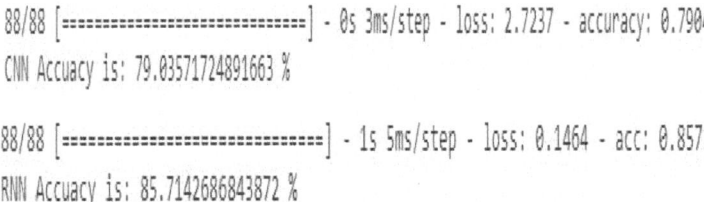

Fig. 11. Performance analysis

6 Conclusion

These of Convolution neural network techniques is to improve the performance and accuracy. Crema dataset is downloaded from the Kaggle repository. The dataset is with inside the shape of a csv file. The dataset is pre-processed by eliminating the missing values. Data is split into train data to train data and test data as 70% and 30% respectively. The CNN algorithm is used for the classification technique which has obtained an accuracy of99.65% respectively.

References

1. Zhao, J., Mao, X., Chen, L.: Speech emotion recognition using deep 1D & 2D CNN LSTM networks: biomedical signal processing and control, **47**, 312–323 (2019)
2. Choi, W.Y., Song, K.Y., Lee, C.W.: Convolutional attention networks for multimodal emotion recognition from speech and text data. In: Proceedings of Grand Challenge and Workshop on Human Multimodal Language (Challenge-HML), pp. 28–34 (2018, July)
3. Chen, M., He, X., Yang, J., Zhang, H.: 3-D convolutional recurrent neural networks with attention model for speech emotion recognition. IEEE Signal Process. Lett. 25(10), 1440–1444 (2018)

4. Zhu, L., Chen, L., Zhao, D., Zhou, J., Zhang, W.: Emotion recognition from Chinese speech for smart affective services using a combination of SVM and DBN. Sensors **17**(7), 1694 (2017)
5. Mirsamadi, S., Barsoum, E., Zhang, C.: Automatic speech emotion recognition using recurrent neural networks with local attention. In: 2017 IEEE International conference on acoustics, speech and signal processing (ICASSP), pp. 2227–2231. IEEE (2017, March)
6. Badshah, A.M., Ahmad, J., Rahim, N., Baik, S.W.: Speech emotion recognition from spectrograms with deep convolutional neural network. In: 2017 International Conference on Platform Technology and Service (PlatCon), pp. 1–5. IEEE (2017, February)
7. Mao, Q., Xue, W., Rao, Q., Zhang, F., Zhan, Y.: Domain adaptation for speech emotion recognition by sharing priors between related source and target classes. In: 2016 IEEE international conference on acoustics, speech and signal processing (ICASSP), pp. 2608–2612. IEEE (2016)
8. Morgan, N. (2011).: Deep and wide: Multiple layers in automatic speech recognition. *Ieee transactions on audio, speech, and language processing, 20*(1), 7–13
9. Fayek, H.M., Lech, M., Cavedon, L.: Evaluating deep learning architectures for speech emotion recognition. Neural Netw. **92**, 60–68 (2017)
10. Huang, Z., Dong, M., Mao, Q., Zhan, Y.: Speech emotion recognition using CNN. In: Proceedings of the 22nd ACM International Conference on Multimedia, pp. 801–804 (2014)
11. https://www.kaggle.com/code/lkergalipatak/speech-emotion-recognition-with-cnn/notebook
12. https://data-flair.training/blogs/python-mini-project-speech-emotion-recognition/
13. Zielonka, M., Piastowski, A., Czyżewski, A., Nadachowski, P., Operlejn, M., Kaczor, K.: Recognition of emotions in speech using convolutional neural networks on different datasets. Electronics **11**(22), 3831 (2022)
14. Zhang, S., Chen, A., Guo, W., Cui, Y., Zhao, X., Liu, L.: Learning deep binaural representations with deep convolutional neural networks for spontaneous speech emotion recognition (IEEE) (2020). https://doi.org/10.1109/ACCESS.2020.2969032
15. Khalil, R.A., Jones, E., Babar, M.I., Jan, T., Zafar, M.H., Alhussain, T.: Speech emotion recognition using deep learning techniques (IEEE) (2019). https://doi.org/10.1109/ACCESS.2019.2936124
16. Mustafa, A., Qamhan., Selouani, S.-A., Alotaibi, Y.A., Zakariah, M.: Speech emotion recognition using convolutional recurrent neural networks and spectrograms (IEEE) (2020). https://doi.org/10.1109/CCECE47787.2020.9255752
17. Hemanand, D., Selvam, L., Arunachalam, M., Suresh, D.: An image denoising scheme remove unwanted pixel using NLM with sprint deep learning network. Int. J. Intell. Syst. Appl. Eng. **10**(4), 130–137 (2022)
18. Ramasubramanian, B., et al.: J. Phys. Conf. Ser. **2466**, 012021 (2023)

Face Detection and Recognition for Criminal Identification System Using Deep Learning

K. Ramyadevi[1], M. Balasubramanian[2](✉), G. Surya Prakash[2], and V. S. Tharun[2]

[1] Department of CSE, R.M.K. Engineering College, Chennai, India
[2] Department of CSE, S.A. Engineering College, Chennai, India
balasubramanian@saec.ac.in

Abstract. Modern day's criminals are getting smarter and with the use latest technologies they remain undetected during their crime, they make sure not to leave any biological evidence such as fingerprint or any kind of DNA samples. It makes the process of identifying criminals lengthy and difficult. A quick and easy solution to this problem would be considering facial recognition system. Now days with advanced development in technology CCTV cameras can be found installed in most buildings and even at traffic control areas for surveillance purpose. Later on footages from these cameras can be used for identifying individuals they can be criminals or suspects, missing persons and more. This proposed paper illustrates how to develop a sophisticated criminal identification system using Machine Learning and training artificial neural networks.

Keywords: LBPH Algorithm · Haar Classifier · ML · deep neural network · CCTV footage · deep learning techniques · image processing

1 Introduction

System presented in this article is a unique fusion of the best facial recognition, feature extraction, and final classification methods available today. Deep learning techniques such as the LBPH (Local Binary Pattern Histogram) algorithm have already proven to be elegant due to its computational simplicity and discriminative accuracy. A face recognition system generally extracts meaningful facial features such as distance between eyes, nose and chin length. These features are useful for classification and database matching. Apart from that it also uses several minute features of face that will also be used for comparison. The system involves two major processes: detection and recognition. Face recognition has two main methods: training and evaluation. Training phase is process of feeding the algorithm with a sample of the images, and the model is trained on the basis of dataset. The Evaluation phase compares the newly recorded test image with an existing database. For years, tracking criminals has been a difficult process. Previously, the entire method consisted of evidence-based clues found at the crime scene. Biological evidence can be easily tracked. But criminals are evolving and smarter than ever in that they cover their footsteps and leave no traceable evidence. This is where face recognition works. Faces are important to human identity, and because of their identifiability, each face is unique. Face recognition system is An exclusive biometric identification technique that employs an individual's facial features to autonomously detect and authenticate their identity from either a video frame or an image/photograph and later on it will be comparing against the faces that are already stored in the database. The facial recognition.

1.1 Significance of the Proposed System

The proposed system consists of 4 crucial steps, first is training the algorithm using real time images, second face detection the algorithm should be able to handle not just one face but multiple faces at same time, third is the comparison where real time images are compared with the dataset in the database and at last the result phase where the output of the comparison is generated. A simple outline would be when a face is detected in surveillance camera, those particular frames are pre-processed which reduces the noise and redundancies present in them. Then the processed image is ready for comparison where it is compared with the dataset which is already available in the database. At last if the comparison satisfies then the criminal's details are sent to mobile phone through firebase.

2 Related Work

1. A face image quality assessment is proposed considering 2 factors of image quality, one visual quality and second mismatch between training and test face images. A face image quality assessor is learned based on discriminate useful faces from unusual faces.
2. Heterogeneous face recognition (HFR) is coming into trend due to its ability to match face images across different domain that may be visible to infrared or visible to thermal. There are many established techniques to achieve it the proposed paper provides a benchmark of the evaluations of these techniques.
3. A autonomous agent is developed whose job is to provide information to teachers about students, it makes use of cameras placed at specific location in classroom and the data is collected by data collector module.
4. Matching faces captured in thermal and visible light images is really tough because they look different. Previous methods used either 2 step process or fuse different channels together but fusing images isn't enough. So this paper proposes a new method that uses 2 parts: a generator that makes visible images from thermal ones and a discriminator that checks how real these generated images look.
5. In heterogeneous face recognition (HFR) face modalities is challenging to overcome it Adversarial discriminative feature learning framework is used which relates raw-pixel to compact feature space.
6. This research focuses on recognizing faces in low light using special near-infrared (NIR) images. These images are often not as clear as regular pictures and can come from different angles. Matching them with high-quality visible images in databases is hard. The paper suggests a new method called "orthogonal dictionary alignment" to help match these images better.

7. CASIA NIR-VIS 2.0 is an updated version of NIR-VIS which overcomes the disadvantage of limited number of subject and also it does not lack certain evaluation protocol.
8. There are techniques that help match a 3D model of a face with a regular 2D picture of a face. This helps analyse the texture of the face. However, these techniques have a problem when parts of the face are hidden. A new method called as Deep Convolutional Neural Network (DCNN) which fills in the missing parts of the face's texture using the 3D model and the picture. To do this, a lot of complete face texture maps from different pictures and videos are gathered.
9. Surveillance Cameras captures data in NIR (near infrared) spectrum whereas law enforcement use (visible light) VIS spectrum. To eliminate this problem NIR images are reconstructed as VIS image and vice versa.
10. Traditional sketch artist take time to draw a sketch and also its expensive and can be errorful. A dual transfer face sketch is proposed which uses framework composed of inter-domain (deals with training photographs) transfer process and an intra-domain (deals with relating photographs to sketches) transfer process.
11. Photorealistic frontal view synthesis uses a single face it may be one's side face or an ill posed face the system would generate a full frontal view of face it is done by a Two-Pathway Generative Adversarial Network (TP-GAN) which simultaneously perceives global structure along with local details of the face.
12. An automated facial recognition system is developed using "Principle Component Analysis" approach which automatically detects and recognize face. It is proven that it can give an accuracy of about 80%.
13. Getting faces from surveillance footage manually can be a lengthy process to eliminate this a automatic face detection system is developed where real-time cropped image is saved which can be later used to track criminals.
14. To reduce crime a Text and face recognition technique is designed. Text identification can be used to identify vehicle's number plate whereas face recognition can be used to identify criminal involved in the crime.
15. A face detection and recognition system where "Haar cascading algorithm" is used. Haar cascading has classifiers which can track faces in openCV platform it makes use of many positive and negative images during training.
16. As crime rate increases it has become a global security issue. Many times criminals are found not guilty due to lack of identification to eliminate this an automated face recognition system has been proposed using Haar feature base cascade classifier.
17. An automated criminal identification system is developed for police department which uses footage from CCTV cameras which detects faces and the detected face is compared with the face in database if it matches it messages the name and place to the system.
18. Typically, a facial recognition system employs two stages: the initial phase involves detecting faces, followed by the subsequent stage of recognizing those faces. The provided document outlines the process of creating and implementing a facial recognition system through the utilization of deep learning techniques alongside openCV. The adoption of deep learning is motivated by its capability to achieve notably high levels of accuracy.

19. Even for police personnel crime prevention and crime detection is a challenging process as property and lives are at stake during crime so they have no other go then to let the criminals flee the crime scene. The potential for escape can be diminished by implementing a Face recognition system. In this context, the Viola-Jones framework is employed to identify the positions of faces and objects within CCTV footage, aiding in the reduction of their likelihood to evade capture.

20. Designing Face recognition system can be a challenging process even for modern days computer vision due to its Accuracy and speed identification issue. The system can even have higher grade issues when humans alter their facial features like wearing wig, contact lens, grow a moustache or beard. The paper provides an analysis of various face detection and recognition techniques and provides a face recognition technique with higher accuracy.

3 Methodology

3.1 System Architecture

The Architecture diagram as shown in Fig. 1 is a visual representation that maps out how the physical components are implemented for a software system. It also depicts the limitations and boundaries between each component. System architecture focuses on scalability, ensuring high availability, fault tolerance, and redundancy to prevent any potential data loss.

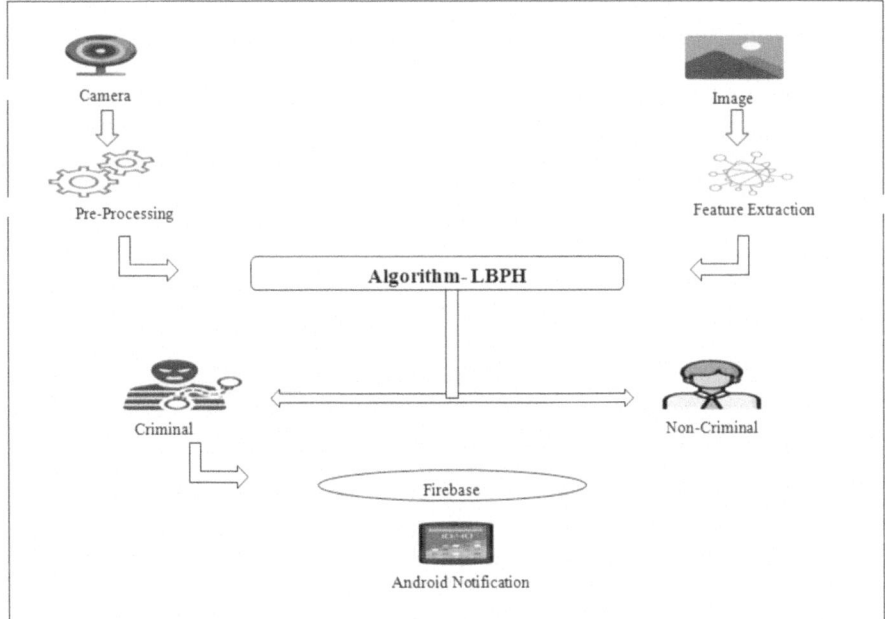

Fig. 1. Architecture Diagram

3.2 Modular Description

3.2.1 Dataset Collection

A dataset is a collection of similar or discrete items of related data which can be accessed either as an individual or as a combination (consist of multiple data). These data samples can be of a image, audio or some text. In this case it is going to be images (faces captured in frames), these collection of real time facial images is used to create a dataset by saving them in a local machine. This collected data will be trained for getting high accuracy result.

3.2.2 Training Algorithm

In this phase a real time face captured by the camera device is used and it will be trained using LBPH (Local Binary Pattern Histogram) Algorithm. This algorithm is capable of detecting multiple faces at same time and it also is able to identify the face of the person from different perspectives. The algorithm uses matrices to store the colour value of each pixels and then assigns them a binary value based upon a threshold key which later gets converted to equivalent decimal and the very next step would be to create a histogram which is done by counting the occurrence of each colour in each square (matrix).

3.2.3 Detection and Notification

Face detection plays a crucial role in the process of face recognition. The algorithm's capacity to autonomously identify faces within video frames is vital. This is achieved through training the algorithm on abundant positive datasets (images containing faces) and negative datasets (images lacking faces). Subsequently, during the comparison phase, the algorithm evaluates the detected face against the faces present in the dataset.

3.2.4 LBPH Algorithm

The algorithm is a combination of 2 procedures.

One is LBP (Local Binary Pattern) is an efficient texture operator that assigns labels to neighbouring pixels within an image based on a specified threshold value, later on it gets converted to binary and then into a decimal equivalent and thus the final result would be a gradient of the original image.

Secondly HOG (Histogram of Gradients) produces histograms based upon the no of occurrence of each colour in the gradient (here it means a pixel).

The algorithm utilizes four primary parameters:

1. Radius: This parameter contributes to the construction of a circular local binary pattern. Typically, its value is set to 1.
2. Neighbours: This parameter signifies the count of sample points used to construct the circular local binary pattern. Typically, it's set to 8.
3. Grid X: This parameter indicates the quantity of cells in the horizontal direction. Increasing the number of cells results in a higher-dimensional feature vector. Usually, it's set to 8.

4. Grid Y: This parameter indicates the quantity of cells in the vertical direction. Similar to Grid X, a higher value leads to a higher-dimensional feature vector. Typically, it's set to 8.

The very first step is to train the algorithm using correct dataset with people's facial images. To perform the computational step, the image is transformed into a matrix.

In Fig. 2, a set of 3 × 3 block is considered by doing so we can point each and every feature that physically exists the face. Each pixel has a range 0–255 and they get converted to grayscale format (B/W) (Table 1).

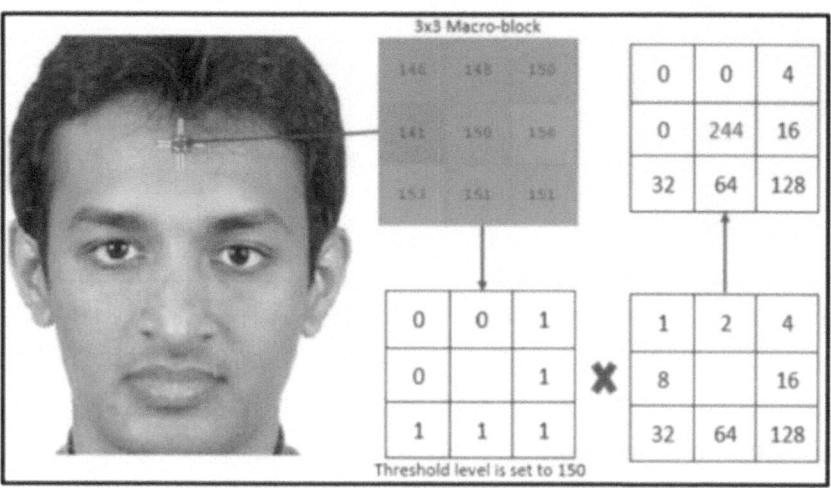

Fig. 2. 3 × 3 Macro Block Representation

Table 1. Accuracy of LBPH Algorithm

Algorithm	Dataset	Detected Faces	Hit Rate	Detection Speed (ms)
LBPH	Color FERET	1004/1127	89%	95.44924
	MIT	1779/2000	89%	101.8864
	Taarlab	674/759	88.8%	104.9335
Overall		**3457/3886**	**89%**	**101**

Next the algorithm generates values for the 3 × 3 block using the colour difference. Which need to be converted into a standard binary pattern. In order to do it the centre pixel value is taken as threshold value and the neighbouring pixel value are set up based upon the threshold for instance any value greater than the threshold will be assigned 1 and the value lower would be assigned 0. After assigning binary value the algorithm concatenates the numbers in a clock wise manner. The binary value is next converted to decimal and will be forwarded to HOG.

HOG accepts the resultant LBP image as shown in Fig. 3 and it is in its gradient form where is equivalent histograms are generated. The algorithm makes use of Grid X and Grid Y to form macro blocks and each block will have its own separate Histogram. The Histogram as shown in Fig. 3 at the end gets concatenated to form a cumulative histogram.

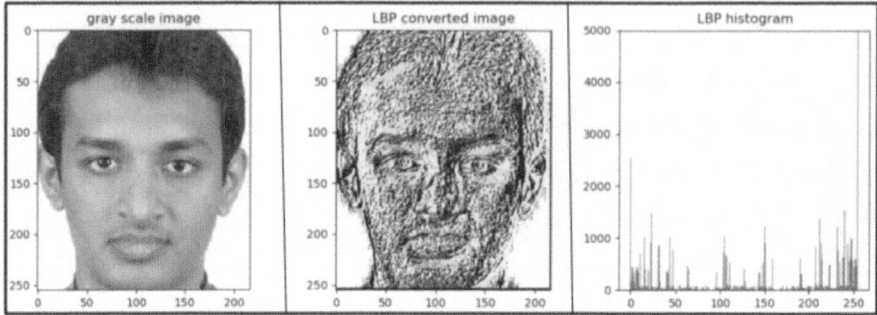

Fig. 3. (Left) gray scale image, (center) LBP image and (right) histogram of the LBP image

4   Result and Discussion

For the performance evaluation it is clearly seen that the model is capable of detecting not one but multiple faces in a single frame and if they are present in the dataset they are eventually recognized.

Here the value of epoch is set to 20 as the accuracy will not get any better even if we increase that value also it will be costing computational resource and training time.

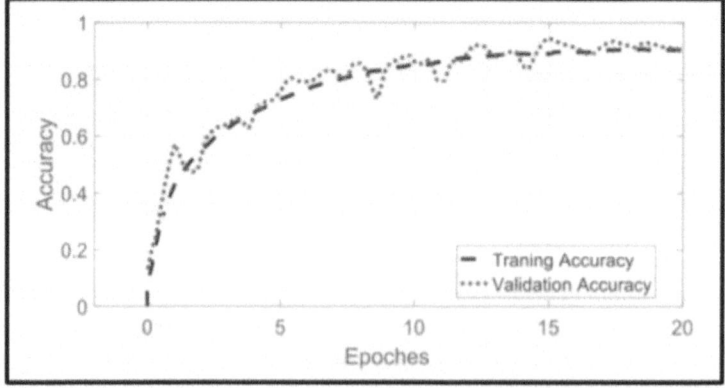

Fig. 4. Trained Dataset vs Validation Accuracy

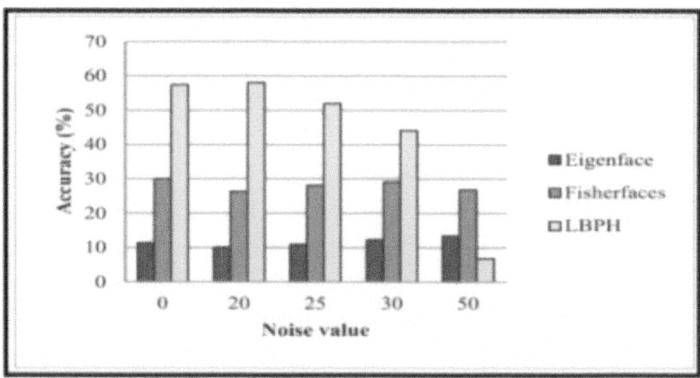

Fig. 5. Comparative of recognition accuracy with different noise values

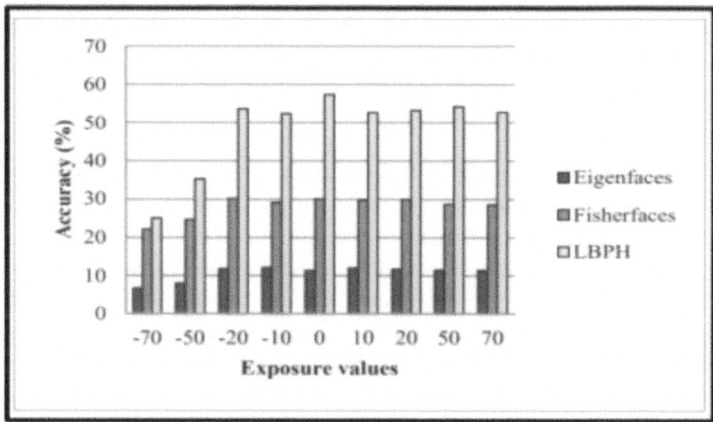

Fig. 6. A comparative of recognition accuracy with different Exposure values

Above graphs as shown in Fig. 4, Fig. 5, Fig. 6, Fig. 7 show a comparative chart between popular face detection algorithms like Eigenface, Fisher faces and LBPH and it is clearly seen that in most of the cases LBPH algorithm surpasses the other face recognition algorithms.

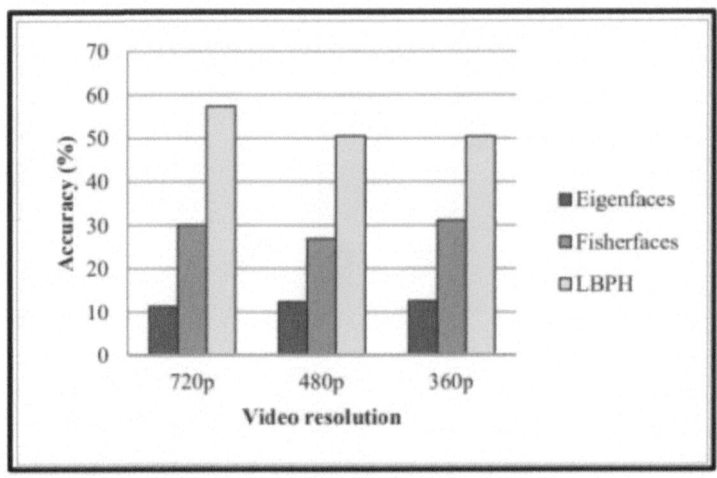

Fig. 7. A comparative of recognition accuracy with different Video resolution.

5   Conclusion and Future Scope

The paper proposes a effective method of facial recognition and how it can be used in criminal identification. The LBPH algorithm used provides a high accuracy result and it also works well in crowded places as it has multiple face detection feature available in it. The LBPH algorithm uses different lighting position to give a better accuracy than most of the other face recognition techniques. In addition to it LBPH can even be trained with a single image. When the application is used in a large scale, say multiple CCTV cameras are installed in multiple locations the criminal or suspect information can be easily gathered. The system can also be used to find runaways, missing persons and lastly a website can also be designed which includes additional details about the criminals such as his age, gender, crime etc.

References

1. Ouyang, S., Hospedales, T., Song, Y.Z., Li, X., Loy, C.C., Wang, X.: A survey on heterogeneous face recognition: sketch, infra-red, 3D and low-resolution. Image Vis. Comput. **56**, 28–48 (2016)
2. Canedo, D., Trifan, A., Neves, A.J.R.: Monitoring students' attention in a classroom through computer vision. In: Bajo, J., et al. (eds.) PAAMS 2018. CCIS, vol. 887, pp. 371–378. Springer, Cham (2018). https://doi.org/10.1007/978-3-319-94779-2_32
3. Lezama, J., Qiu, Q., Sapiro, G.: Not afraid of the dark: NIRVIS face recognition via cross-spectral hallucination and low-rank embedding. In: IEEE Conference on Computer Vision and Pattern Recognition (2017)
4. Song, L., Zhang, M., Wu, X., He, R.: Adversarial discriminative heterogeneous face recognition. In: AAAI Conference on Artificial Intelligence (2018)
5. Mudunuri, S.P., Venkataramanan, S., Biswas, S.: Dictionary alignment with re-ranking for low-resolution NIR-VIS face recognition. IEEE Tran. Inf. Forensics Secur. **14**(4), 886–896 (2019)

6. Li, S.Z., Yi, D., Lei, Z., Liao, S.: The CASIA NIR-VIS 2.0 face database. In: IEEE Conference on Computer Vision and Pattern Recognition Workshops, pp. 348–353 (2013)
7. Deng, J., Cheng, S., Xue, N., Zhou, Y., Zafeiriou, S.: UV-GAN: adversarial facial UV map completion for pose-invariant face recognition. In: IEEE Conference on Computer Vision and Pattern Recognition (2018)
8. Juefei-Xu, F., Pal, D.K., Savvides, M.: NIR-VIS heterogeneous face recognition via cross-spectral joint dictionary learning and reconstruction. In: IEEE Conference on Computer Vision and Pattern Recognition Workshop (2015)
9. Zhang, M., Wang, R., Gao, X., Li, J., Tao, D.: Dual-transfer face sketch photo synthesis. IEEE Trans. Image Process. 28(2), 642–657 (2019)
10. Huang, R., Zhang, S., Li, T., He, R.: Beyond face rotation: global and local perception GAN for photorealistic and identity preserving frontal view synthesis. In: IEEE International Conference on Computer Vision, pp. 2458–2467 (2017)
11. Wang, K., He, R., Wang, L., Wang, W., Tan, T.: Joint feature selection and subspace learning for cross-modal retrieval. IEEE Trans. Pattern Anal. Mach. Intell. 38(10), 2010–2023 (2016)
12. Zhang, K., Zhang, Z., Li, Z.: Joint face detection and alignment using multi-task cascaded convolutional networks. IEEE Sig. Process. Lett. 23(10), 1499–1503 (2017)
13. Yang, B., Yan, J., Lei, Z., Li, S.Z.: Aggregate channel features for multi-view face detection. In: IEEE International Joint Conference on Biometrics (2014)
14. Viola, P., Jones, M.J.: Robust real-time face detection. Int. J. Comput. Vis. 57(2), 137–154 (2004)
15. Yang, S., Luo, P., Loy, C.C., Tang, X.: From facial parts responses to face detection: a deep learning approach. In: IEEE International Conference on Computer Vision (2015)
16. Kan, M., Shan, S., Zhang, H., Lao, S., Chen, X.: Multi-view discriminant analysis. IEEE Trans. Pattern Anal. Mach. Intell. 38(11), 188–194 (2016)
17. Gui, J., Li, P.: Multi-view feature selection for heterogeneous face recognition. In: IEEE International Conference on Data Mining, pp. 983–988 (2018)
18. Shao, M., Fu, Y.: Cross-modality feature learning through generic hierarchical hyperlingual-words. IEEE Trans. Neural Netw. Learn. Syst. 28(2), 451–463 (2016)
19. Gong, D., Li, Z., Huang, W., Li, X., Tao, D.: Heterogeneous face recognition: a common encoding feature discriminant approach. IEEE Trans. Image Process. 26(5), 2079–2089 (2017)
20. Peng, C., Wang, N., Li, J., Gao, X.: Re-ranking high-dimensional deep local representation for NIR-VIS face recognition. IEEE Trans. Image Process. 28, 4553–4565 (2019)
21. Wu, X., He, R., Sun, Z., Tan, T.: A light CNN for deep face representation with noisy labels. IEEE Trans. Inf. Forensics Secur. 13(11), 2884–2896 (2018)
22. Jin, Y., Lu, J., Ruan, Q.: Coupled discriminative feature learning for heterogeneous face recognition. IEEE Trans. Inf. Forensics Secur. 10(3), 640–652 (2015)
23. Reale, C., Nasrabadi, N.M., Kwon, H., Chellappa, R.: Seeing the forest from the trees: a holistic approach to near-infrared heterogeneous face recognition. In: IEEE Workshop on Perception Beyond the Visible Spectrum, pp. 54–62 (2016)
24. Abdullaha, N.A., Saidi, M.J.: Face recognition for criminal identification: an implementation of principal component analysis for face recognition. In: The 2nd International Conference on Applied Science and Technology (ICAST 2017) (2017)
25. Patil, S.P., Yadav, G.S.: Criminal identification for low resolution surveillance. VIVA-IJRI 1(4), 1–6 (2021). Article 6

Automated Essay Grading System for IELTS Using Bi-LSTM

Chandan Kumar Sangewar, Chinmay Pagey, Aman Kumar, and R. Krithiga(⊠)

Vellore Institute of Technology, Chennai, India
krithiga.r@vit.ac.in

Abstract. Using machine learning and natural language processing methods, this paper intends to create an automated essay grading system. Essays submitted for the IELTS and TOEFL tests are evaluated using Long Short-Term Memory (LSTM) and Support Vector Machine (SVM) models. The essays are graded on a scale from 1 to 5 based on the system's analysis of many factors, such as cohesion, syntax, vocabulary, phraseology grammar and conventions. The model learns the associations between language characteristics and evaluation scores by studying a dataset of essays that have already been graded. To evaluate the system's performance, it is first applied to a new pool of essays. Since the suggested approach does not rely on human subjectivity or prejudice, it has the potential to enhance the efficiency and uniformity of essay grading. As a whole, this work illustrates how machine learning and NLP may be used to create useful tools in the field of language evaluation and instruction.

Keywords: Automatic Essay Evaluation (AES) · TOEFL (Test of English as a Foreign Language) · IELTS (International English Language Testing System) · Word embedding · Score

1 Introduction

Automatic Essay Evaluation, often known as AEE, is a method that may quickly and objectively assess student compositions. In recent years, this method has gained greater popularity. There is need for development in both the accuracy and the scalability of AEE systems that are based on methods of machine learning. These approaches have shown promise in increasing the precision of grading, but there is still room for improvement in both of these areas [1]. The purpose of this research is to increase the accuracy and scalability of AEE by integrating the most beneficial aspects of several machine learning methods that are already in use. The suggested AEE system makes use of deep learning methods to extract important features from text as well as Learning Vector Quantization in order to classify essays according to varied degrees of difficulty (LVQ) [3]. The LVQ technique is a kind of neural network that has been found to be successful in a variety of natural language processing tasks. It is especially well-suited for tasks that require categorization. The suggested AEE system makes use of a multi-model approach, which blends the best parts of multiple different machine learning approaches, in order

R. Geetha et al. (Eds.): AAIMB 2023, CCIS 2202, pp. 280–291, 2025.
https://doi.org/10.1007/978-3-031-73065-8_23

to increase the system's accuracy and scalability. Since there are no specific writing tasks included into the system, it may easily be adapted for use in a broad range of contexts requiring writing [5]. It is hoped that the proposed AEE method would reduce the amount of work required of teachers while simultaneously increasing the level of impartiality and effectiveness of essay grading.

1.1 Statement

An automated essay grading system using natural language processing (NLP) is a promising approach to address the challenges of grading essays manually. This approach can provide several benefits, including increased efficiency, consistency, and objectivity in grading, individualized feedback to students, improved student engagement and performance, cost-effectiveness, and improvement in the overall educational system. However, developing a robust automated essay grading system using NLP requires addressing challenges such as the lack of training data, ambiguity and subjectivity in grading, understanding context and intent, handling non-standard language, dealing with plagiarism, and adapting to different grading scales. By overcoming these challenges, an automated essay grading system can offer a more efficient and personalized grading solution, ultimately enhancing the learning experience of students and supporting the work of teachers and professors. Our approach is to use LSTM and SVR for our research.

1.2 Motivation

Developing an automated essay grading system using natural language processing (NLP) can have several motivations. Firstly, it can increase efficiency as grading essays can be a time-consuming task for teachers and professors. By automating the grading process, it can help save time and increase productivity. Secondly, it can ensure consistency and objectivity in grading. Human grading can be influenced by factors such as mood, bias, and fatigue, leading to inconsistent and subjective grading. An automated grading system can help ensure more consistent and objective grading. Additionally, an automated essay grading system can provide individualized feedback to each student based on their writing strengths and weaknesses, allowing for a more personalized learning experience. It can also help improve student engagement and performance by providing timely and actionable feedback, creating a more personalized learning experience, and reducing the cost of grading essays, as it eliminates the need for hiring additional staff for grading purposes. Lastly, it can improve the overall educational system by providing consistent and objective grading, enabling teachers to focus on other important tasks and creating a more personalized learning experience for students.

1.3 Challenges

Some potential challenges in developing an automated essay grading system using natural language processing (NLP) are as follows:

- Lack of training data: Developing a robust automated essay grading system requires a large corpus of well-annotated essays. However, such data may be scarce, and even when available, it may not be diverse enough to capture the full range of writing styles and topics.

- Ambiguity and subjectivity: Grading essays often involves subjective judgment, and there may be multiple valid interpretations of a given essay. This ambiguity makes it challenging to develop an automated grading system that is consistent and accurate.
- Understanding context and intent: To accurately grade an essay, an automated system needs to understand not just the surface-level meaning of the text but also the underlying context
- to students that can help them improve their writing skills. This requires not only accurate grading but also the ability to provide targeted feedback based on specific writing and intent. This requires advanced NLP techniques such as semantic analysis and deep learning models.
- Handling non-standard language: Essays may contain non-standard language, such as slang, dialects, or specialized terminology, that can be difficult for an automated system to parse and analyzed.
- Dealing with plagiarism: Automated essay grading systems need to be able to detect plagiarism and accurately attribute authorship. This requires advanced techniques for text similarity analysis and authorship attribution.
- Providing meaningful feedback: Automated grading systems need to be able to provide meaningful feedback strengths and weaknesses.
- Adapting to different grading scales: Different educational institutions and programs may use different grading scales, making it challenging to develop a one-size-fits-all grading system. An automated grading system may need to be adapted to different grading scales and criteria to be useful across different contexts.

2 Exploratory Data Analysis

a. Dataset
The dataset which we use is taken from English language learning for grades. The essay dataset used in the study includes several features that are essential for developing an accurate automated essay grading system. The features include text_id, full_text, cohesion, syntax, vocabulary, phraseology, grammar, and conventions. The full_text feature includes the entire essay text, while cohesion measures how well the essay is structured and organized. Syntax evaluates the use of grammatical structures and sentence construction, while vocabulary measures the range and complexity of the words used in the essay. The phraseology feature evaluates the essay's idiomatic expression and figurative language, while grammar measures the essay's adherence to grammatical rules. Finally, conventions evaluate the essay's adherence to formatting and stylistic guidelines, such as the use of punctuation, capitalization, and spelling.

b. Pre-processing
As part of our effort to automate the grading of essays using NLP methods, we subjected the essay dataset to a number of preprocessing processes. Since non-printing characters like /t, /r, \n might introduce noise into the data and divert the model's attention from the most relevant features, we first eliminated them from the essays. Second, we reduced the dataset's dimensionality by eliminating often used, meaningless terms known as "stop-words." To further clean up the information, we also used stemming and lemmatization to

break down the words into their elemental forms. In addition, we scrubbed the dataset of extraneous information to make sure the model only considers what is really important. The repetitive nature of the preparation stages allowed us to try out many approaches before settling on the one that worked best for our dataset [2]. In order to guarantee that the model could train properly and effectively, we preprocessed the dataset extensively. This allowed the model to provide insightful feedback to students and aid educators in their daily tasks.

We also employed feature engineering, a kind of preprocessing, to increase the model's accuracy by generating additional features from the current data. For instance, we tallied up the word count of each essay and used it as a criterion, as lengthier essays have been shown to have a greater degree of complexity and maybe call for a more advanced command of the English language. Since longer sentences could signify more complicated concepts, we also factored in the average amount of words used in a sentence.

We also determined the essay's readability score, which indicates how straightforward it is to read and comprehend. We were able to improve the model's assessment of the essays' quality by adding in this additional data point [7, 11]. To further ensure the model could comprehend the text input, we additionally used tokenization and vectorization. For the model to be able to assess and interpret the data, the text must first be tokenized, or split up into individual words or phrases. However, vectorization includes transforming the text into numerical vectors that reflect the frequency of individual words and phrases. In order to minimize the dimensionality of the data and provide the model with the necessary information for learning, we employed vectorization, which included methods like one-hot encoding and TF-IDF.

c. Visualization

We used data visualization to gain insights into our dataset's most frequent n-grams and each metric's average score. We found that the most frequent n-grams in the dataset were 2-grams and 3-grams, indicating that the essays contain a lot of phrases and collocations. By visualizing these n-grams, we could see the common patterns and structures used in the essays, which can help in identifying specific features that contribute to high or low scores. We also found that the essays' average score varied significantly for each metric, with cohesion and syntax receiving higher scores on average than vocabulary and phraseology. By visualizing these differences, we could identify the areas where the model may need improvement and adjust our approach accordingly. Furthermore, by visualizing the distribution of scores for each metric, we could see the patterns and trends in the dataset. For example, we found that the grammar and conventions metrics had a higher concentration of lower scores, indicating that the model may have difficulty identifying errors in these areas. We also found that the vocabulary metric had a more evenly distributed score distribution, indicating that the model was better at identifying the range of vocabulary used in the essays. By visualizing these trends, we could identify the areas where the model needs improvement and adjust our approach accordingly. Overall, data visualization was a valuable tool for gaining insights into our dataset and identifying areas for improvement in our automated essay grading system using NLP techniques (Fig. 1 and 2).

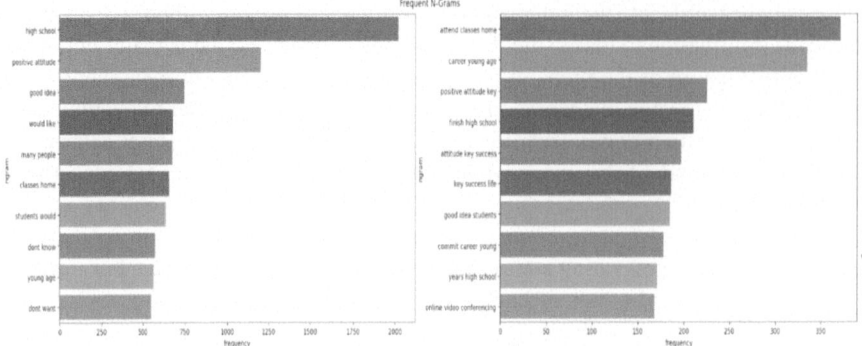

Fig. 1. Shows the Frequent N-Grams

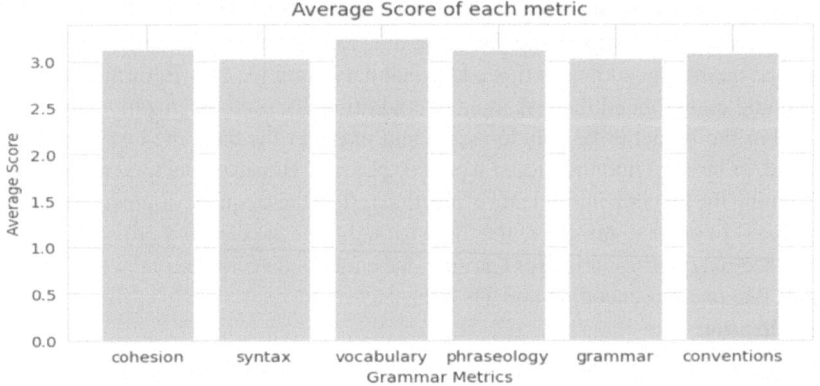

Fig. 2. A visualization of the Average Score of each metric in a Grammar Metrics VS Average Score Graph

3 Model Building

3.1 LSTM Model

The first step is to bring in the required libraries and begin data reading using pandas. Next, we use the Tensor Flow library's Text-vectorization layer to tokenize the textual input. As a result, we may use the numbers to represent the text in the model. We restricted the output sequence length to 800 tokens and the maximum number of tokens to 37,000. Constructing the LSTM model comes next. After the input layer is defined, the textual data is vector zed using the Text-vectorization layer. The numerical information is then embedded into a dense vector representation using an Embedding layer. This is followed by three LSTM layers, each with 'return_sequences' set to TRUE and a different number of units (128, 128, and 64). This indicates that the results from one LSTM layer will be used as input for the following layer. In order to avoid over fitting, we add two Dense layers, each with 128 units, after the LSTM layers. Finally, six output layers are added, each consisting of a single unit and activated by the ReLU function. The model is built using the Adam optimizer and the mean squared error loss. Over fitting is avoided by fitting the model in batches of 32 and quitting early. We next build a function to assess how well the model does on test data and another to generate predictions on fresh text data. We also add a new text data point for prediction and load in the test data. Using the response function, we feed this fresh text input into the model and get back the projected scores for each measure. At each time step, the LSTM layer's output is a function of both the current input and the state stored in its memory. Three LSTM layers, each with a different number of units, are used in our model to account for varying degrees of generalization in the text input. The output of one LSTM layer may be used as input for the following layer if **'return_sequences' = TRUE** is set. As a result, the model may discover subtler connections between the input data and the final ratings.

3.2 Multi Output Regressor and SVR

In scikit-learn, multi-output regression is handled via the Multi Output Regressor wrapper. It's meant to work with non-native multi-output regression models to add such capability. We begin by characterizing the characteristics and the end points. There are six different aspects of written communication that make up these features: (i) cohesiveness; (ii) syntax; (iii) vocabulary; (iv) phraseology; (v) grammar; and (vi) norms. A data frame containing these feature values from the training dataset constitutes the target variable. Then, we create a single variable called "text" that contains both the training and testing text data. We then created two new variables, 'X' and 'X_test,' where 'X' would hold the training data and 'X_test' the test data. The scikit-learn library's TFIDF-Vectorizer is used for the textual data processing. Machine learning models need a numerical representation of the data, and this vectoriser provides that. The stop words were set to "english," so that common English terms like "the," "a," and "an" would be deleted from the text before vectorisation. Terms that occur more than 50% of the time in the documents are disregarded, while terms that occur less than 1% of the time are disregarded. We also set the minimum document frequency to 0.01%. The 'fit_transform' function is used to optimize the vectoriser for the training set, whereas the 'transform' function is used to normalize the test set. Then, the regression model will be established. When there are many possible outcomes to be predicted, we employ the Multi Output Regressor. For this purpose, we use a special kind of Support Vector Machine (SVM) known as the Support Vector Regression (SVR) method. To assess the model's efficacy, we utilise the scoring method and SVR to fit the training data. Finally, we use the trained model to generate predictions on the test data and save those predictions in the 'predictions' variable. For the first data point in the test set, the 'predictions' variable stores an array containing predicted values for all six attributes. We achieve better accuracy when the second approach/method was followed.

4 Results

To begin, we used a dataset of essays to train an LSTM model, which ultimately achieved an accuracy of 65.65%. Although this is a solid outcome, we had hoped to find a better way to forecast essay grades. We then applied the Multi Output Regressor and SVR models to the same dataset and discovered that they achieved a much higher accuracy (86.69%) than the LSTM model. This proves that these models are superior to the LSTM model and can accurately predict essay grades. The Multi Output Regressor and SVR models may have performed better because they are better equipped to handle the intricate, multidimensional nature of the essay grading problem. In contrast to more basic models like LSTM, which may fail to capture the subtleties of the many grading criteria, these models are built to handle numerous output variables. Our findings indicate that Multi Output Regressor and SVR are promising alternatives to conventional LSTM models for automating the grading of essays. However, further study is required to establish the best method for this endeavor and to isolate the elements that boost these models' efficiency (Fig. 3).

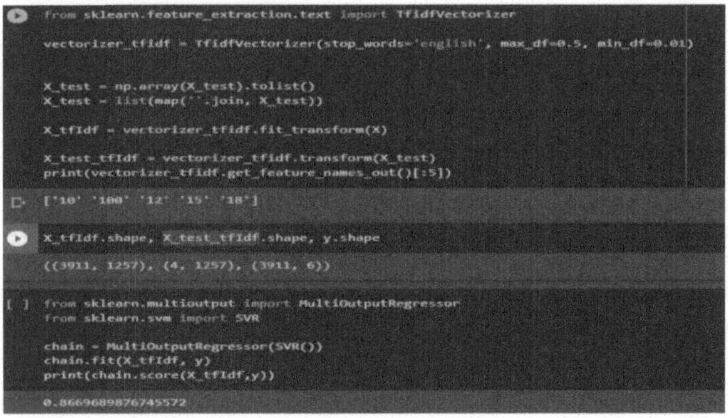

Fig. 3. The snapshot of accuracy of the results along with the score based on different criteria "'cohesion', 'syntax', 'vocabulary', 'phraseology', 'grammar', 'conventions'" respectively of Multi Output Regressor and SVR Model.

5 Discussion

This review compares essay-grading studies. NLP-based writing feature analysis tools and a neural network grading engine that assesses student answers based on pre-graded essays were examined in the article. Intensively examined. Datasets from Mansoura University computer and information science students were assessed. Linear regression, clustering, classification, 5-fold cross validation, forward feature selection, and Quadratic Weighted Kappa error assessment trained the model. These were used in forward feature selection. N-Grams and content testing may improve the article's final model, which had a kappa of 0.73 across all eight essay sets.

A Two-Stage Learning Framework (TSLF) that uses feature-engineered and end-to-end Automated Essay Score (essay grading) methods offered a third alternative for automated essay grading. Two-stage learning framework [29]. on five-eighths of the prompts and achieved new state-of-the-art average performance with no negative samples. "Taking an essay grading system to its logical conclusion" included task-independent features [17, 18]. Our study showed that our task-independent feature collection boosts model transfer across tasks, especially similar tasks. Several studies suggest that quickly assessing many student works with comments is difficult. It's academic.

AES is popular because to technology and the necessity for faster, more accurate grading. Faster, more objective, and consistent essay grading isn't perfect. Language may be misjudged in essay grading. AES technology should support human assessors. Essay grading is popularized by natural language processing and machine learning. AES systems may misinterpret student writing. The technologies provide instructors and students quick, accurate, and objective feedback. The authors argue training data quality, evaluation criteria, and essay topic influence essay grading system performance. Criterion uses NLP and ML to evaluate student writing's grammar, mechanics, organization, and content. The study suggests using AEE systems like Criterion to quickly and objectively evaluate student writing, identify needs, and reduce teacher labour. In the study, AEE systems' failure to assess creative and critical thinking, gender, racial, and language discrimination, dependability, and equality are discussed.

Our AEE system uses NLP and sentiment analysis to evaluate student writing. The system verifies academic articles. To evaluate their AEE system, human assessors reviewed the algorithm's score for bias and inaccuracy [24]. This review examines rule-based, machine learning, and hybrid AES. Rule-based methods are simpler yet inaccurate. Rules-based systems are simpler. Machine learning enhances statistical and NLP-based student text analysis [19]. Essay grading systems help instructors grade papers faster and more equitably, yet they may favour specific student groups [9]. For fairness, the authors suggest revising automated essay scoring (AES) systems and improving machine learning algorithms to account for student writing differences. Automatic essay grading improves objectivity, student responses, and teacher stress, but it makes judging advanced writing skills and impartiality harder. Rule-based, machine learning, and hybrid AEE methods are reviewed here. Rule-based systems may not match writing. Machine learning algorithms can learn from massive human-graded essay datasets. AEE systems employ rules and NLP.

Another popular essay grader is the TOEFL. Ancient methods. These tools provide writers and students frequent feedback. However, essay grading by machine is flawed [12]. When editing English writings, they ignore regional Indian languages. Frameworks for limits are needed. Article analysis will use deep learning framework layers like LSTM and dense. LSTM and thick layers evaluate essays.

The framework handles pre-processing, language translation, feature extraction, and grade assignment [4]. The recommended technique captures human language complexity, reduces local Indian language effect on English writings, and is more accurate than automated essay grading approaches. Introduction AES speeds and improves essay scoring. Early approaches struggled with score prediction, user perspective, and unwanted content. These problems demand a scalable, adaptive writing style and fast relevance assessment mechanism. To match human ratings, this initiative automates essay grading. In our study, LSTM had an accuracy of 65.65%, whereas Multi Output Regressor with SVR had 86.69% accuracy. Compared to other models' past investigations.

5.1 Novelty of the Study

Our study focused on evaluating the essays that are presented for IELTS and TOEFL so that the evaluator's work load is reduced and the results can be posted much prior than compared to the time they take to evaluate the essays now. There were study done on the same or similar field which gave the following results (Table 1).

Table 1. Comparative Study of Accuracy of Results with Different Models

Model	Accuracy
The research evaluates rule-based and machine learning automated essay scoring techniques for assessing the moral content of student writings	The tested approaches' accuracies range from around 60% for the worst to 76% for the best
The research investigates the potential of Latent Semantic Analysis (LSA) in essay evaluation by comparing student and expert-generated reference writings for semantic similarities	LSA-based essay grading is said to have an accuracy of 70–80%
In this study, the authors use essay feature extraction to probe the feasibility of combining NLP with machine learning methods for AEE	The research shows that using established rubrics for grading essays may result in an accuracy rate of between 78% and 82%
Using natural language processing methods, the study suggests a way for assessing persuasive writings by locating viewpoints and their intended recipients	It has been estimatedthat opinion andtarget identification accuracy is between 70% and 80%
The research evaluates rule-based and machine learning automated essay scoring techniques for assessing the moral content of student writings	The tested approaches' accuracies range from around 60% for the worst to 78% for the best

From the similar studies, we can see how different methods have been used and the level of accuracy they have obtained. There are several other studies which gives more or less the same level of accuracy with different models and datasets. Our study gives the best accuracy score of 86.69% to 93.42% to evaluate cohesion, syntax, vocabulary, phraseology grammar and conventions of any given essay.

5.2 SOTA

You probably already know that our model was trained using data from the top essays in our dataset. Inputs to our model consist just of essay IDs; for instance, if we wish to submit an essay for grading, we simply ask for the essay's unique ID. The essay is checked and sent on for grading after some time has elapsed. Here, the best model ('Multi Output Regressor and SVR') is used to assess the essay and provide a score (out of 5) based on factors like the essay's coherence, syntax, vocabulary, phraseology, grammar, and norms. The essay's grade will reflect how well you do.

6 Conclusion

In conclusion, we have developed a model for automated essay grading using two different approaches: LSTM and Multi Output Regressor with SVR. The LSTM model showed promising results with an accuracy of 65.65%. However, we observed that the Multi Output Regressor with SVR model performed significantly better with an accuracy of 86.69%. As for output, we get the score of the essay based on different criteria that are required to evaluate an IELTS and TOEFL essay which includes cohesion, syntax, vocabulary, phraseology grammar and conventions. We believe that the reason behind the improvement in accuracy is that the Multi Output Regressor with SVR model is a more robust and flexible approach to predict the scores of multiple criteria used for essay grading. This approach considers the interdependence between these criteria, and hence, provides better results. Overall, our study provides a feasible solution for automating the essay grading process based on cohesion, syntax, vocabulary, phraseology grammar and conventions of the essay. It can be utilized by educators and students to evaluate their writing skills and identify areas of improvement. Moreover, it can help reduce the workload of teachers and enable them to focus on other aspects of teaching. We hope that our study will encourage further research in this area and pave the way for more sophisticated automated essay grading systems in the future.

7 Future Works

Finally, further research is needed to determine how well they work and what limits they have. We suggest more study be conducted to determine effective methods of assessing student writing quality and to enhance the validity and reliability of automated essay evaluation (AES) systems. These method has the potential to enhance the effectiveness and impartiality of essay evaluation in educational contexts, but further study is needed to alleviate fears and enhance the validity of essay grading systems before they are used in classrooms. The authors acknowledge the limitations of AES systems and the concerns that have been raised about their usage in the classroom, but stress the need for continuing research and development to increase the reliability and validity of essay grading systems.

Data Availability Statement. The data supporting the findings and the reported results, including analyzed archived datasets, are available from the corresponding author upon reasonable request.

Conflicts of Interest. The authors declare no conflict of interest.

Code. The code for the following study could be found here in the following GitHub repository https://github.com/cnd-sw/Automatic-Essay-Evaluation.

References

1. Ramesh, D., Sanampudi, S.K.: An automated essay scoring systems: a systematic literature review. Artif. Intell. Rev. **55**, 2495–2527 (2022). https://doi.org/10.1007/s10462-021-10068-2
2. Ifenthaler, D.: Automated essay scoring systems. In: Zawacki-Richter, O., Jung, I. (eds.) Handbook of Open, Distance and Digital Education, pp. 1057–1071. Springer, Singapore (2023). https://doi.org/10.1007/978-981-19-2080-6_59
3. Susanti, M.N.I., Ramadhan, A., Warnars, H.L.H.S.: Automatic essay exam scoring system: a systematic literature review. Procedia Comput. Sci. **216**, 531–538 (2023). ISSN: 1877-0509. https://doi.org/10.1016/j.procs.2022.12.166
4. Srivastava, K., Dhanda, N., Shrivastava, A.: An analysis of automated essay grading systems. Int. J. Recent Technol. Eng. (IJRTE) **8**(6), 5438–5441 (2020). ISSN: 2277-3878
5. Sadanand, V.S., Guruvyas, K.R.R., Patil, P.P., Acharya, J.J., Suryakanth, S.G.: An automated essay evaluation system using natural language processing and sentiment analysis. Int. J. Electr. Comput. Eng. **12**(6), 6585–6593 (2022). ISSN: 2088-8708
6. Doewes, A., Saxena, A., Pei, Y., Pechenizkiy, M.: Individual fairness evaluation for automated essay scoring system. International Educational Data Mining Society (2022)
7. Liu, J., Xu, Y., Zhu, Y.: Automated essay scoring based on two-stage learning. arXiv preprint arXiv:1901.07744 (2019)
8. Williams, R.: Automated essay grading: an evaluation of four conceptual models. In: New Horizons in University Teaching and Learning: Responding to Change, pp. 173–184. Centre for Educational Advancement, Curtin University (2001)
9. Burstein, J., et al.: Automated scoring using a hybrid feature identification technique. In: Proceedings of 36th Annual Meeting of the Association for Computational Linguistics, Montreal, Canada, pp. 206–210 (2017)
10. Aluthman, E.S.: The effect of using automated essay evaluation on ESL undergraduate students' writing skill. Int. J. Engl. Linguist. **6**(5), 54–67 (2016)
11. Ding, Y., et al.: Don't take for an answer–the surprising vulnerability of automatic content scoring systems to adversarial input. In: Proceedings of the 28th International Conference on Computational Linguistics (2020)
12. Dong, F., Zhang, Y.: Automatic features for essay scoring–an empirical study. In: Proceedings of the 2016 Conference on Empirical Methods in Natural Language Processing, pp. 1072–1077 (2016)
13. Horbach, A., Zesch, T.: The infuence of variance in learner answers on automatic content scoring. Front. Educ. **4**, 28 (2019). https://doi.org/10.3389/feduc.2019.00028
14. Knill, K., Gales, M., Kyriakopoulos, K., et al.: Impact of ASR performance on free speaking language assessment. In: Interspeech 2018, Hyderabad, India, 02–06 September 2018. International Speech Communication Association (ISCA) (2018)
15. Kopparapu, S.K., De, A.: Automatic ranking of essays using structural and semantic features. In: 2016 International Conference on Advances in Computing, Communications and Informatics (ICACCI), pp. 519–523 (2016)
16. Kumar, Y., Aggarwal, S., Mahata, D., Shah, R.R., Kumaraguru, P., Zimmermann, R.: Get it scored using autosas—an automated system for scoring short answers. In: Proceedings of the AAAI Conference on Artifcial Intelligence, vol. 33, no. 01, pp. 9662–9669. (2019)
17. Loukina, A., et al.: Feature selection for automated speech scoring. In: BEA@NAACL-HLT (2015)
18. Loukina, A., et al.: Speech-and text-driven features for automated scoring of English-speaking tasks. In: SCNLP@EMNLP (2017)
19. Loukina, A., et al.: The many dimensions of algorithmic fairness in educational applications. In: BEA@ ACL (2019)

20. Lun, J., Zhu, J., Tang, Y., Yang, M.: Multiple data augmentation strategies for improving performance on automatic short answer (2020)
21. Salim, Y., Stevanus, V., Barlian, E., Sari, A.C., Suhartono, D.: Automated English digital essay grader using machine learning. In: 2019 IEEE International Conference on Engineering, Technology and Education (TALE), pp. 1–6. IEEE (2019)
22. Wu, X., Knill, K., Gales, M., Malinin, A.: Ensemble approaches for uncertainty in spoken language assessment (2020)
23. Xia, L., Liu, J., Zhang, Z.: Automatic essay scoring model based on two-layer bi-directional long-short term memory network. In: Proceedings of the 2019 3rd International Conference on Computer Science and Artificial Intelligence, pp. 133–137 (2019)
24. Zhu, W., Sun, Y.: Automated essay scoring system using multi-model achine learning. In: CS & IT Conference Proceedings, vol. 10, no. 12 (2020)

Automized Quick Prediction of Skin Cancer Diagnosis by Enhanced Deep Convolutional Neural Network

V. S. Jeyalakshmi[1]([✉]), N. Bala Shunmugam[2], M. Kavitha[3], and D. Paulin Diana Dani[4]

[1] Centre for Information Technology and Engineering, Manonmaniam Sundaranar University, Abhisekapatti, Tirunelveli 627012, India
`vsjeyalakshmiap@gmail.com`
[2] Department of CSE, Manonmaniam Sundaranar University, Abhisekapatti, Tirunelveli 627012, India
[3] Department of Computer Science, S.A. Engineering College, Chennai, India
[4] Department of IT, Sri Ram Nallamani Yadava College of Arts and Science, Manonmaniam Sundaranar University, Tirunelveli, India

Abstract. The most common and possibly lethal type of cancer, skin cancer, will be most people's first encounter with it. Information technology must also be used to make a skin cancer diagnosis. This emphasizes how important it is to create and use extremely efficient deeplearning techniques for quick and accurate identification and prevention of skin cancer. In this paper, the Deep Convolution Neural Network (DCNN) is recommended for automated skin cancer diagnosis. The original contribution of this study is the application of a deep convolution neural network with 12 stacked processing layers to increase the precision of skin cancer detection and diagnosis. The results of this study have led researchers to the conclusion that deep learning methods are more effective than machine learning for detecting skin cancer. As a result, automated evidence-based skin cancer identification may increase the accuracy and proficiency of pathologists. In this study, we introduce a deep convolution neural network (DCNN) network that uses deep learning to successfully identify between malignant and benign skin lesions. The results of this study have led researchers to the conclusion that deep learning methods are more effective than machine learning for detecting skin cancer. As a result, automated evidence-based skin cancer identification may increase the accuracy and proficiency of pathologists. In this study, we introduce a deep convolution neural network (DCNN) network that uses deep learning to successfully identify between malignant and benign skin lesions.

Keywords: Deep Convolution Neural Network · skin cancer · Gaussian Filter · classification · accuracy

1 Introduction

Several illnesses that may affect a person's health are referred to as "cancer" interchangeably. Cancer, which is an uncontrolled multiplication of cells, spreads to other organs through the process of metastasis, when it invades nearby cells. Cancer is the second-leading cause of death globally, taking the lives of around 9.6 million people each year [1].

© The Author(s), under exclusive license to Springer Nature Switzerland AG 2025
R. Geetha et al. (Eds.): AAIMB 2023, CCIS 2202, pp. 292–302, 2025.
https://doi.org/10.1007/978-3-031-73065-8_24

The most prevalent and possibly lethal kind of human cancer in recent years is skin cancer. Squamous and carcinomas of basal cells are additional kinds of skin cancer, though moles are the most challenging to identify. Over the past few decades, both non-melanoma and malignant skin cancer rates have increased. With between 2.5 and 3.5 trillion instances of other types of cancer skin tumors detected each year worldwide, there are currently 132,000 instances of skin cancer melanoma. One in five Americans may get skin cancer at some point in their lives, according to statistics published by the Skin Cancer Foundation Statistics (SCFS). The 19th most common kind of cancer among men and women is melanoma. In 2018, almost 300,000 new cases of cancer of the skin were identified. A thousand men and women are diagnosed with nonmelanoma cancer each year, making it the sixth most common kind of cancer [2]. Identification and diagnosis are particularly challenging since cancer and other illnesses are similar to one another.

Skin cancer is invisible to the naked eye in its earliest stages because it resembles a benign mole [3]. A method for processing data IT's ability to do diagnostics is growing quickly. As information technology advances, medical care may become more effective in terms of sickness diagnosis, patient support, and treatment. Improved decision support tools that take evidence-based therapies into account can aid physicians in the diagnosing process [4].

There have been several studies on the use of IT in identifying and diagnosing skin cancer. Many of them included machine learning methods, including Naive Bayes, Random Forest, Agent Technologies, and Neural Networks [5]. Convolution neural networks (CNN), probabilistic neural networks (PNN), and artificial neural networks (ANN) have shown the most promise in identifying and diagnosing instances of skin cancer. The training classifier must be supplied with an illustration of protruding structures according to the conventional method for categorizing pictures. In the case of skin cancer, it might be difficult to extract characteristics and categorize them based on an image; the structures of the features could be color, shape, texture, or color images. The researchers are considering attempting to use CNNs to determine the structural properties. Recent research has demonstrated that CNNs can extract either low- or high-level details thanks to their layer-based abstraction capacity. Despite this, research indicates that CNNs will offer the most precise outcomes for the identification and evaluation of skin cancer.

2 Related Works

The most helpful and pertinent research articles that have been created to detect and diagnose have been presented, and they have been thoroughly examined. [6] earlier described a hybrid melanoma skin cancer method for evaluating worrisome spots.

Convolutional neural networks (CNN) and two separate machine classifiers are among the three methods the suggested system employs to produce predictions. Data on the border, texture, and color of skin lesions are utilized to train the machine-learning classifier. Then, these techniques are blended to restore their respective displays by employing the majority rule. The effectiveness of the suggested approach has been evaluated using the 640-image International Skin Imaging Collaboration (ISIC) dataset. The results demonstrate that combining all three strategies results in the highest accuracy (88.4%).

The most effective and efficient approach for identifying skin cancer, according to their study [7], is CNN. Considerable attention is also placed on an SVM-based skin cancer classification system and an error-correcting output coding (ECOC) method. From KAGGLE, images of skin cancers in RGB (red, green, blue) format were downloaded and used for performance assessment. According to the comparison tools used, there is background noise in the collection of images. The best result is obtained when these images are cropped. A total of 3753 images of four different types of skin cancer were used to evaluate the recommended algorithms. The implementation's findings show that a 95.1% average maximum accuracy was attained.

An illustration of this is the visualization of a dataset for the classification of skin cancer [8] using an enhanced CNN classifier for trained network models. Although there are several methods for optimizing neural networks, deep learning-based neural networks have a limited supply of resources. In this study, we use a novel strategy and adjust the weights and biases of a CNN model using the monster optimization technique. The Demurest Database and the Dermi S Digital Database, two datasets of counts of skin cancer, have been utilized to evaluate the proposed method with ten generic classifiers. In comparison to conventional algorithms, our method achieves a 91% success rate.

It provided a more sophisticated method for identifying three different types of skin cancer [9]. With the recommended method, a skin lesion could be entered as a skin cancer lesion and the classifier could assess if it was malignant or not. Fuzzy C techniques are used in image segmentation to separate the homogenous parts of the image. Several filters are applied during preprocessing to enhance the quality of the images; moreover, the RGB color space, the local binary pattern (LBP), and the graylevel co-occurrence matrix (GLCM) are used to assess other features. A fictitious neural network is trained to utilize several development methodologies for subsequent classification. The differences between the PH2 and HAM10000 skin cancer data sets have been well calculated.

The earlier study tried to assess CNN's effectiveness in classifying skin lesion comparisons they had not previously assessed using the approach outlined in the Stanford

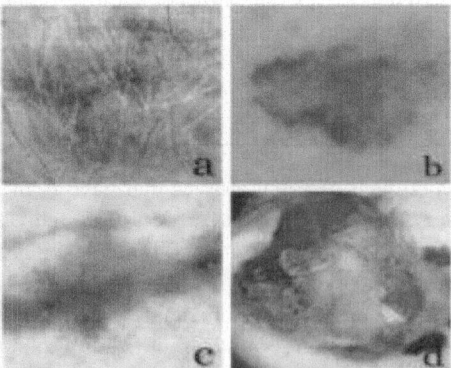

Fig. 1. Hairstyle artifacts (a), poor contrast (b), uneven borders (c), and color lighting (d) are all obstacles to spotting skin lesions.

publication. Melanoma vs. solar lentigo and melanoma vs. seborrheic keratosis are two examples of novel binary classification use cases [10, 11].

Inception v3 was prepared for a range of skin lesions using transfer learning. There were a total of 16 different courses used to train CNN. Using CNN, three-way classification was confirmed with an accuracy of 68.3%. Similar to the Stanford study, we discovered that melanoma and nevus comparisons had an accuracy rate of 71%, whereas comparisons of seborrheic keratosis with basal and squamous cell carcinoma had an accuracy rate of 91%. Seborrheic keratosis and melanoma could be distinguished 84% of the time, however, solar lentigo and melanoma could not be distinguished 84% of the time. Digital photographs and their accompanying metadata have been used to develop a method for identifying skin cancer. Tests on the ISIC 2019 dataset were done to determine the effectiveness of the suggested strategy.

The dataset also includes nine lectures, however, one of them is an outlier since it is outdated. To address the issue, we employ a collection of classifiers made up of 13 CNNs, two approaches for dealing with outliers, and a basic way for using meta-data with images. The obtained findings are consistent with those of the previous challenges, and it appears that the methods for locating the outlier class and managing the meta-data are effective. The proposed approach operates effectively with an accuracy rate of 91%.

A. Problems in skin cancer diagnosis

Accurately detecting skin cancer is difficult due to the broad range of image forms and sources. Due to individual variations in skin tone, detecting skin cancer may be difficult and time-consuming [12, 15]. These challenges are illustrated visually in Fig. 1, and the most salient characteristics of images of skin lesions are explored below.

1. The fact that different-sized and shaped photos don't provide the same findings is the largest obstacle to correctly detecting skin cancer. This viewpoint contends that substantial pre-processing is necessary for accurate analysis.
2. It could be essential to give up a small number of unneeded signals that weren't initially included in the image to achieve the desired result. In the pre-processing steps, these artifacts and noise should be filtered away.
3. In the absence of contrast from surrounding tissues, it may be difficult to make an accurate diagnosis of skin cancer.
4. In addition to the difficulties already discussed, color illumination creates additional difficulties because of things like color texture, light beams, and reflections.
5. Even though certain moles on a person's body may never develop into cancer, they provide difficulties when trying to accurately identify skin cancer from pictures of cancerous moles [16–20].
6. The inherent bias in the available data presents another challenge in the detection of skin cancer and makes it more difficult for models to produce reliable findings.

3 Methodology

In this section, we go into great depth on our method, including the procedures seen in Fig. 2.

Deep convolution neural networks (DCNNs) and learning through transfer are used to represent the difficulty of distinguishing between benign and malignant tumors. Our proposed DCNN model uses certain layers.

A. Dataset

The collection solely comprises images showing various types of skin cancer. To properly apply DL techniques, a lot of information is required. However, the collection of skin cancer images is far more significant. A significant barrier to employing DL algorithms is the lack of training data. We used the HAM10000 dataset, a collection of 10015 dermoscopy, photos gathered from patients in Australia and Austria, to solve these problems. The collection includes 6705 pictures of benign lesions, 1113 pictures of cancer, and 2197 pictures of unidentified lesions. The findings were validated by confocal microscopy, an expert agreement, or pathology. The rigorously assembled HAM10000 dataset that we used in this study consisted of melanocytic lesions that had been biopsy-proven and described.

B. Preliminary Data

"Data preparation" refers to the process of converting unstructured data into a more usable format. We have incorporated in our collection photos or photographs of the same lesion on other individuals to enable comparisons. Too blurry or distant images were removed from the training set, but not from the test or validating sets. Before training can start, processes like these in the preparation process—data maintenance, selecting features, data modifications, engineering features, dimensionality reduction, etc.—bring the dataset to its final form.

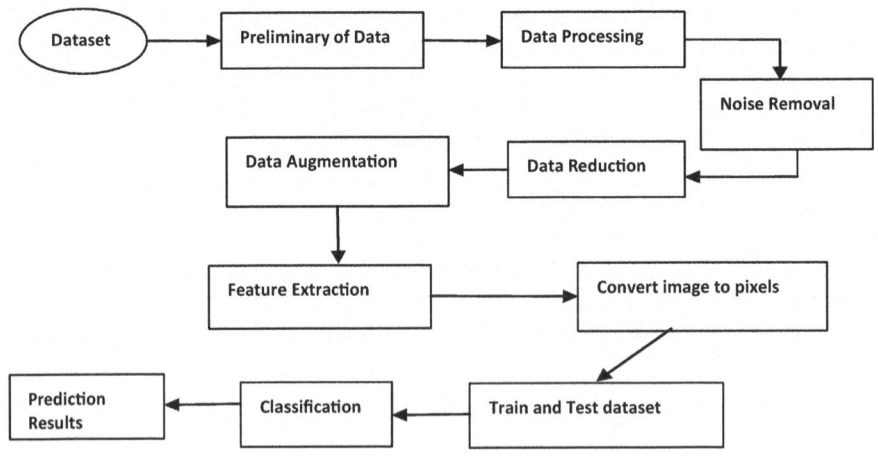

Fig. 2. Proposed Workflow

C. Data Preprocessing

"Data preparation" refers to the first and most crucial step in preparing data for a machine-learning model. Our dataset has been preprocessed to enhance the original medical images by removing air bubbles, noise, and artifacts from the gel application procedure before capture. We have removed noise and artifacts from the photographs to attain good performance in classification in the suggested research. Additionally, we have applied feature extraction, data normalization, and label text to numeric data conversion.

D. Reduction of Data

In imaging datasets, data reduction refers to the process of condensing more images into fewer. It is quite challenging to extract the most effective classification results from the entire dataset due to a large number of photographs with noises and irregularities, some blur views, some images with low contrast, some images with molecular structure near the wounds, and some photos with color illumination. We removed some images with similar characteristics from the DCNN model we manually displayed. After removing the photographs, we had 6136 images of benign lesions and 979 images of malignant lesions.

E. Data Normalization

"Data normalization" is the process of arranging a database in a way that minimizes duplication, assures the correctness of the data, and gets rid of undesired elements such as abnormalities in the total amount of insertions, updates, and deletions. The min-max normalization, z-score normalization, and decimal scaling normalization are some of the current normalization techniques. Data normalization seeks to accomplish two main objectives: 1) It enhances the uniformity and uniformity of input types, which in turn improves the accuracy of the data, lead generation, and segmentation; and 2) It makes all of the data in a collection look consistent and comparable across all records and fields.

F. Extraction of Features

Feature extraction is used to divide up processed pictures into more manageable pieces. We extract a huge number of attributes that help us recognize the patterns in the different datasets we analyze.

Additionally, it picks and chooses variables to extract features from the raw data that consume fewer resources without losing any information.

G. Augmentation of Data

Rather than by collecting new data, data enhancement is a strategy for artificially growing a dataset. The training dataset's size may be manipulated by data warping and sampling too much to keep the model from overfitting. To improve the variety of our incoming data and hence decrease the possibility of overfitting, several augmentation variables, such as movement, random trimming, replicating, and colorshifting, were applied in conjunction with PCA.

H. DCNN Classification

Our proposed DCNN model uses several tuning variables. The primary ones we used in our experiment, By adding modified versions of the initial database to the training

database along with the values we chose for each, are mentioned below. This configuration is contrasted with several other experimental settings and deep neural network
models as a benchmark.

- To reduce the size of the input photographs and speed up training, we use the pooling
  layer (MaxPooling2D).
- Iteration size: the entire number of photographs to be analyzed in just one batch (in
  this example, 128)

.

We divided the collection of data into two stages: either 70% as the initial training
set, 20% as the set used for validation, and 10% as the set used for testing, or 80%
as the training set, 10% as the validation set, and 10% as the testing set. This allowed
for a more precise contrast between existing transfer learning models and our proposed
DCNN model. The model was created on a Jupyter Notebook and Google Colab (12 GB
RAM) using a PC with an Intel Core i5-8250U CPU, 8 GB of RAM, and an NVIDIA
GeForce MX130 GPU card. Each new model's accuracy, precision, recall, F-measure,
and execution durations are contrasted with those of the existing neural network shown
in Fig. 3.

- True Positives (TP) are times when the outcome matched our expectations.
- True Negative (TN): Examples where both predicted and actual yields were incorrect.
- False positive (FP) is a circumstance in which the anticipated outcome differs from
  the actual outcome.
- False negative (FN) occurs when we estimate a low yield wrongly, but the actual
  outcome is larger than anticipated.

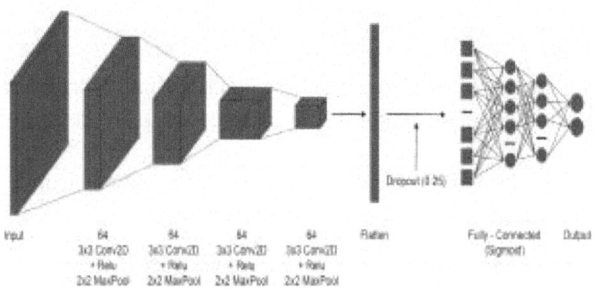

Fig. 3. DCNN Architecture

Divide the total number of accurate positive finds by the overall number of accurate
positive predictions generated by the classifier to determine precision. The learning
process is started by adjusting the "Initial Learning Rate" to 0.001 (Table 1).

The validation frequency, shown by the symbol, is the number of cycles between
validity metric assessments. The Adam optimizer, an alternative for the use of stochastic
gradient descent in the optimization process, assists in minimizing the loss function
when training DL models. After careful consideration, Standard Gradient Descent with
a Momentum of 0.999 was found to be the most effective technique.

Table 1. Illustrates the accuracy chart of the existing and propose model

Algorithms/Results	F1 Score	Precision	Recall	Accuracy
Existing Algorithm (CNN)	0.87	87.4%	81.1%	86%
Proposed Algorithm (DCNN)	0.95	95.8%	92.4%	93%

4 Results and Discussions

The major objective of our proposed model is to classify skin lesions as benign or malignant using the DCNN-extracted dataset. After data reduction, the conclusions are based on the total number of photos we took, which were obtained from the Internet Scientific Image Collection and Archive (http://www.isic-archive.com), a freely accessible online picture collection. Figure 4 displays several illustrations of both malignant and healthy tissue (Fig. 5).

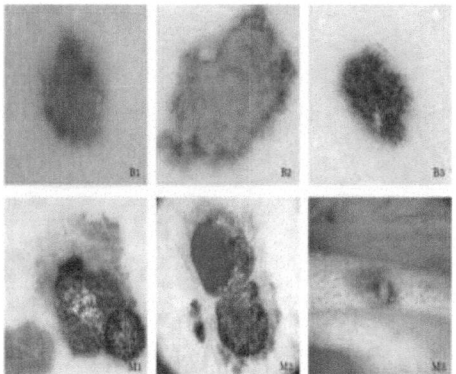

Fig. 4. Skin cancers are categorized as B1, B2, or B3, or M1, M2, or M3 for benign, malignant, or both

In this study, we evaluated our proposed DCNN model twice: once with just 70% of the training images, and again with 80% of the training images to get the best accuracy. On the same dataset, we also tried a few more DNN models; nonetheless, our proposed DCNN model produced the greatest classification accuracy of above 92%.

Figures 6 and 7 display the outcomes of our proposed DCNN model in terms of accuracy and loss between the training and testing phases.

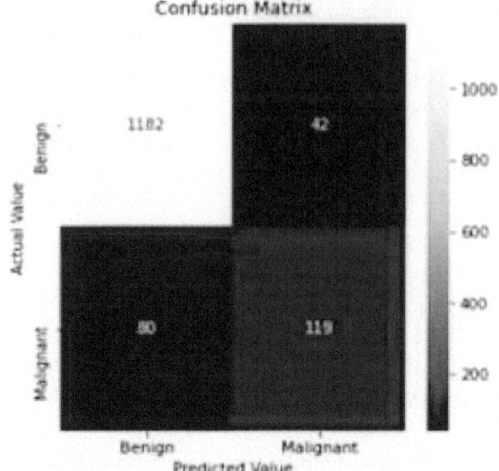

Fig. 5. Depicts the confusion matrix

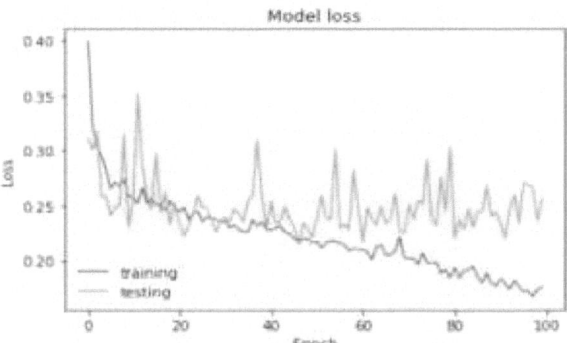

Fig. 6. Train and Test the Accuracy Model

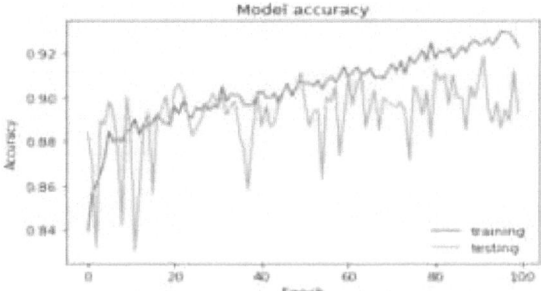

Fig. 7. Train and Test the Loss Model

5 Conclusion

In terms of classification accuracy, the DCNN model presented in this work beats rival transfer learning methods. The proposed method may discriminate between benign and malignant skin lesions by changing the output activation layer to a sigmoid activation layer. Additionally, we put the recommended strategy to the test on the HAM10000 dataset, where we outperformed previous transfer learning models in terms of training and testing accuracy. Additionally, the dataset's imbalance and the lack of a significant number of photographs hindered the model's efforts to increase accuracy. Therefore, to maximize classification accuracy, we ensured that the dataset was spread equally throughout both tiers. We also trained many transfer learning models on the same dataset, however, their performance lagged below that of our recommended DCNN model.

References

1. Al-Masni, M.A., Al-Antari, M.A., Choi, M.-T., Han, S.-M., Kim, T.-S.: Skin lesion segmentation in dermoscopy images via deep full-resolution convolutional networks. Comput. Methods Programs Biomed. **162**, 221–231 (2021)
2. Alfed, N., Khelifi, F., Bouridane, A., Seker, H.: Pigment network-based skin cancer detection. In: 2015 37th Annual International Conference of the IEEE Engineering in Medicine and Biology Society (EMBC), pp. 7214–7217. IEEE (2015)
3. Aljanabi, M., Özok, Y.E., Rahebi, J., Abdullah, A.S.: Skin lesion segmentation method for dermoscopy images using artificial bee colony algorithm (2018)
4. Biessmann, F., Salinas, D., Schelter, S., Schmidt, P., Lange, D.: Deep learning for missing value imputation in tables with non-numerical data. In: Proceedings of the 27th ACM International Conference on Information and Knowledge Management, pp. 2017–2025 (2018)
5. Targ, S., Almeida, D., Lyman, K.: Resnet in Resnet: generalizing residual architectures (2016). arXiv preprint arXiv:1603.08029
6. Tschandl, P., Rosendahl, C., Akay, B.N., Argenziano, G., Blum, A., Braun, R.P., et al.: Expert-level diagnosis of nonpigmented skin cancer by combined convolutional neural networks. JAMA Dermatol. **155**(1), 58–65 (2019)
7. World Health Organization (WHO) (2019). How common is skin cancer? https://www.who.int/uv/faq/skincancer/en/index1.html. Accessed 30 Dec 2020
8. Kadampur, M.A., Al Riyaee, S.: Skin cancer detection: applying a deep learning-based model-driven architecture in the cloud for classifying dermal cell images. Inform. Med. Unlocked **18**, Article 100282 (2021)
9. Salvi, M., Acharya, U.R., Molinari, F., Meiburger, K.M.: The impact of preand post-image processing techniques on deep learning frameworks: a comprehensive review for digital pathology image analysis. Comput. Biol. Med. **128**, 104129 (2021)
10. Pandey, S., Ramesh, S.P., Singh, A.: Real Time Object Detection and Recognition, 7 May 2021. SSRN: https://ssrn.com/abstract=3866094 or https://doi.org/10.2139/ssrn.3866094ICA ST-2021: 4th International Conference on Advances in Science and Technology
11. Kumar, A., Ramesh, S.P., Sharma, A., Tiwari, P.K.: An optimized image caption generator. J. Huazhong Univ. Sci. Technol. **50**(4) (2021). ISSN-1671-4512
12. Mehr, R.A., Ameri, A.: Skin cancer detection based on deep learning. J. Biomed. Phys. Eng. 559–568 (2022)
13. Wu, Y., Chen, B., Zeng, A., Pan, D., Wang, R., Zhao, S.: Skin Cancer Classification with Deep Learning: A Systematic Review, July 2022

14. Senthil Kumar, R., Singh, A., Srinath, S., Thomas, N.K., Arasu, V.: Skin cancer detection using deep learning. In: International Conference on Electronics and Renewable Systems (2022)
15. Rajarajeswari, S., Prassanna, J., Quadir, M.A., Jackson, C.J., Sharma, S., Rajesh, B.: Skin Cancer Detection using Deep Learning, Volume-15, Issue-10 (2022)
16. Naqvi, M., Gilani, S.Q., Syed, T., Marques, O., Kim, H.C.: Skin Cancer Detection using Deep Learning - A Review, vol. 13, no. 11, p. 1911 (2023)
17. Jojoa Acosta, M.F., Caballero Tovar, L.Y., Garcia-Zapirain, M.B., Percybrooks, W.S.: Melanoma diagnosis using deep learning techniques on dermatoscopic images, 06 January 2021
18. Neema, M., Nair, A.S., Joy, A., Menon, A.P., Haris, A.: Skin Lesion/Cancer Detection using Deep Learning, vol. 15, Number 1 (2020). ISSN 0973-4562
19. Agrahari, P., Agrawal, A., Subhashini, N.: Skin cancer detection using deep learning. In: Sivasubramanian, A., Shastry, P.N., Hong, P.C. (eds.) Futuristic Communication and Network Technologies. VICFCNT 2020. LNEE, vol. 792, pp. 179–190. Springer, Singapore (2022). https://doi.org/10.1007/978-981-16-4625-6_18
20. Dildar, M., et al.: Skin Cancer Detection: A Review Using Deep Learning Techniques, vol. 18, p. 5479 (2021)

Clustering Based Demand Prediction Using Long Short-Term Memory (LSTM) in Retail Supply Chains

S. Praveena and S. Prasanna Devi$^{(\boxtimes)}$

Department of CSE, SRM Institute of Science and Technology, Chennai, India
ps4851@srmist.edu.in

Abstract. Demand prediction is a significant component of managing supply chains because it can increase profit and efficiency by identifying supply channels with predicted consumer demand. The data appropriately chosen to enhance corporate sales and profits and also to identify the customer buying patterns and behavior. This paper proposes an integrated data-driven cluster-based demand forecasting approach for successful retail businesses, segmenting customers based on Recency, Frequency, and Monetary (RFM). After the segmentation of customers, the prediction of demand for each segment is carried out by Long Short-Term Memory (LSTM). The results show that consumer segmented prediction gives better accuracy of MAPE = 8.23, MAE = 17.45, RMSE = 145.8. Since accurate and precise demand forecasting is crucial for efficient inventory and supply chain management, the current study tackles one of the most severe problems in retail management. The suggested model performs better than other popular existing demand forecasting techniques.

Keywords: Demand forecasting · LSTM · Clustering · Customer Segmentation · KMeans Clustering · Silhoutte Score · Root Mean Square Error (RMSE)

1 Introduction

The retail supply chain is the end-to-end process that assures retailers have appropriate products in stock at the right moment and in sufficient quantities. It starts from the suppliers/manufacturers and continues through shipment centers until the product reaches the store. The reliable demand forecasting is essential for optimal management of stocks, which helps to maintain supply-demand equilibrium. Due to competition and globalization, Sales forecasting is becoming an increasingly significant component of the commercial sector [1]. Understanding the market better and developing prospects for better pricing and marketing strategies, in the long run, may be possible by identifying the products with comparable prices and sales patterns over a specific period [2]. Time series clustering can assist in locating the goods whose sales volumes and prices fluctuate similarly over a particular time frame in this context. For massive time-series data, an improved and reliable demand estimates can be successfully done by the machine learning and deep learning approaches. However, when dealing with single-step forecasting

© The Author(s), under exclusive license to Springer Nature Switzerland AG 2025
R. Geetha et al. (Eds.): AAIMB 2023, CCIS 2202, pp. 303–314, 2025.
https://doi.org/10.1007/978-3-031-73065-8_25

issues with univariate datasets, conventional approaches such as Seasonal Autoregressive Integrated Moving Average (SARIMA) and exponential smoothing (ES) might outperform [3, 4]. As a result, before investigating into more data-intensive methods, it is necessary to comprehend and evaluate traditional time-series forecasting approaches. Because traditional methodologies may lag in terms of enabling feature engineering techniques. In the past, it was the traditional procedure of incorporating all training data into the model creation stage at once when building prediction models. The cluster-based hybrid prediction model is frequently used to enhance predictions since using data with various features in the model is likely to give inaccurate prediction results. Clustering also enhances prediction accuracy by lowering the degree of data heterogeneity [5].

This paper aims to propose the clustering-based forecasting model, which employs a data clustering technique that divides the entire featured input set into several groups. The mathematical prediction model is then built for each group to generate the desired result. To get the final forecasting results for the prediction target, locate the cluster to which it belongs and then utilise the trained forecasting method connected to that cluster. The proposed framework is intended to help decision-makers in demand forecasting, especially when encountering a lack of historical data and demand unpredictability.

The remainder of the paper is structured as follows. Section 2 discusses similar works on demand forecasting. Section 3 describes our suggested demand forecasting model and the underlying methodology. Section 4 discusses the study's data processing and experimental results. Finally, Sect. 5 brings the conclusion of the work.

2 Related Works

Many real-time application areas need forecasting to improve business strategies and replenish inventory. This related work section includes studies that apply demand forecasting in similar circumstances. In the retail sector, demand forecasting is a crucial and complex problem [1]. Numerous strategies have been put out in recent years to enhance forecasting performance. The two main demand forecasting techniques are machine learning-based forecasting models and traditional statistics-based forecasting approaches [7, 8].

In Tourism applications, time series analysis techniques were applied to forecast tourist demand. There were Five different machine learning (ML) models compared for tourism forecasting LR, SVM, GRNN, SARIMA, and SARIMAX [9]. This comparative study helps identify the performance of traditional statistics-based approaches and machine learning techniques for forecasting future demand. Clustering based Hotel demand forecasting model [10] proposed to predict weekly hotel demand for tourists by clustering hotel demand data and determining hotel tags by training K-Means embedded hotel data. The Attention-LSTM outperforms well and produced error values MAE of 4.61 and MAPE = 23.97 comparatively with existing methods such as RNN, LSTM, and GBR. Some of the clustering methods investigated in existing work concentrated only on price and sales patterns; the numerical distance between the data points and metrics should not be capable of capturing the dynamic behavior of time series patterns.

Nevertheless, [11] proposed alternatives for clustering the products according to price trends and sales numbers. In this work, they have suggested a novel distance metric that considers the relationship between the movement of product pricing and sales rather than determining the distance using numerical data. The electrical demand dataset was analyzed for time series forecasting by [12]. The time series data were grouped and labelled using k-means clustering before time series forecasting. Using the silhouette index, (Davies-Bouldin) DB index, and (Dunn) DU index, three well-known internal validity indices, for assessing the clustering algorithm's performance.

Buyers are commonly separated into segments based on their buying power, which e-commerce enterprises can use for predicting demand from each group for their products. [13, 14]. Each segment's data is associated with a specific group of clients that are similar to one another, creating forecasting models simpler and efficiently to identify trends in the data and elevate the accurateness of forecasting [14]. The use of data-mining methods for customer segmentation has been the primary focus of many research studies in the marketing field. In order to analyse Recency, Frequency, and Monetary value (RFM), segmented Clustering, logistic regression (LR), decision trees (DT), and neural networks (NN) are used [15–17].

3 Proposed Methodology

The primary objective of the study is to develop an improved demand forecasting method in order to provide more accurate forecasting results for optimizing inventory. The suggested approach is divided into two parts: buyer segmentation and multivariate time-series demand forecasting.

- Initially, Data is pre-processed, and outliers are imputed. Next, the target and predictor variables are chosen. Sales volume (demand) has been considered as the target variable.
- Demand forecasting and Buyer segmentation are part of the training phase. The segmentation focuses on grouping buyers.
- The Buyer's Data is initially clustered using the KMeans Technique. The resultant groups are then connected to form three segments that represent Low, medium, and High according to their financial liablity. Following buyer segmentation, demand predictions are generated for each buyer category using the LSTM and SARIMA algorithms.
- The results are compared with Traditional methods such as Holtwinter, Auto Regressive Integrated Moving Average (ARIMA), and with machine learning models such as simple RNN and prophet. Our clustered based approach outperforms well and predicting the demand efficiently compared with existing techniques (Fig. 1).

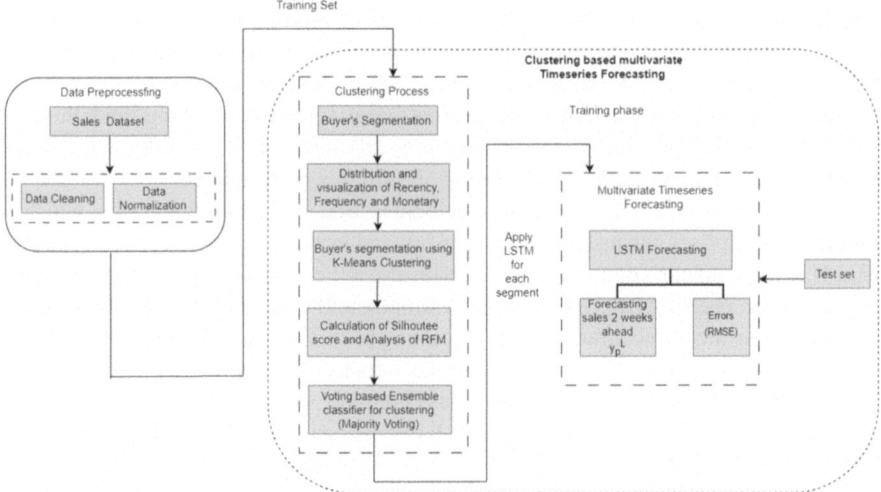

Fig. 1. Proposed work flow

3.1 RFM Analysis

RFM study [18] is a highly well-liked database marketing strategy for consumer segmentation and identification. This can be done by looking up recent transactions, consumer spending patterns, and amounts. It is essential, particularly for the retail industry, for each consumer RFM assigns a score based on Recency, Frequency, and monetary criteria.

- Recency: The term recency refers that the customer bought the product most recently or interacting with the product recently. The greater the rate of recency, the greater the client visits a store.
- Frequency: The interval between a customer's subsequent two purchases is known as Frequency. Customers visit the company more frequently, the higher the value of Frequency.
- Monetary: The amount of money a consumer spends over a given time frame, and the firm will make more money when monetary value is high.

3.2 Choosing of Clustering Algorithm

A conceptually significant group of things with standard features is called a cluster. For further analysis, the Buyer's segmentation can be done using clustering. Many related works suggested using K-Means to segment and analyze customer buying behavior [14–16] effectively. The K-Means clustering technique determines the number of clusters that the user specifies, as represented by their centroids. In contrast, K-Means outperforms other clustering approaches on large datasets and is computationally faster. A further benefit of utilizing k-Means over other algorithms is that it minimizes the rate of data misclassification by employing the parameter K. The most common applications of k-Means are Buyer's segmentation. Finally, this paper uses the k-Means clustering used to segment Buyer's purchasing behavior.

k-Means

K-Means alternates between these to obtain the best centroids. (1) clustering data points according to the current centroids p1, p2, p3....pk $\in$ Rn randomly (2) selecting centroids (the points at the center of a cluster) by the existing clustering of data points. To achieve this until convergence, for every i, set

$$c^{(i)} = arg\ arg\left\|x^{(i)} - p_j\right\| \tag{1}$$

The Mathematical Model for Clustering Algorithm

Most commonly used data analysis techniques for unsupervised learning is K-Means clustering algorithm. This method distributes non-overlapping data points to the closest clusters. After identifying 'K' centroids from the dataset 'D.' In the K-means method, the intra-cluster distance is the greatest in comparison to the inter-cluster distance. Data points are iteratively shifted to distinct clusters based on the centroid's computation. The products in this case study are categorized based on consumer transaction data for recency vs. monetary and Frequency vs. monetary.

$$k = \{\{(x_1, y_1), (x_2, y_2), \ldots\ldots(x_a, y_a)\}, \{(x_1, y_1), (x_2, y_2)\ldots\ldots(x_b, q_b)\}, \ldots$$
$$\{(x_1, y_1), (x_2, y_2)\ldots\ldots(x_z, y_z)\} \tag{2}$$

where K = cluster number, (x, y) = data point in a cluster. Each data point in the cluster represents RFM Values. Assume $\{x_1, y_1\}$ in cluster 1, compute the mean distance from $\{x_1, y_1\}$ of the similar cluster (intra-distance value a_i). We can identify the mean distance from $\{x_1, y_1\}$ to the other data points of further groups by the specified equation below (3)

$$d = \sqrt{\sum_{i=1}^{n}(x_1 - x_i) + (y_1 - y_i)^2} \tag{3}$$

The shortest mean distance from $\{x_1, y_1\}$ from 1 to n clusters has been calculated and Evaluation of optimal clusters can be identified using silhouette coefficient s(i) by the Eq. (4)

$$s_i = \frac{b_i - a_i}{max(b_i, a_i)} \tag{4}$$

After the evaluation, calculate the recency and frequency using sales amount should be done with one cluster to next and investigates the comparative results with both Recency and Frequency of sales amount. It assists in identifying the customer group with the uppermost sales Recency, Frequency, and the Monetary.

3.3 Long Short-Term Memory (LSTM)

LSTM is a type of recurrent neural network that exhibits sequence component order dependency [7]. LSTMs are quite good at patten extraction from input features, even when the input data is extensive. Because of their gated architecture and ability to alter memory state, LSTMs are appropriate for time series challenges. The formula employed by the forget gate to figure out the difference between the previous moment and current moment of the unit state.

$$f_t = \sigma \left(w_f x \left[h_{t-1}, x_t \right] + b_f \right) \tag{5}$$

where the weight matrix of forget gate is denoted as w_f; $\left[h_{t-1}, x_t \right]$ is the present moment's input coupled with the previous moment's output, offset term and sigmoid function is mentioned as b_f and σ and its output value f_t is between 0 to 1. The input gate chooses how much input is currently preserved to the unit state. The numerical computations for the input gate is denoted as follows:

$$i_t = \sigma \left(w_i x \left[h_{t-1}, x_t \right] + b_i \right) \tag{6}$$

$$\tilde{c}_t = tanh \left(w_c . \left[h_{t-1}, x_t \right] + b_c \right) \tag{7}$$

$$\tilde{c}_t = f_t x \left[c_{t-1} + i_t \right] + \tilde{c}_t) \tag{8}$$

(See Fig. 2).

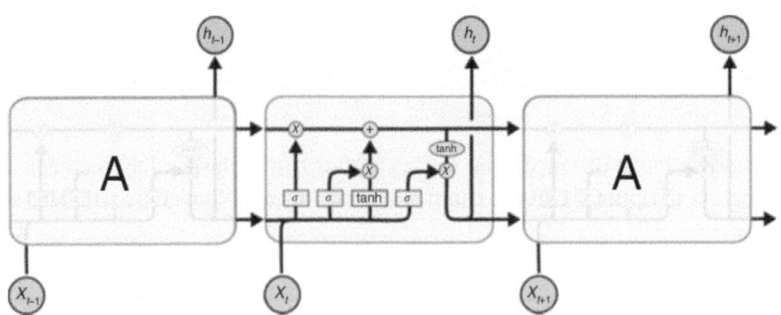

Fig. 2. Basic Structure of LSTM (source of Image: Christopher [19])

4   Experiments, Results, and Discussions

4.1   Retail Dataset

The suggested methodology applied to the 3-year customer Transactions and supply chain store data from an online retail store from 1-1-2010 to 09-12-2012. The dataset consists of 9 features, as shown in Table 1.

Table 1. Online Retail Store Dataset Description

S. No	Feature Name	Type	Description
1	Invoice No	Nominal	5 Digit Number for every Transaction
2	Stock Code	Nominal	4 Digit number for each Item
3	Item Details	Nominal	Item description
4	Quantity	Numeric	Quantity sold
5	Invoice Date	Numeric	Transaction's Date and Time
6	Unit Amount	Numeric	Per unit Item Amount
7	Customer ID	Numeric	Unique Id given for each customer
8	On Promotion	Numeric	No of promotions for each product
9	Holiday Event	Categorical	Represents the sales if any holidays in calendar

4.2 Feature Engineering

Most of the features are not needed for predicted the demand so that we can comprehend this data set's data distribution and extract useful features through visual analysis. By using data visualization, it can be clearly identifying holiday events, on promotion, commodity sales data, month and weeks are the effective features. The data is then cleaned for the training model is built based on 3 criteria. Once effective features have been identifying we can cleanse the data using subsequent steps:

- Handling of missing values – Missing values in the retail sales data can be replaced with 0 and remaining with –1
- Treating anomalous values – Kalman smoothing is used to identify anomalous data sales peak value
- Reducing Dimensionality-Remove the irrelevant features in the dataset by performing some of the dimensionality reduction techniques.
- The training data which consists of quantity of item sales with on promotion from January 1, 2010 to December 31, 2012. The test data which includes same features but it covers from December 15, 2012 to December 31, 2012.

4.3 Clustering with k-Means Clustering

For each transaction, we deducted the present day—from the date the transaction was completed in order to compute recency. The quantity of transactions that each consumer made was also factored into the frequency calculation. We counted up all of a customer's transactions to determine their financial value. To guarantee that the within-cluster variation of each cluster is minimized, it is crucial to detect the ideal clusters count in order to construct the aforementioned properties.

Above shown Fig. 3 clearly depicts the segmented customers with respect to recency, frequency and revenue and clusters were based on RFM. This result helps us to give the next day purchase of each customer (Figs. 4 and 5).

	CustomerID	NextPurchaseDay	Recency	RecencyCluster	Frequency	FrequencyCluster	Revenue	RevenueCluster	OverallScore	Segment
0	13313.0	104.0	26	3	64	0	1242.64	0	3	Mid-Value
1	18097.0	999.0	0	3	78	0	1823.93	0	3	Mid-Value
2	16656.0	24.0	16	3	36	0	4103.04	0	3	Mid-Value
3	14210.0	14.0	5	3	71	0	1292.11	0	3	Mid-Value
4	13397.0	16.0	10	3	11	0	542.69	0	3	Mid-Value
...										
2783	14298.0	33.0	23	3	692	2	20847.62	1	6	High-Value
2784	17735.0	999.0	8	3	555	2	10651.29	1	6	High-Value
2785	14096.0	10.0	1	3	823	2	10395.18	1	6	High-Value
2786	17841.0	4.0	3	3	3877	3	18880.19	1	7	High-Value
2787	18102.0	24.0	18	3	289	1	132044.36	3	7	High-Value

Fig. 3. Segmented Buyers based on recency, frequency and revenue in data frame

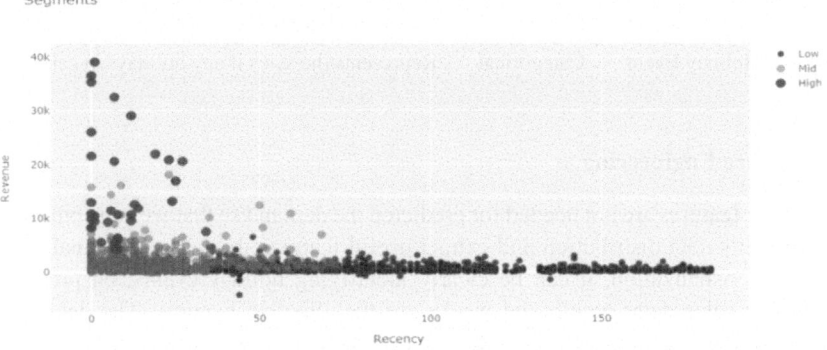

Fig. 4. Visualization of Buyers Segmentation using K-Means Recency Vs Revenue

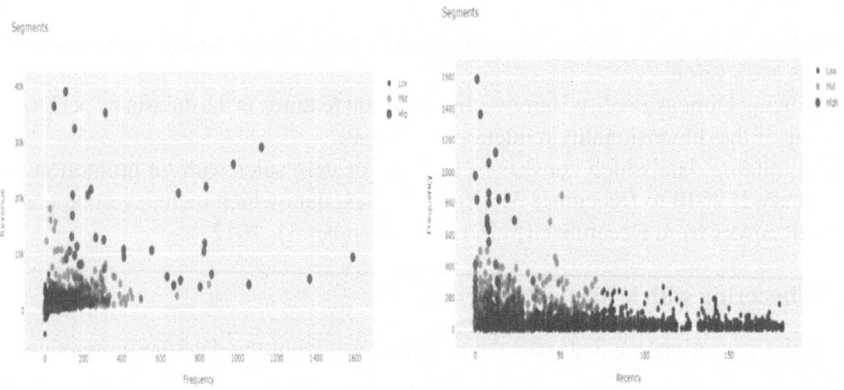

Fig. 5. Visualization of Buyers segmentation using K-Means a) Recency vs Frequency b) Recency vs Revenue

K-Means Clustering can be validated using silhouette score with which RFM analysis has been done by assigning k = 3 and k = 5. It is noted that cluster 5's silhouette score is less ideal than cluster 3's because, in comparison to other clusters, an ideal value is closer to +1. After applying clustering techniques, three categories are used to classify buyers into low, medium, and high value ranges. We aggregated the results using majority voting classifier for the further future prediction of demands (Table 2).

Table 2. Silhouette Score (optimal) k = 3 and k = 5 for segmented clusters

Clusters count	Recency	Frequency	Monetary
0	252.191	25.76	182.85
1	22.373	118.371	4427.800
2	31.32	39.87	983.32
Silhouette Score for number of Cluster 3 is 0.36			
0	321.33	21.26	281.59
1	31.30	86.24	1912.82
2	21.77	371.086	7424.39
3	87.29	31.23	645.01
4	3.78	43.65	709.62

4.4 Predictions with LSTM

The next phase involved by combining everyday demands and mean sale prices to each consumer segment. We subsequently utilized this segmented input to predict the everyday demand. Days that involve holidays were also collected in addition to daily demand and sales because they have a big impact on demand in the future. For non-holiday days (362), as well as holidays (364), the average daily demand was computed. This demonstrates that consumers were more likely to purchase goods on holidays than on non-holidays. The LSTM algorithm were used to as explained in the section [3.2] to forecast the future daily demand. The ideal parameters selected for our LSTM were identified using an exhaustive search strategy as follows: 50 hidden layers, 50 epochs, and 72 batch sizes were used, with adam being chosen as the accuracy optimizer function.

4.5 Accuracy Test

The proposed Cluster Based LSTM forecasting model is evaluated using three different scenarios. The prediction was applied to the training set directly without clustering and it is evaluated by performance metrices (MAPE, MAE and RMSE) (Tables 3, 4 and 5).

The results showed that the accuracy considerable improved and performance of the prediction is also improved while segmenting customers. In this scenarios LSTM achieved MAPE = 8.23, MAE = 17.45, RMSE = 145.8 during univariate analysis. By adoption with multivariate forecasting, we integrated the daily demand estimates could incorporate the effects of additional characteristics, improving their accuracy by MAPE = 8.09, MAE = 27.10, RMSE = 98.45 in comparison to univariate forecasting.

Table 3. Prediction performance- Without Clustering

ML/DL Models for prediction	MAPE%	MAE	RMSE
LSTM	9.47	19.82	67.54
RNN	10.81	10.3	86.1
ARIMA	13.34	12.53	72.34
Holt winter	7.89	8.30	67.9
SARIMA	9.23	20.56	73.34

Table 4. Prediction performance- With Clustering

Clustering Type	Forecasting Techniques	MAPE%	MAE	RMSE
K- Means Clustering	LSTM	7.37	29.56	56.12
	RNN	12.71	18.6	45.23
	ARIMA	15.54	23.12	72.37
	Holt winter	10.23	78.23	68.24
	SARIMA	15.48	78.56	73.34

Table 5. Prediction performance – Majority voting in clustering

ML/DL Models for prediction	MAPE%	MAE	RMSE
LSTM	9.23	17.45	145.8
RNN (univariate)	8.70	21.6	224.6
ARIMA	8.46	27.45	330.46
Holt winter	9.96	30.4	203.201
SARIMA	9.1	26.25	351.815
LSTM(Multivariate)	8.09	21.23	104.56
RNN(Multivariate)	8.120	27.10	98.23

5   Conclusion and Future Work

Demand forecasting has a significant impact on enterprise choices, such as planning for production, inventories, finances, and sales. In the context of retail supply chains, this investigation created an approach for estimating demand using a two-phase technique with the intent to improve prediction accuracy. We have employed segmentation and forecasting as an ensemble technique by utilizing time-series forecasting techniques such as LSTM and K- Means Clustering algorithms. Various forecasting circumstances, including those with and without grouping of clients, were used to analyze the suggested framework. The prediction accuracy of LSTM achieved for segmented univariate analysis of MAPE = 8.23, MAE = 17.45, RMSE = 145.8 and for multivariate analysis of MAPE = 8.09, MAE = 27.10, RMSE = 98.45. The proposed methodology outperforms well when compared with RNN, ARIMA, SARIMA, Holtwinter etc. This proposed model could be extended by inventory optimization combined with retail product segmentation in response to anticipated demand scenarios in the future.

References

1. Nguyen, D.T., et al.: Enhancement of Sales Forecast Using Hybrid Sarima and Extreme Machine Learning: A Case for a Jewelry Retailer in Viet Nam. SSRN 4330707
2. Bozanta, A., et al.: Time Series Clustering for Grouping Products Based on Price and Sales Patterns (2022). arXiv preprint arXiv:2204.08334
3. Fattah, J., Ezzine, L., Aman, Z., El Moussami, H., Lachhab, A.: Forecasting of demand using ARIMA model. Int. J. Eng. Bus. Manag. **10**, 1847979018808673 (2018)
4. Arunraj, N.S., Ahrens, D.: A hybrid seasonal autoregressive integrated moving average and quantile regression for daily food sales forecasting. Int. J. Prod. Econ. **170**, 321–335 (2015)
5. Chen, I.F., Lu, C.J.: Sales forecasting by combining clustering and machine-learning techniques for computer retailing. Neural Comput. Appl. **28**, 2633–2647 (2017)
6. Dai, W., Chuang, Y.Y., Lu, C.J.: A clustering-based sales forecasting scheme using support vector regression for computer server. Procedia Manuf. **2**, 82–86 (2015)
7. Dave, E., Leonardo, A., Jeanice, M., Hanafiah, N.: ScienceDirect forecasting Indonesia exports using a hybrid model ARIMA- LSTM. Procedia Comput. Sci. **179**(2020), 480–487 (2021)
8. Hong, W.-C: Application of chaotic ant swarm optimization in electric load forecasting. Energy Policy **38**(10), 5830–5839 (2010)
9. Khatibi, A., Belém, F., da Silva, A.P.C., Almeida, J.M., Gonçalves, M.A.: Fine-grained tourism prediction: Impact of social and environmental features. Inf. Process. Manag. **57**(2), Article 102057 (2020)
10. Kaya, K., Yılmaz, Y., Yaslan, Y., Öğüdücü, Ş.G., Çıngı, F: Demand forecasting model using hotel clustering findings for hospitality industry. Inf. Process. Manag. **59**(1), 102816 (2022)
11. Bozanta, A., et al.: Time Series Clustering for Grouping Products Based on Price and Sales Patterns (2022). arXiv preprint arXiv:2204.08334
12. Alvarez, F.M., Troncoso, A., Riquelme, J.C., Ruiz, J.S.A.: Energy time series forecasting based on pattern sequence similarity. IEEE Trans. Knowl. Data Eng. **23**(8), 1230–1243 (2010)
13. Murray, P.W., Agard, B., Barajas, M.A.: Forecasting supply chain demand by clustering customers. IFAC-PapersOnLine **48**(3), 1834–1839 (2015). https://doi.org/10.1016/J.IFACOL.2015.06.353
14. Kashwan, K.R., Velu, C.M.: Customer segmentation using clustering and data mining techniques. Int. J. Comput. Theory Eng. **5**(6), 1–6 (2013). https://doi.org/10.7763/IJCTE.2013.V5.811
15. McCarty, J.A., Hastak, M.: Segmentation approaches in data-mining: a comparison of RFM, CHAID, and logistic regression. J. Bus. Res. **60**, 656–662 (2007). https://doi.org/10.1016/j.jbusres.2006.06.015
16. McCarty, J.A., Hastak, M.: Segmentation approaches in data-mining: a comparison of RFM, CHAID, and logistic regression, p. 60
17. Yesil, E., Kaya, M., Siradag, S.: Fuzzy forecast combiner design for fast fashion demand forecasting. In: International Symposium on Innovations in Intelligent Systems and Applications, IEEE, 2012

18. Christy, A.J., Umamakeswari, A., Priyatharsini, L., Neyaa, A.: RFM ranking–an effective approach to customer segmentation. J. King Saud Univ.-Comput. Inf. Sci. **33**(10), 1251–1257 (2021)
19. Christopher, O.: "LSTM structure" (2015). http://colah.github.io/posts/2015-08-Understanding-LSTMs

Early Detection of Diabetic Retinopathy Using Deep Convoulutional Neural Network

K. Vijay, P. Krithiga$^{(\boxtimes)}$, S. Kavirakesh, S. Swetha, and B. Vishal

Department of CSE, Rajalakshmi Engineering, Chennai, India
krithiga.p.2019.cse@rajalakshmi.edu.in

Abstract. If overlooked diabetic retinopathy (DR), a degenerative eye condition that affects people with diabetes, can cause serious vision loss. Early DR detection is essential for prompt management and action. Deep convolutional neural networks (DCNNs) have become effective tools for automated processing of retinal pictures in recent years, making it possible to diagnose DR earlier. To learn and extract distinguishing features from retinal fundus images, the suggested method makes use of a DCNN architecture. For the training of models and evaluation, a dataset containing quite a few of retinal pictures from diabetic patients—including both normal and DR cases is utilised. Combining methods of supervised learning, the DCNN model undergoes training with a focus on improving performance indicators including accuracy, sensitivity, and specificity. To intentionally enhance the quantity and diversity of the training dataset, methods like image rotation, flipping, scaling and random cropping are used. This successfully simulates changes in the circumstances of collecting images and the clinical traits of DR. The efficiency of the suggested DCNN-based technique for early DR detection is shown by experimental findings. The trained model performs exceptionally well in spotting early indications of DR, such as microaneurysms, haemorrhages, and exudates, with high accuracy. An independent test dataset is used to assess the model's performance, highlighting its potential for use in practical situations. The suggested approach advances automated DR detection, enabling immediate action and greater control of this illness that threatens eyesight. The suggested DCNN-based approach needs to be further improved and validated, and the resulting system needs to be integrated into the current medical system to facilitate widespread early DR detection and treatment. The model achieved validation accuracy of 76.80% and training accuracy of 99.58%. A training loss of 0.0132 and a validation loss of 1.9230 have also been attained.

Keywords: Automated processing · DCNN · Supervised learning

1 Introduction

Many people across the world suffer from diabetic retinopathy (DR), a degenerative eye disease brought on by long-term diabetes mellitus. Preventing serious vision loss and blindness from this illness requires prompt diagnosis and treatment. Retinal fundus images may now automatically detect early DR thanks to recent advancements

R. Geetha et al. (Eds.): AAIMB 2023, CCIS 2202, pp. 315–327, 2025.
https://doi.org/10.1007/978-3-031-73065-8_26

in deep learning, namely deep convolutional neural networks (DCNNs). In this publication, we give a systematic analysis of the research on deep convolutional neural networks (DCNNs) for DR detection. Successful treatment approaches and improved patient outcomes depend on an accurate early diagnosis of DR. Manual evaluation of retinal pictures by competent ophthalmologists is the gold standard for DR screening, although this method can be time consuming and prone to inter-observer variability. [7] DCNNs, with their superior capacity to develop hierarchical representations directly from raw image data, have emerged as potent methods for automated DR detection. Microaneurysms, haemorrhages, exudates, and aberrant blood vessel patterns are some of the earliest signs of DR that can be detected by these techniques.

In many cases, deep convolutional neural networks (DCNNs) have outperformed more conventional machine learning algorithms. Their deep architecture, which consists of many convolutional and pooling layers, successfully models the complex structures and patterns seen in retinal pictures by capturing both low-level and high-level data. [5] The generalisation and resilience of DR detection models can be further enhanced by using DCNNs because of their ability to automatically learn and adapt to fluctuations in image properties. DCNNs are effective and flexible because their end-to-end learning strategy eliminates the need for human feature engineering across a variety of DR datasets.

The purpose of this study is to examine DCNNs and their potential use in early DR detection from a number of angles. Popular models like VGGNet, ResNet, and InceptionNet are just some of the examples of DCNN architectures and variants investigated. It also looks into how different hyperparameter tuning, transfer learning, and data augmentation strategies affect DCNN model performance for DR early detection. In addition, the benefits and drawbacks of DCNN-based DR detection are discussed, including the need for vast and varied datasets, the interpretability of model predictions, and the incorporation of DCNNs into clinical processes. [10] It also emphasises the need for DCNN-based DR detection systems to undergo extensive testing and refinement in collaboration with ophthalmologists and other healthcare professionals before being implemented into clinical practise.

This evaluation will help to clarify the current state of DCNN-based early DR detection and highlight areas where further study is needed. Researchers and clinicians can take advantage of DCNN developments to improve the precision, speed, and availability of early DR screening and diagnosis by being familiar with the technology's strengths, weaknesses, and potential future directions. Finally, deep convolutional neural networks (DCNNs) provide promise as a novel tool in the battle against diabetic retinopathy-related blindness by early detection of the disease. This study seeks to shed light on the current state, obstacles, and potential of DCNN-based early DR detection by conducting a thorough literature review. Insights from this analysis will be useful for researchers, clinicians, and policymakers interested in using DCNNs for early DR identification and bettering healthcare outcomes for diabetic patients.

2 Architecture

The proposed DCNN architecture for early detection of diabetic retinopathy comprises several key components. The input layer receives retinal fundus images as input, which are resized to a fixed size, such as 224 × 224 pixels, to ensure compatibility with the DCNN architecture. The subsequent convolutional layers form the core of the model, employing multiple filters or kernels to extract local features through convolution operations. Non-linear activation functions, such as ReLU or its variants, are applied to introduce non-linearity and facilitate the learning of complex patterns. To reduce spatial dimensions and retain important information, pooling layers, such as max pooling or average pooling, are strategically placed within the architecture. These layers downsample the feature maps, improving computational efficiency and promoting translation invariance. Dropout layers are introduced as a regularization technique to prevent overfitting by randomly deactivating a fraction of the input units during training. This promotes the learning of robust features and reduces the co-adaptation of neurons.

After the 2D feature maps have been flattened into a 1D vector by the flatten layer, the data is ready to be used by the subsequent fully linked layers. The retrieved features from the convolutional layers are used by the fully connected layers to learn global features and engage in high-level reasoning. The predicted probabilities for the different classes are generated by the output layer, which is typically a fully connected layer with a softmax activation function, and are then used to display the severity degree of DR or to distinguish images as normal or DR. The optimal loss function to use is task-specific. Categorical cross-entropy and binary cross-entropy are two popular loss functions used in DR detection. During training, the model's weights are modified via optimising algorithms such stochastic gradient descent (SGD) or more sophisticated approaches like Adam, RMSprop, or AdaGrad. The training process involves iterative adjustments of the model's weights through backpropagation, optimizing the chosen loss function. Figure 1 Explains the overall system architecture.

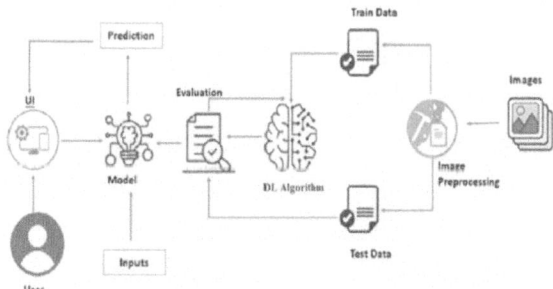

Fig. 1. Overall architecture system

The efficacy of the trained model is measured by its performance on a testing dataset. Some measures used to evaluate the model's performance at correctly classifying DR circumstances and distinguishing them from normal cases include accuracy, sensitivity, specificity, precision, and F1-score. These readings demonstrate the model's ability to detect DRR in a timely manner. The model can detect diabetic retinopathy early on because of the built and optimised DCNN architecture, which allows it to train and extract pertinent properties. The architecture must be tailored to the individual needs of the dataset and the desired DR severity levels. Improvements in patient treatment and outcomes may be possible thanks to the suggested architecture's ability to further the study of DR detection.

3   Methodology

For the purpose of early diagnosis of diabetic retinopathy, the methodology comprises a systematic approach to developing a DCNN-based model. Images of the retinal fundus are collected and preprocessed using techniques including scaling and data augmentation, and then the dataset is partitioned into training, validation, and testing subsets. Model parameters are then pre-trained, and a DCNN architecture informed by transfer learning is selected. The outcome can be enhanced by modifying the model's hyperparameters. The model is fine-tuned using loss functions on the training set before being put to the test on the testing set. Using these interpretability techniques, it is feasible to make out crucial elements in the photos. Ethics are closely upheld, especially in the areas of data protection and regulatory compliance.

3.1   Data Collection

The data collection process for the development of an early detection model for diabetic retinopathy involved multiple steps. Initially, a diverse range of retinal fundus images was obtained from reputable sources, including hospitals, clinics, and publicly available datasets. Ethical approvals and informed consent were acquired to ensure the privacy and confidentiality of patients' data. In order to ensure an inclusive representation of DR cases, clear criteria were established to identify eligible images. The selection process

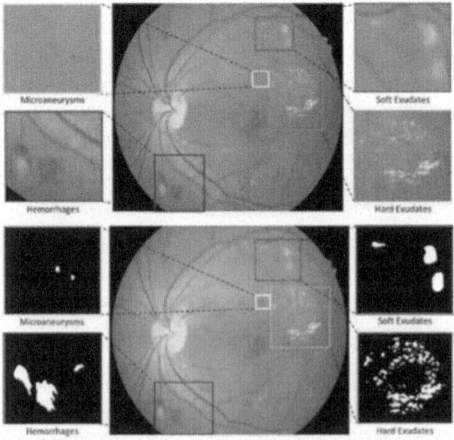

Fig. 2. Different retinal lesions and their masks

took into account factors such as image quality, availability of relevant annotations, and clinical relevance. Figure 2 Represents the different retinal lesions and their masks.

Experts then painstakingly annotated the collected photos with markers for microaneurysms, haemorrhages, and exudates, all of which are hallmarks commonly linked with DR. A thorough data preprocessing stage was performed to improve the dataset's quality and trustworthiness. Noise was reduced, the resolution was harmonised, and the pixel values were normalised. The dataset was split into training, validation, and testing sets to facilitate model construction and evaluation.

3.2 Data Preprocessing

The preprocessing of data plays a crucial role in developing an accurate early detection model for diabetic retinopathy. The retinal fundus images were subjected to a series of preprocessing steps to enhance their quality and prepare them for effective model training. For consistency and compatibility with the deep learning model architecture, the photos were first downsized to a standard resolution of 224×224 pixels. This resizing step aimed to maintain consistent image dimensions throughout the dataset. The pixel values were then scaled to a common range, between 0 and 1, for example. The normalisation procedure ensured that the model was trained with consistent data by eliminating the effect of photos with widely varied pixel intensities. Artefacts and distortions were also reduced through the use of noise reduction methods including Gaussian smoothing and median filtering. The purpose of this process was to clean up the photos and make them more usable for training the model. These preprocessing methods were applied to the dataset, guaranteeing uniform image sizes, regularised pixel values, and improved image quality. These preprocessing steps were essential to facilitate optimal model performance and generalization during the subsequent training and evaluation stages of the research. Figure 3 Represents the pipeline for the pre-processing stage.

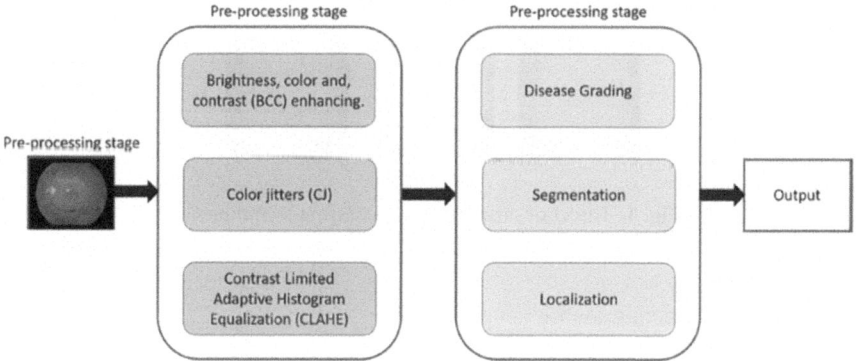

Fig. 3. Pipeline for the pre-processing stage

3.3 Data Augmentation

A robust and transferable early detection algorithm for diabetic retinopathy requires data augmentation. Data was added to retinal fundus images to improve diversity and quality. Augmentation methods included rotation, flipping, zooming, shifting, and adding random noise to create augmented versions of the original images. Rotation involved randomly rotating the images by a certain degree to simulate different orientations. Flipping horizontally or vertically reflected the images to introduce variations in their spatial arrangement. Zooming in or out on the images created different scales and perspectives. [3] Shifting the images horizontally or vertically introduced translations in their position. Additionally, random noise was added to the images to mimic real-world imperfections and enhance model robustness. These augmentation techniques resulted in an augmented dataset with a larger number of images, each exhibiting variations in orientation, spatial arrangement, scale, position, and noise levels. The augmented data provided the model with a more diverse training set, enabling it to generalize better and improve its ability to detect early signs of diabetic retinopathy in various scenarios. The goals of this research are to increase the model's accuracy, decrease its susceptibility to overfitting, and make it more robust against artefacts brought on by variations in clinically-relevant retinal fundus images (Fig. 4).

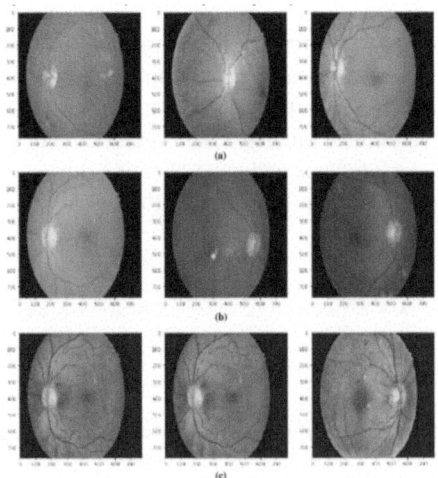

Fig. 4. Effect of different pre-processing techniques

3.4 Model Training

There were a number of changes made to the early detection model for diabetic retinopathy as it was being trained. We split the dataset into a training set, a validation set, and a testing set after cleaning, enriching, and preprocessing it. Approximately 80% of the data in the training set was used throughout the iterative optimisation process that was used to train the model's parameters. The final 10–15% of the training data was utilised to fine-tune the model's hyperparameters and assess its performance. A testing set distinct from the training and validation sets was used to assess the final model's generalisation capabilities. [4] At the outset of training, random weights were seeded into the DCNN architecture of the deep neural network. The model was then trained with the training set images and their associated validated labels. Backward propagation involved refining the model's parameters based on the loss that was estimated by gradient descent, while forward propagation involved feeding input images into the network. Using a loss function, such as binary cross-entropy or categorical cross-entropy, the discrepancies between the predicted and ground-truth labels were measured. The model's efficiency was improved by tuning its hyperparameters with the help of the validation set. We conducted experiments with the model's hyperparameters, including its learning rate, batch size, number of layers, and regularisation processes, to discover the ideal configuration that minimised loss and maximised accuracy. Grid search and random search were used to try out various settings for the hyperparameters and find the optimal one. It was common practise for the training procedure to span numerous epochs, during which the full training dataset was repeated multiple times.

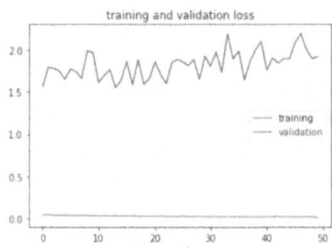

Fig. 5. Training and validation loss   **Fig. 6.** Training and validation accuracy

The model's performance on the training set and the validation set was evaluated using accuracy, precision, recall, and F1-score. In order to address this problem, we implemented early stopping, which stops training before a fixed number of iterations if the model's performance does not increase on the validation set. The final model was put through its paces using data from a different source than the training set. The model's capacity to detect early-stage diabetic retinopathy was evaluated using standard performance criteria such as accuracy, sensitivity, specificity, and area under the receiver operating characteristic curve (AUC-ROC). Through repeated training, validating, and testing, the DCNN model's parameters and hyperparameters were fine-tuned. Because early detection is crucial in the diagnosis and management of diabetic retinopathy, this method set out to develop a trustworthy and precise model for doing so. The accuracy during training was 99.58%, whereas during validation it dropped to 76.80%. The loss during validation was 1,9230, but during training it was only 0.0132. Since the model's training accuracy and validation loss are both inflated relative to reality, it has been deemed overfit. Figures 5 and 6 depict the gain and loss in accuracy during training and validation, respectively.

4 Model Evaluation and Results

The built DR detection model's performance and efficacy can only be gauged through a thorough examination of the model. It allows for measuring the model's accuracy, benchmarking against existing methods, evaluating generalization and robustness, establishing clinical relevance, and identifying areas for future improvement. Through performance assessment, the model's ability to accurately detect DR at an early stage can be determined, aiding in timely intervention and treatment. [6] Comparisons with benchmarks validate the model's efficacy, while evaluation of generalization and robustness ensures its reliability in real-world clinical settings. The clinical relevance of the model is established by assessing its performance using clinically relevant metrics, demonstrating its potential impact on healthcare. Model evaluation provides valuable insights for future improvements and refinements, enabling iterative advancements in the field of early DR detection.

4.1 Evaluation Methods

Train-Test Split Evaluation: The dataset must be partitioned into a training set and a test set before this technique can be applied. The model is trained on one set of data, and then its accuracy is measured using an other set of data (the "test set"). Accuracy, sensitivity, specificity, precision, recall, and F1-score are just few of the evaluation measures that can be used to define the model's performance on unlabeled data. The train-test split is a quick and simple method for gauging a model's scalability.

$$\text{Accuracy} = (\text{TP} + \text{TN})/(\text{TP} + \text{TN} + \text{FP} + \text{FN})$$

where: TP = True Positives (number of correctly predicted positive cases) TN = True Negatives (number of correctly predicted negative cases) FP = False Positives (number of incorrectly predicted positive cases) FN = False Negatives (number of incorrectly predicted negative cases)

$$\text{Sensitivity} = \text{TP}/(\text{TP} + \text{FN})$$

$$\text{Precision} = \text{TP}/(\text{TP} + \text{FP})$$

$$\text{F1 - Score} = 2 * (\text{Precision} * \text{Sensitivity})/(\text{Precision} + \text{Sensitivity})$$

K-Fold Cross-Validation: In k-fold cross-validation, the dataset is divided into a number of subsets ("folds") that are then tested independently. During training, the first fold is utilised as a test set, while the subsequent folds are used for actual training. The performance is averaged when each fold's evaluation criteria are established. A more precise assessment of the model's performance can be obtained using k-fold cross-validation, which removes the bias introduced by a single train-test split.

Receiver Operating Characteristic (ROC) Curve Analysis: ROC curve analysis is a graphical representation of the sensitivity and specificity (sensitivity minus specificity) for a particular classification threshold. One common method for illustrating a model's discriminatory power is to plot the area under the receiver operating characteristic (AUC) curve. A higher area under the curve (AUC) indicates better classification performance, and the ROC curve can shed light on the sensitivity-specificity trade-off of a model.

4.2 Hyperparameter Tunning

With the use of hyperparameter tuning, a deep learning model could be made more effective at spotting diabetic retinopathy in its early stages. Several key hyperparameters were explored and fine-tuned to improve the model's accuracy and generalization. One of the crucial hyperparameters was the learning rate, which determined the step size in adjusting the model's parameters during optimization. A range of learning rates, such as 0.001, 0.01, and 0.1, were considered to find the optimal value that balanced convergence speed and stability. Another important hyperparameter was the batch size, which determined the number of samples processed in each iteration during training. Different batch sizes, such as 16, 32, or 64, were experimented with to strike a balance between computational efficiency and convergence quality. The number of layers and the complexity of the deep convolutional neural network (DCNN) architecture were also hyperparameters to be tuned. Overfitting was reduced with the help of regularisation strategies like dropout and L2 regularisation. The complexity of the model was kept in check and generalisation was enhanced by experimenting with various regularisation strengths and dropout rates.

To systematically explore the hyperparameter space, techniques such as grid search or random search were employed. Grid search involved exhaustively trying out all possible combinations of hyperparameters from predefined ranges. Random search randomly sampled hyperparameter configurations and evaluated their performance. The optimum hyperparameter configurations were identified by applying evaluation metrics like accuracy, precision, recall, and F1-score to a validation set. Minimising the amount of parameters and employing regularisation to prevent overfitting due to a two-layer reduction in the neural network have had no influence on the difference between training and validation loss.

4.3 Outcomes and Discussion

The input images used in the study undergo a preprocessing step, where they are classified into different skin tone labels, namely dark skin, mild skin, and fair skin. To facilitate this classification, the RGB input images are converted to the YCbCr color space. Images after transformation along with their class labels make up the dataset required to train the deep learning model. The trained model is subsequently put to use for testing image detection. These test images, representative of real-world scenarios, are fed into the model, which processes them and predicts their respective classes. The model's ability to accurately classify these test images is a critical aspect evaluated in the research.

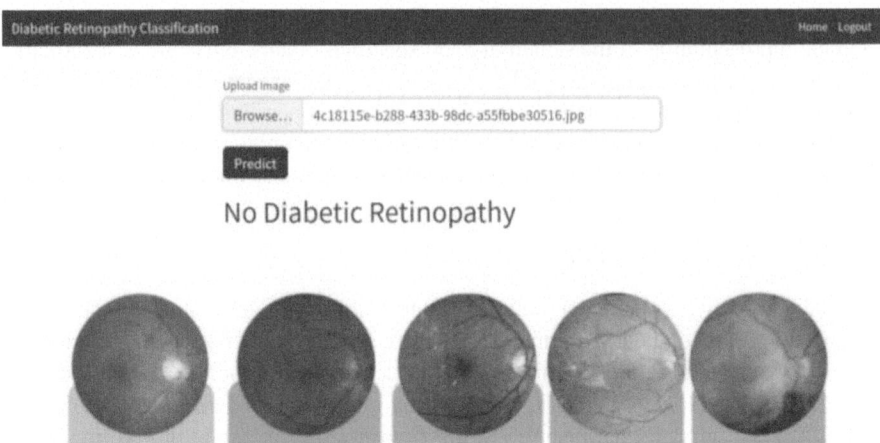

Fig. 7. Detection of No diabetic retinopathy

In addition, random picture URLs are used for detection to evaluate the model's generalisation capabilities. This method can be used to examine the model's adaptability to different inputs and evaluate its performance on previously unseen photos. Indicating the skin tone classification of the input image, the model's output matches the anticipated class. It is observed that the developed deep learning model demonstrates successful prediction capabilities for the input images. Figure 7 represents the detection of no diabetic retinopathy for the given image.

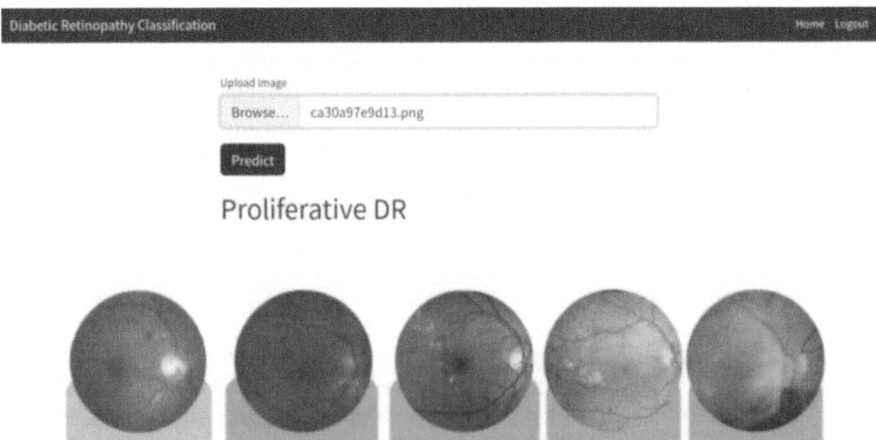

Fig. 8. Detection of proliferative DR

By incorporating the YCbCr conversion and training the model on the collected dataset, it achieves accurate classification results. The utilization of random images for testing further confirms the model's effectiveness in handling real-world scenarios. Overall, the study showcases the potential of the developed model for early detection of diabetic retinopathy by effectively classifying skin tones based on input images. The results of this study benefit from the combination of preprocessing methods, training with a deep learning model, and testing with both test and random images. Figures 8 and 9 show the results of the identification of proliferative DR and severe NPDR, respectively, for the same image.

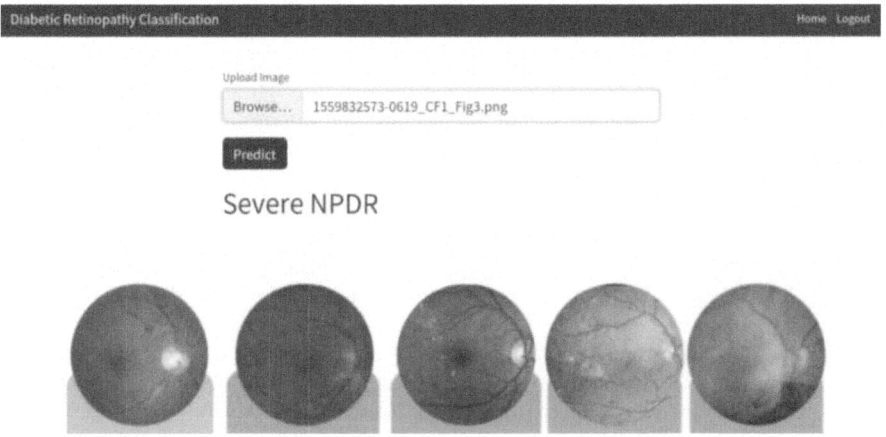

Fig. 9. Detection of Severe NPDR

5 Conclusion

In conclusion, the DCNN algorithm has shown promising results in detecting diabetic retinopathy (DR) at an early stage. The model's validation accuracy is 76.80%, whereas the training accuracy is 99.58% when classifying images of the retina. However, there is a slight indication of overfitting, as evidenced by the higher validation loss compared to the training loss. Nevertheless, the achieved accuracy surpasses the baseline and suggests the model's potential in assisting healthcare professionals with early DR detection. Further improvements can be made by implementing regularization techniques to address overfitting. Overall, the results highlight the effectiveness of the DCNN algorithm in aiding early DR detection, offering valuable insights for improved diagnosis and management of this condition.

References

1. Rahhal, D., Alhamouri, R., Albataineh, I., Duwairi, R.: Detection and classification of diabetic retinopathy using artificial intelligence algorithms. In: 13th International Conference on

Information and Communication Systems (ICICS), Irbid, Jordan, pp. 15–21 (2022). https://doi.org/10.1109/ICICS55353.2022.9811197

2. Das, D.,Biswas, S.K.,Bandyopadhyay, S., Laskar, R.H.: Deep learning techniques for early detection of diabetic retinopathy: recent developments and techniques. In: 5th International Conference on Computing, Communication and Security (ICCCS), Patna, India, pp. 1–7 (2020). https://doi.org/10.1109/ICCCS49678.2020.9276781

3. Shelar, M., Gaitonde, S., Senthilkumar, A., Mundra, M., Sarang, A.: Detection of diabetic retinopathy and its classification from the fundus images. In: International Conference on Computer Communication and Informatics (ICCCI), Coimbatore, India, pp. 1–6 (2021). https://doi.org/10.1109/ICCCI50826.2021.9402347

4. Bhatkar, A.P., Kharat, G.U.: Detection of diabetic retinopathy in retinal images using MLP classifier. In: IEEE International Symposium on Nanoelectronic and Information Systems, Indore, India, pp. 331–335 (2015). https://doi.org/10.1109/iNIS.2015.30

5. Hasan, D.A., Zeebaree, S.R.M., Sadeeq, M.A.M., Shukur, H.M., Zebari, R.R., Alkhayyat, A.H.: Machine learning-based diabetic retinopathy early detection and classification systems-a survey. In: 1st Babylon International Conference on Information Technology and Science (BICITS), Babil, Iraq, pp. 16–21 (2021). https://doi.org/10.1109/BICITS51482.2021.9509920

6. Kumari, C.U., Hemanth, A., Anand, V., Kumar, D.S., Naga Sanjeev, R., Sri Harshitha, T.S.: Deep learning based detection of diabetic retinopathy using retinal fundus images. In: Third International Conference on Intelligent Computing Instrumentation and Control Technologies (ICICICT), Kannur, India, pp. 1312–1316 (2022). https://doi.org/10.1109/ICICICT54557.2022.9917709

7. Elzennary, A., Soliman, M., Ibrahim, M.: Early deep detection for diabetic retinopathy. In: International Symposium on Advanced Electrical and Communication Technologies (ISAECT), Marrakech, Morocco, pp. 1–5 (2020). https://doi.org/10.1109/ISAECT50560.2020.9523650

8. Jayakumari, C., Lavanya, V., Sumesh, E.P.: Automated diabetic retinopathy detection and classification using ImageNet convolution neural network using fundus images. In: International Conference on Smart Electronics and Communication (ICOSEC), Trichy, India, pp. 577–582 (2020). https://doi.org/10.1109/ICOSEC49089.2020.9215270

9. Sugasri, M., Vibitha, V., Paveshkumar, M., Bose, S.S.: Screening system for early detection of diabetic retinopathy. In: 6th International Conference on Advanced Computing and Communication Systems (ICACCS), Coimbatore, India, pp. 760–762 (2020). https://doi.org/10.1109/ICACCS48705.2020.9074436

10. Gunawardhana, P.L., Jayathilake, R., Withanage, Y., Ganegoda, G.U.: Automatic diagnosis of diabetic retinopathy using machine learning: a review. In: 5th International Conference on Information Technology Research (ICITR), Moratuwa, Sri Lanka, pp. 1–6 (2020). https://doi.org/10.1109/ICITR51448.2020.9310818

11. Vijay, K., Vijayakumar, R., Sivaranjani, P., Logeshwari, R.: Scratch detection in cars using mask region convolution neural networks. Adv. Parallel Comput. **37**, 575–581 (2020)

12. Bhavani, M., Ravikumar, S., Prithi, S., Arockia Raj, Y., Rajendiran, B.: Methodical Tamil character recognition using fabricated CNN model. In: International Conference on Computer Communication and Informatics (ICCCI), Coimbatore, India, pp. 1–6 (2023). https://doi.org/10.1109/ICCCI56745.2023.10128316

13. Nilaiswariya, R., Manikandan, J., Hemalatha, P.: Improving scalability and security medical dataset using recurrent neural network and blockchain technology. In: International Conference on System, Computation, Automation and Networking (ICSCAN), pp. 1–6. IEEE (2021)

14. Carmel, M.B.M.J., Ravikumar, S., Muhammad, A., Dhilip, K.V., Antony, K.K., Arulkumaran, G.: Linguistic analysis of Hindi-English mixed tweets for depression detection. J. Math. (2022). https://doi.org/10.1155/2022/3225920

15. Oh, K., Kang, H.M., Leem, D.., et al. : Early detection of diabetic retinopathy based on deep learning and ultra-wide-field fundus images. Sci. Rep. **11**, 1897 (2021). https://doi.org/10.1038/s41598-021-81539-3

Author Index

A

Abinaya, J.   I-3
Abraham, Daniel   II-20
Achyutha Gowda, CP   II-72
Adaikkammai, A.   II-150
Adlin Layola, J.A.   II-303
Ahmed, A. Hammadh   I-111
Akshitha, N.   I-197
Algumalai, Vasudevan   I-32
Allagadda, NagaHemanth Murari   I-81
Anbukkarasi, S.   II-25
Aranidhi, A.   II-20
Archana, M.   I-257
Arumugam, S. S.   II-126
Arumugam, Vengatesan   I-32
Aruna, K. B.   II-139
Arunadevi, J.   I-234, II-314
Arunesh, K.   II-234

B

Bala Shunmugam, N.   I-292
Balachandran, S.   II-250
Balaji, V. S.   II-37
Balakrishnan, C.   I-225
Balasubramanian, M.   I-270
Belshia Jebamalar, G.   II-303
Bharath, P.   I-147
Bhavani, M.   I-21, I-68
Bhuvaneswaran, B.   I-45
Boral, Pushpita   I-56

C

Celina Rekha, J.   II-61

D

Deepa, A.   I-81
Devi, D.   II-150
Dhanalakshmi, K.   I-246
Durgadevi, M.   II-84
Dutta, Aritra   I-56

G

Ganapathy, P. S. Anirudh   II-37
Ganesh, A.   I-111
Ganesh, S.   I-111
Gangadhar, P. Rohit   I-68
Geetha, R.   I-181
Gopirajan, P. V.   II-220
Gowtham, Andene   II-20
Gowthami, S.   II-165
Gunavathie, M. A.   II-61
Gupta, Pranav   I-56

H

Hameed, P. T. S. Shahul   I-21
Hemakirthiga, K.   II-314
Hemalatha, K.   II-165

J

Jaeyalakshmi, M.   I-21, I-68
Jaison, B.   II-176
Janakiraman, R.   I-197
Jayanthi, S.   II-126
Jayaprakash, J.   II-20
Jebamani, Anitha   II-150
Jegadeesan, R.   I-197
Jegatha, R.   I-158
Jeyalakshmi, J.   I-158, I-171
Jeyalakshmi, V. S.   I-292
Jhansi, M.   I-197
Judith, J. E.   I-124

K

Kabilamani, P.   I-213
Kaliraj, V.   I-213
Kamalanaban, E.   II-20
Kamalesh, P.   II-37
Karthika, S.   II-250
Kathambari, P.   I-81
Kausthuban, M.   II-303
Kavirakesh, S.   I-315
Kavitha, M.   I-292

Keerthana, R.   I-3
Kousalya, S.   II-61
Krishna, Alla Vamsi   II-72
Krithiga, P.   I-315
Krithiga, R.   I-280
Kumar, Aman   I-280
Kumar, Leti Manish   II-72
Kumar, M. Selva   I-147
kumar, V. S. Prassana   I-225

M

Madhan Kumar, A.   II-195
Mahalakshmi, D.   II-61
Mahalakshmi, K.   II-176
Mahalakshmi, S.   I-135
Mahesh, Vijayalakshmi G. V.   II-72
Mahi, A. R.   I-135
Mani, A.   I-97
Manickavasagan, V.   II-220
Meena, R.   I-135
Murugesan, Shravanthi   II-45
Muthukumarasamy, S.   I-97
Muthurasu, N.   II-292

N

Nandhan, S. Harish   I-135
Nanmaran, R.   II-195
Nareshraj, R.   II-303
Navaz, K.   II-116
Nethra, M.   II-3
Niranjana, G.   I-56
Nivedha, S.   I-97

O

Oliviya, K. Jency   I-3

P

Pagey, Chinmay   I-280
Pal, Subha Bal   I-56
Paulin Diana Dani, D.   I-292
Pillai, N. Muthuvairavan   II-116
Pragadeesh, M.   I-45
Prasanna Devi, S.   I-303
Prasanth, K.   I-213
Praveena, S.   I-303
Priya, V.   II-205
Purushotham, N.   II-116

R

Radhika, R.   II-150
Raj, S. Lakshman   I-147
Rajalakshmi, J.   II-303
Rajasekar, V.   I-213, II-292
Rajeashwari, S.   II-234
Ram, N. Sankar   I-197
Rameshraja, M.   I-234
Ramyadevi, K.   I-270
Rasikaranjani, S.   II-61
Reddy, A. Sai Simha   I-81
Rekha, K. S.   II-220
Rethi, M.   II-25
Rifana Fathima, A.   I-246
Roobika, R.   II-276

S

Sabitha, E.   II-84
Saicharan, C. H.   I-197
Sambath, M.   I-111
Sangar, G.   I-213
Sangeetha, K.   II-37
Sangewar, Chandan Kumar   I-280
Sanjana, C. H.   I-197
Santhiya, M.   I-158, I-171
Santhya, S.   II-3
Sarala, R.   I-147
Saraswathi, V.   II-150
Sathya Sofia, A.   II-205
Sazzad, K. H. Mohammed   II-3
Selvarani, P.   II-20
Selvi, C. P. Thamil   II-105
Selvi, M.   I-81
Selvi, P. Francis Antony   II-105
Senthil, G. A.   I-3
Shanthalakshmi, M.   II-105
Shanthi, D. Sorna   I-45
Shanthi, D.   I-257, II-263, II-276
Sharanyaa, S.   I-111
Shiney, S. Arumai   I-181
Shobana, M.   I-171
Shriram, M. S.   II-45
Sindhuja, R.   II-195
Sivakamy, N.   II-276
Sivankalai, S.   II-250
Sivasekaran, K.   II-25
Sofia, A. Sathya   II-105
Sridevi, S.   II-165
Srinivas, Dava   I-197
Srinivas, M. D. Harish   I-147

Srivatsan, M.   I-68
Sudha, L.   II-139
Suganya, S.   II-105
Sujatha, E.   II-220
Suntheya, A. K.   II-139
Sureka, V.   II-139
Surya Prakash, G.   I-270
Suseela, G.   I-56
Sushmitha, S.   II-45
Swetha, S.   I-315
Sylevester, A. Arnold   II-3

T
Tharun, V. S.   I-270
Thilagam, T.   II-303
Thomas, Ansu Liz   I-124

U
Umarani, V.   II-220

V
Vadrevu, Pavan Kumar   I-257
Vaishnavi, M. G.   II-263
Veena, K.   I-81
Vijay, K.   I-21, I-315
Vijayakumar, R.   I-45
Vishal, B.   I-315

Y
Yazhinian, S.   II-116

Z
Zean, J. Remoon   I-135

SPRINGER NATURE

GPSR Compliance

The European Union's (EU) General Product Safety Regulation (GPSR) is a set of rules that requires consumer products to be safe and our obligations to ensure this.

If you have any concerns about our products, you can contact us on ProductSafety@springernature.com

In case Publisher is established outside the EU, the EU authorized representative is:

Springer Nature Customer Service Center GmbH
Europaplatz 3
69115 Heidelberg, Germany

The manufacturer's authorised representative in the EU is Springer
Nature Customer Service Centre GmbH, Europaplatz 3, 69115 Heidelberg,
Germany. If you have any concerns regarding our products, please
contact ProductSafety@springernature.com

Printed and bound by CPI Group (UK) Ltd, Croydon, CR0 4YY
29/04/2026
02099541-0002